Universitext

Universitext

Universitext is a series of textbooks that presents material from a wide variety of mathematical disciplines at master's level and beyond. The books, often well class-tested by their author, may have an informal, personal even experimental approach to their subject matter. Some of the most successful and established books in the series have evolved through several editions, always following the evolution of teaching curricula, to very polished texts.

Thus as research topics trickle down into graduate-level teaching, first textbooks written for new, cutting-edge courses may make their way into Universitext.

More information about this series at http://www.springer.com/series/223

Fernando Q. Gouvêa

p-adic Numbers

An Introduction

Third Edition

 Springer

Fernando Q. Gouvêa
Department of Mathematics and Statistics
Colby College
Waterville, ME, USA

ISSN 0172-5939   ISSN 2191-6675   (electronic)
Universitext
ISBN 978-3-030-47294-8   ISBN 978-3-030-47295-5   (eBook)
https://doi.org/10.1007/978-3-030-47295-5

Mathematics Subject Classification (2010): 11S80, 11-01, 12J25, 13F30

This Springer imprint is published by the registered company Springer Nature Switzerland AG
The registered company address is: Gewerbestrasse 11, 6330 Cham, Switzerland

Contents

Introduction

In the course of their undergraduate careers, most mathematics majors see little beyond "standard mathematics:" basic real and complex analysis, abstract algebra, some differential geometry, etc. There are few adventures in other territories, and few opportunities to visit some of the more exotic corners of mathematics. The goal of this book is to offer such an opportunity, by way of a visit to the p-adic universe. Such a visit offers a glimpse of a part of mathematics which is both important and fun, and which also is something of a meeting point between algebra and analysis.

Over the last century, p-adic numbers and p-adic analysis have come to play a central role in modern number theory. This importance comes from the fact that they afford a natural and powerful language for talking about congruences between integers, and allow the use of methods borrowed from calculus and analysis for studying such problems. More recently, p-adic numbers have shown up in other areas of mathematics, and even in physics.

Despite their strangeness, p-adic numbers are not an extremely difficult concept; in fact, they are quite accessible to an undergraduate audience. The goal of this book is to realize this possibility, taking its readers for a short promenade along the p-adic path. Our aim is sightseeing, rather than a scientific expedition, so we will not worry too much if we fail to emphasize a subtle point here and there, nor if our theorems are less general than they could be, nor, in fact, if we do not learn all there is to know. Rather, our goal is to introduce the reader to the rather strange world of the p-adic numbers, and to begin to make it feel familiar. What we will cover will not be sufficient for those students which will need to use p-adic numbers as a research tool. For them, a lot more reading will be necessary (and in an appendix we discuss some of the texts that are available for further reading). Instead, we try to touch a lot of bases, and set the stage for further study.

There are many ways to begin our task. Of the available options, I chose to go with the theory of absolute values on fields, and to view the p-adic numbers as directly analogous to the real numbers. In this approach, the main ingredient is a change of attitude about absolute values. It starts with the observation that from an *algebraic* point of view there is no reason to view the usual absolute value on the field $\mathbb{Q}$ of rational numbers as a given. Rather, any function satisfying the same basic properties should be just as good. If we start with the usual absolute value and look for a completion[1] of $\mathbb{Q}$ as a metric space, we get the real numbers; starting with a different absolute value, we get something else. What that something else is, and why it is interesting, is the subject of this book.

Besides its importance, the study of p-adic numbers is attractive because it blends together so many parts of mathematics. While it is certainly a part of number theory, its language is often the language of analysis, and its theorems are often analogous to, but slightly different from, those found in calculus textbooks. Both the analogy and the differences are fascinating, so that at times one gets the feeling that things are slightly out of whack, and p-adic analysis seems like classical analysis in a distorting mirror. I have tried to include many examples of this sort of thing, and I hope they are convincing.

I have done much less to convince the reader that p-adic numbers are actually useful. For the most part, I have limited myself to stating that certain things are true or that certain methods are fruitful. In every case, developing the details of the application would make this book much harder than it is supposed to be. Once again, a lot can be learned from other texts, and the student who wants to know should go to the references mentioned in the text and in the appendix.

Some business: the pre-requisites for reading this book are a basic knowledge of algebra and number theory, and a few courses in calculus or analysis. To be a bit more precise, the reader should be familiar with the language of congruences, with the basic theory of fields and rings, and with basic concepts about point-set topology, continuity, and infinite series. I have tried to provide as many definitions (and also informal descriptions) as I could, consistent with the requirement that the result not be too ungainly. I hope that this approach may be useful both to refresh students' memories of other subjects and to display the unity and interconnectedness of mathematics in a dramatic way.

The use of the topics mentioned above as pre-requisites is not uniform. Most students will know enough to read the first few chapters without needing to run back to their textbooks from other courses. The analysis requirements become more serious beginning in Chapter 5, and the algebraic requirements come in more strongly in Chapter 6. Even so, the whole book remains[2] well within the reach of undergraduate mathematics majors.

There are many kinds of books about mathematics, from encyclopedic treatises to brief surveys, from dry as dust to boringly chatty. This book is closer to being a survey than to being encyclopedic, and is intended to be easy to read, but not as bed-time reading: the reader is expected to do some work. (Maybe even a lot of work.) To this end, I have included a great

---

[1] If you're wondering what a "completion" is, the definition will be met later, in full gory detail. Don't worry about it yet.

[2] I hope!

many problems throughout. The problems are meant to be solved, or at least attempted, at about the time when they are met in the text.[3] Most of them offer an opportunity to work with concepts that have just been introduced, and it is the author's fond hope that such problems will help create familiarity with the material. The majority of these problems ask the reader to work out the details of arguments which have been only sketched in the text, or to supply the proofs for statements given in the text (for the most part, this is only done when the proof is straightforward, and even then hints are often given). Other problems stretch out to mention matters not touched upon in the text, to indicate to the reader that there are many themes we have not had time to discuss. Finally, many are intended to prepare the reader for the discussion to follow. Such problems will often become trivial in the light of what comes later (they may be special cases or simple corollaries of theorems we will prove); leaving them for later will only render them boring.

Besides offering practice and a chance of active interaction with the material, the many problems are intended to stimulate the reader to read in a certain way. In many mathematics textbooks, one finds proofs that are "left to the reader" or dismissed as "clear" and throwaway lines mentioning interesting sidelines to the material being discussed. The experienced mathematical reader knows that these are signals to dig out pencil and paper and verify what has been said or to find a reference and verify things that way. In this book, I have tried to make sure that most such signals are followed by explicit problems. My hope is that this will help my less experienced readers gain experience of how to interact with mathematical texts.

I have provided hints and comments on all the problems and complete solutions for most of the harder problems. These should be used only after some meditation on the problem, or they may spoil the fun.

Every writer creates in his or her mind an imaginary audience for his or her text. In the case of this book, what I imagined was an upper-level undergraduate course for mathematics majors. It would include honest-to-goodness undergraduates and not only graduate-level students who just happen not to have finished their undergraduate degrees yet. (In other words, this is not only for hot-shots, though hot-shots should be welcome too.) The course would very likely use an approach where students are asked to read the text, attempt the problems, and discuss the results with each other and with their instructor. The many problems asking the reader to "make a conjecture," or to attempt something ("Can you...") presuppose such a situation.

A note to the specialists: this book is intended as a pedagogical tool. It is *not* intended as a replacement for the standard references nor as a model of an elegant or detailed treatment of this (or any other) subject. Rather, I have tried to make it fun to work with, demanding, and ample. I have often spent time discussing interesting mathematics (the point-set topology,

---

[3]I realize this is very different from what most of my readers are used to.

for example, or the various definitions of the field norm) just because it was interesting. I welcome any comments, and ask students in particular to tell me their reactions.

Note on the first edition: This book grew from a set of notes for a mini-course given at the "17° Colóquio Brasileiro de Matemática," the 1989 edition of the bi-annual congress of Brazilian mathematicians. It has since been used in a course (much like the one described above) at Colby College. I would like to thank the organizers of the "Colóquio" for their invitation, and also to thank the students who sat through preliminary versions of this material for their interest and for their patience with its shortcomings. Many shortcomings will undoubtedly remain, and I would like to hear about them (who knows, there may even be a second edition someday). Please drop me a note if you have any comments.

During the final stages of the writing of this book, the author's research was partially supported by NSF grant number DMS–9203469. The writing was done in three phases, at the Universidade de São Paulo, at Queen's University at Kingston, Ontario, and at Colby College. I would like to thank NSF and all three universities for their support; Colby College, where most of the work was done, and whose computer equipment is responsible for the physical existence of this book, deserves special thanks for providing pleasant and fruitful working conditions.

This book was typeset in LaTeX using several different kinds of computers and a large number of standard macro packages. It depends, thus, on the work of many people who have given of their talents to the community of TeX users. I thank you all.

Finally, I would also like to thank César Polcino, of the Universidade de São Paulo, who first put a book on p-adic numbers in my hands, and Noriko Yui, of Queen's University, who insisted that I develop the original notes into this book; the project would not have been undertaken without them.

Note on the second edition: I am grateful to Springer-Verlag for giving me the opportunity to revise the book for this new printing. The largest changes happened in chapter four. I'd like to thank the various people who made comments and suggestions, including Silvio Levy, Alain Robert, and especially Keith Conrad.

Note on the second printing of the second edition: The need for a new printing has given me the opportunity to correct several typos, update references, and make a few small changes in the text.

Note on the third printing of the second edition: The main change for this new printing was to correct the numbering of the solutions to the problems, which was incorrect in the previous printing. I apologize to those who were inconvenienced by that mistake. Other than that, I have only made a few minor changes.

I'd like to thank the many people who found typos, made suggestions and comments, and generally gave me useful feedback. You are all encouraged to keep at it!

τι ποιεῖτε, πάντα εἰς δόξαν θεοῦ ποιεῖτε

On the Third Edition

Almost twenty years have passed since the second edition of this book. One of the privileges of mathematicians is, of course, that the passage of time does not change the truth of theorems nor invalidate correct proofs. But things do change. One major change in our mathematical environment has to do with software: the availability of Sage and GP means that all mathematicians can have access to powerful computational engines. The major change in the third edition reflects this: I have added quite a bit of material on how to compute, both by hand and using a computer. Readers who have never used these computer tools before should read Appendix A for a brief introduction.

Beyond that, there are many small changes. Often I added more pointers to where the theory ends up going or an extra theorem that seemed interesting. I also added, at the suggestion readers, short sections on visualizing $\mathbb{Z}_p$ and $\mathbb{Q}_p$ and on integration. The bibliography has been updated. I have updated Appendix C accordingly.

There has also been a change that I noticed while revising the text: I find that I am no longer comfortable with telling readers to "look this up in your real analysis textbook." (I'm not sure why, but there it is.) I have tried, in each instance, to give a precise reference instead. While it is often true that the results I need are in standard textbooks, I felt I owed my readers at least one specific place to look.

There are some structural changes. I decided to split the old third chapter into two parts, one giving the construction of $\mathbb{Q}_p$ and the other exploring its properties. I also broke off the final section of the old chapter six, creating a new chapter with suggestions for further exploration. Several chapters have new sections or have had long sections broken up into smaller ones.

Two students at Colby College, Shuofeng Xu and Qidong He, read through the book and made many comments that have been very useful. As a result, I have ended up clarifying a few arguments, adding and removing things, and generally playing around with the text. I read through the entire book and found many passages where it seemed possible to be clearer. In many cases I have made the solutions to the problems more detailed. Many readers had sent me questions or found typos, which I have attempted to correct. Springer's reviewers also made several useful suggestions, most of which I have followed.

Revising a book sometimes involves replacing known errors by new and subtler errors. I hope that the errors I have corrected will be more numerous than those I may have introduced. In particular, I have corrected a bad mistake (on the range of the p-adic logarithm) that has been around since the first edition and was caught by one of Springer's referees. I am very grateful to everyone who pointed out mistakes, unclear passages, and other issues.

Finally, the (mostly) number theorist author of 1999 is now (mostly) a historian of mathematics. In particular, I now know much more about the history of the p-adic numbers than I did then. As a result, many of the historical comments have been updated, usually in subtle ways that matter to me but may not matter to anyone else.

The overall effect is that there have been few global changes but there are changes locally everywhere.

Warnings: In my world, "x is positive" means $x \geq 0$, with "strictly positive" reserved for $x > 0$. A similar convention is in place for "increasing," but luckily that word doesn't play much of a role in p-adic analysis. I know this may annoy my Anglo-American readers, and I apologize in advance.

In this book, all rings have a multiplicative identity. Quotients of $\mathbb{Z}$ are denoted $\mathbb{Z}/m\mathbb{Z}$.

Acknowledgments: I am immensely grateful for all the comments I have received. The two referees made many helpful comments; I have followed most of their suggestions. Shuofeng Xu and Qidong He did a fantastic job of reading through the book and pointing out the places where I had missed the target. Thank you!

1 Apéritif

The idea of considering new ways to measure the "distance" between two
rational numbers, and then of considering the corresponding completions,
did not arise merely from some desire to generalize, but rather from several
concrete situations involving problems from algebra and number theory. Each
of the new metrics on $\mathbb{Q}$ will be connected to a certain prime, and they will
codify a great deal of arithmetic information related to that prime. This
point of view, however, arrived after the fact, as a way to justify what Hensel
had done.

The goal of this first chapter is to offer an *informal* introduction to these
ideas. Thus, we proceed without worrying too much about mathematical
rigor,[1] but rather emphasizing the ideas that are behind what we are trying
to accomplish. This was in fact Hensel's original approach. Then, in the next
chapter, we will begin to develop the theory in a more formal way.

1.1 Hensel's Analogy

The p-adic numbers were first introduced by the German mathematician Kurt
Hensel. Hensel's starting point was the analogy between the ring of integers
$\mathbb{Z}$, together with its field of fractions $\mathbb{Q}$, and the ring $\mathbb{C}[X]$ of polynomials with
complex coefficients, together with its field of fractions $\mathbb{C}(X)$. He learned the
analogy from his doctoral adviser, Leopold Kronecker, who even attempted
to develop a single theory that covered both cases.

To be specific, let's use X as an indeterminate, saving x to stand for a
number. An element of $f(X) \in \mathbb{C}(X)$ is a "rational function," i.e., a quotient
of two polynomials:

$$f(X) = \frac{P(X)}{Q(X)},$$

with $P(X), Q(X) \in \mathbb{C}[X], Q(X) \neq 0$; we can always require that $Q(X)$ is
monic, i.e., its leading coefficient is 1. Similarly, any rational number $x \in \mathbb{Q}$
is a quotient of two integers:

$$x = \frac{a}{b},$$

with $a, b \in \mathbb{Z}, b \neq 0$; we can always require that $b > 0$. Furthermore, the
properties of the two rings are quite similar: both $\mathbb{Z}$ and $\mathbb{C}[X]$ are rings where

---

[1] which always runs the risk of becoming mathematical *rigor mortis*...

© Springer Nature Switzerland AG 2020

F. Q. Gouvêa, *p-adic Numbers*, Universitext, https://doi.org/10.1007/978-3-030-47295-5_1

there is *unique factorization*: any integer can be expressed uniquely as ± 1 times a product of primes, and any polynomial can be expressed uniquely as

$$P(X) = a(X - \alpha_1)(X - \alpha_2)\ldots(X - \alpha_n),$$

where a and $\alpha_1, \alpha_2, \ldots \alpha_n$ are complex numbers. This gives us the main point of the analogy Hensel explored: *The primes $p \in \mathbb{Z}$ are analogous to the linear polynomials $X - \alpha \in \mathbb{C}[X]$.*

The analogy extends to solutions of equations. Given a polynomial with coefficients in $\mathbb{Z}$, any root is an *algebraic number*; if a function is a root of a polynomial with coefficients in $\mathbb{C}[X]$, it is an *algebraic function*. So $\sqrt{2}$, which is a root of $Y^2 - 2$, is an algebraic number, while $f(X) = \sqrt{X^3 - 3X + 1}$, which is a root of $Y^2 - (X^3 - 3X + 1)$, is an algebraic function.

Hensel was studying a specific problem about algebraic numbers. Pursuing the analogy, he considered the identical problem in the context of algebraic functions; that problem turned out to be easy to solve, because he could expand the algebraic functions into power series.

Suppose we are given a polynomial $P(X) \in \mathbb{C}[X]$ and a particular $\alpha \in \mathbb{C}$. Then it is possible (for example, using a Taylor expansion) to write the polynomial in the form

$$P(X) = a_0 + a_1(X - \alpha) + a_2(X - \alpha)^2 + \cdots + a_n(X - \alpha)^n$$

$$= \sum_{i=0}^{n} a_i(X - \alpha)^i$$

with $a_i \in \mathbb{C}$. This gives very precise information on how the polynomial behaves near α.

Can we do something like this for integers? For positive integers, we can, and indeed we do it every day when we write them down:

$$321 = 1 + 2 \times 10 + 3 \times 10^2$$

is in that form. The annoying thing is that 10 is not a prime, while $(X - \alpha)$ is a prime in $\mathbb{C}[X]$. But we can fix that: choose a prime number p and write our number in base p: given a positive integer m, we can write it in the form

$$m = a_0 + a_1 p + a_2 p^2 + \cdots + a_n p^n = \sum_{i=0}^{n} a_i p^i$$

with $a_i \in \mathbb{Z}$ and $0 \le a_i \le p - 1$. For example, if $p = 7$ we can write

$$320 = 5 + 3 \times 7 + 6 \times 7^2.$$

We can even record this as a string of digits "635" (as in base ten, we record the digits backwards: lowest powers[2] come last). To keep the distinction

---

[2]In the p-adic context it would perhaps be better to reverse this, and put lowest powers first. The downside is that doing that would require us to "do arithmetic backwards," which can be confusing. So we have decided to stick to the usual convention.

between standard (base ten) notation and base p, let's use red for the latter, so $320 = \mathbf{635}$ as long as it's understood that we are working with $p = 7$.

How do we find the expansions? Well, to find the last base seven digit of 320 we use division with remainder: $320 = 45 \times 7 + 5$. Then we take the quotient, 45, and divide it as well: $45 = 6 \times 7 + 3$. And finally $6 = 0 \times 7 + 6$. The main rule is that the remainder *must* be one of the numbers $0, 1, \ldots,$ $p - 1$.

Such expansions are already interesting in that they give "local" information: the expansion in powers of $(X - \alpha)$ will show, for example, if $P(X)$ vanishes at α, and to what order. Similarly, the expansion "in base p" will show if m is divisible by p, and to what order. For example, expanding 72 in base 3 gives

$$72 = 0 + 0 \times 3 + 2 \times 3^2 + 2 \times 3^3 = \mathbf{2200},$$

which shows at once that 72 is divisible by 3^2.

Now, for polynomials and their quotients, one can in fact push this much further. Taking $f(X) \in \mathbb{C}(X)$ and $\alpha \in \mathbb{C}$, there is always an expansion

$$f(X) = \frac{P(X)}{Q(X)} = a_{n_0}(X - \alpha)^{n_0} + a_{n_0+1}(X - \alpha)^{n_0+1} + \ldots$$
$$= \sum_{i \geq n_0} a_i(X - \alpha)^i.$$

This is just the Laurent expansion from complex analysis, but in our case it can be very easily obtained either by doing long division with the expansions of $P(X)$ and of $Q(X)$ or by using division with remainder as before. Notice that it is a much more complicated object than the preceding expansion:

- We can have $n_0 < 0$, that is, the expansion can begin with a negative exponent; this would signal that α is a root of $Q(X)$ and not of $P(X)$ (more precisely, that its multiplicity as a root of $Q(X)$ is bigger). In the language of analysis, we would say that $f(X)$ has a *pole* at α of order $-n_0$. This is not much of a problem: we first remove the pole by multiplying by $(X - \alpha)^{|n_0|}$, expand the result into powers of $(x - \alpha)$, then divide again at the end.

- The expansion will usually not be finite. In fact, it will only be finite if when we write $f(X) = P(X)/Q(X)$ in lowest terms with $Q(X)$ monic, then $Q(X)$ happens to be a power of $(X - \alpha)$ (can you prove it?). In other words, this is usually an infinite series, and it can be shown that the series $f(\lambda)$ that we get when we replace X by $\lambda \in \mathbb{C}$ will converge whenever λ is close enough (but not equal) to α. However, since we want to focus on the algebraic structure here, we will treat the series as a *formal* object: it is just there, and we do not care about convergence.

Here's an example. Take the rational function

$$f(X) = \frac{X}{X - 1},$$

and let's look at the expansions for different α. If $\alpha = 0$, we get

$$\frac{X}{X-1} = -X - X^2 - X^3 - X^4 - \dots$$

which shows that $f(0) = 0$ with multiplicity one. For $\alpha = 1$, we get

$$\frac{X}{X-1} = \frac{1+X-1}{X-1} = (X-1)^{-1} + 1$$

which highlights the pole of order one at $\alpha = 1$ (and also gives an example of an expansion that is finite). Finally, if we take, say, $\alpha = 2$, where there is neither pole nor zero, we get

$$\frac{X}{X-1} = \frac{2+(x-2)}{1+(x-2)} = 2 - (X-2) + (X-2)^2 - (X-2)^3 + \dots$$

Problem 1 Refresh your calculus memory and check these three equalities. Can you find the regions of convergence? (Hint: all you need to remember is the geometric series.)

Problem 2 Suppose $f(X) = P(X)/Q(X)$ is in lowest terms, so that $P(X)$ and $Q(X)$ have no common zeros. Show that the expansion of $f(X)$ in powers of $(X - \alpha)$ is finite if and only if $Q(X) = (X - \alpha)^m$ for some $m \geq 0$.

The punchline is that any rational function can be expanded into a series of this kind in terms of each of the "primes" $(X - \alpha)$. (The quotes aren't really necessary, since the ideals generated by the elements of the form $(X - \alpha)$ are exactly the prime ideals of the ring $\mathbb{C}[X]$, so that $(X - \alpha)$ is a rightful bearer of the title of "prime." But all that comes later.) On the other hand, not all such series come from rational functions. In fact, we have already met examples in our calculus courses: the series for $\sin(X)$, say, or the series for e^X, which cannot be expansions of any rational function (calculus exercise: why not?).

Here's how to read the situation from an algebraic point of view. We have two fields: the field $\mathbb{C}(X)$ of all rational functions, and another field which consists of all Laurent series in $(X - \alpha)$. (The next exercise asks you to check that it is indeed a field.) Let's denote the second by $\mathbb{C}((X - \alpha))$. Then the function

$$f(X) \mapsto \text{expansion around } (X - \alpha)$$

defines an *inclusion* of fields

$$\mathbb{C}(X) \hookrightarrow \mathbb{C}((X - \alpha)).$$

There are, of course, infinitely many of these (one for each α), and each one contains "local" information about how rational functions behave near α.

Problem 3 Let $\mathbb{C}((X - \alpha))$ be the set of all finite-tailed Laurent series (with complex coefficients) in $(X - \alpha)$

$$f(X) = \sum_{i \geq n_0} a_i (X - \alpha)^i.$$

Define the sum and product of two elements of $\mathbb{C}((X - \alpha))$ in the "obvious" way, and show that the resulting object is a field. Show that one may in fact take the coefficients to be in any field, with the same result.

Since such power series expansions were so useful in studying rational and algebraic functions, Hensel's idea was to extend the analogy between $\mathbb{Z}$ and $\mathbb{C}[X]$ to include the construction of such expansions. Recall that the analogue of choosing α is choosing a prime number p. As we have already seen, we already know the expansion for a positive integer m: it is just the "base p" representation of m:

$$m = a_0 + a_1 p + a_2 p^2 + \cdots + a_n p^n,$$

with $a_i \in \mathbb{Z}$, $0 \leq a_i \leq p - 1$. As in the case of polynomials, this is a finite expression.[3]

In the case of positive rational numbers whose denominator is a power of p, it's easy: take the expansion of the numerator and divide by a power of p. So, since $320 = 5 + 3 \times 7 + 6 \times 7^2 = \mathbf{635}$ in base 7, we see that

$$\frac{320}{49} = 5 \times 7^{-2} + 3 \times 7^{-1} + 6 = \mathbf{6.35},$$

where we are using the dot in analogy to the "decimal point" in base ten, to mark the place where we move to negative powers of 7.

To pass to more general positive rationals, we need to allow infinitely long expansions. What we do is expand both numerator and denominator in powers of p, and then either *divide formally* or use *repeated division with remainder*. The only thing one has to be careful with is that one may have to "carry." The sum of two of our a_i, for example, may be larger than $p - 1$, and one has to do the obvious thing. It's probably easier to go straight to examples.

Let's start with an easy one, $1/2$, and take $p = 5$. We divide $1/2$ by 5, like this:

$$\frac{1}{2} = 5 \times \frac{-1}{2} + \mathbf{3},$$

so the last digit is $\mathbf{3}$. (Because 5 is prime, it can be shown that quotient and remainder are unique; for now, take that as given.) Now we divide again:

$$\frac{-1}{2} = 5 \times \frac{-1}{2} + \mathbf{2}.$$

---

[3] The condition $0 \leq a_i \leq p - 1$ may seem to break the analogy with the complex case. But not so! The point is that the quotient of $\mathbb{C}[X]$ by the ideal generated by $(X - \alpha)$ is isomorphic to $\mathbb{C}$, and the constants in $\mathbb{C}[X]$ give a "canonical" choice of coset representatives. Similarly, the numbers between 0 and $p - 1$ are a choice of coset representatives for the quotient of $\mathbb{Z}$ by the ideal generated by p.

The new quotient is $-1/2$ again, so from now on all the divisions are the same:

$$\frac{-1}{2} = 5 \times \frac{-1}{2} + 2, \text{ forever!}$$

So we get

$$\frac{1}{2} = 3 + 2 \times 5 + 2 \times 5^2 + \cdots = \ldots 22223.$$

Seems crazy, but notice that if we multiply by 2 it works: $2 \times 3 = 6 = 11$ in base 5, so we write 1 and carry a 1, then $2 \times 2 + 1 = 5 = 10$, so we write 0 and carry a 1, and then we repeat forever to get $\ldots 00001 = 1$.

Let's do a harder one: take $p = 3$, and consider the rational number $24/17$. Then we have

$$a = 24 = 0 + 2 \times 3 + 2 \times 3^2 = 220 = 2p + 2p^2$$

and

$$b = 17 = 2 + 2 \times 3 + 1 \times 3^2 = 122 = 2 + 2p + p^2.$$

(Though of course $p = 3$, it's probably less confusing to write p because one is less tempted to "add it all up." The point is to operate *formally* with our expansions.)

Suppose we want to get an expansion for $a/b = 24/17$. One way to do it is to set

$$\frac{a}{b} = \frac{24}{17} = \frac{2p + 2p^2}{2 + 2p + p^2} = a_0 + a_1 p + a_2 p^2 + a_3 p^3 + \ldots$$

then multiply through by $2 + 2p + 2p^2$. This gives infinitely many equations involving the coefficients a_n, but they are easy to solve. (See the next section for more on how to compute.) We can also do "long division" or use repeated divisions by 3 as before. However we do it, we get

$$\frac{24}{17} = \frac{2p + 2p^2}{2 + 2p + p^2}$$

$$= p + p^3 + 2p^5 + p^7 + p^8 + 2p^9 + \ldots$$

$$= \ldots 2110201010.$$

Let's check that this is correct by multiplying it by (the expansion of) 17, remembering that $p = 3$:

$$(2 + 2p + p^2)(p + p^3 + 2p^5 + p^7 + p^8 + 2p^9 + \ldots)$$

$$= 2p + 2p^2 + \underbrace{p^3 + 2p^3} + 2p^4 + p^5 + 4p^5 + 4p^6 +$$

$$+ 2p^7 + 2p^7 + 2p^8 + 2p^8 + p^9 + 2p^9 + 4p^9 \ldots$$

since $p = 3$, we get $p^3 + 2p^3 = 3p^3 = p^4$, so

$$= 2p + 2p^2 + \underbrace{p^4 + 2p^4} + p^5 + 4p^5 + 4p^6 + 2p^7 + 2p^7 +$$
$$+ 2p^8 + 2p^8 + p^9 + 2p^9 + 4p^9 + \ldots$$

$$= 2p + 2p^2 + \underbrace{p^5 + p^5 + 4p^5} + 4p^6 + 2p^7 + 2p^7$$
$$+ 2p^8 + 2p^8 + p^9 + 2p^9 + 4p^9 + \ldots$$

$$= 2p + 2p^2 + \underbrace{2p^6 + 4p^6} + 2p^7 + 2p^7 + 2p^8 + p^9 + 2p^8 + 2p^9 + 4p^9 + \ldots$$

$$= 2p + 2p^2 + \underbrace{2p^7 + 2p^7 + 2p^7} + 2p^8 + 2p^8 + p^9 + 2p^9 + 4p^9 + \ldots$$

$$= 2p + 2p^2 + \underbrace{2p^8 + 2p^8 + 2p^8} + p^9 + 2p^9 + 4p^9 + \ldots$$

$$= \ldots$$

$$= 2p + 2p^2$$

so that the higher powers of p disappear "to the right," leaving us with $2p + 2p^2 = 24$! We could also have done it with the usual multiplication algorithm as above, working in base three, as long as we remember that $3 = 10$, $4 = 11$, etc., and we must carry as necessary. Indeed, there will be an infinite amount of carrying. (The reader will probably feel something has been shoved under the rug, and in fact there is something to prove here. But the point is that, at least formally, it works.)

In our example, the denominator was not divisible by p. If it was, we would factor out a negative power of p to get a fraction without p in the denominator, expand, then multiply back by the power of p we factored out. The effect is just to move the dot. For example, from the computation above we get

$$\frac{8}{17} = 1 + p^2 + 2p^4 + p^6 + p^7 + 2p^8 + \cdots = \ldots 211020101$$

and

$$\frac{8}{51} = p^{-1} + p + 2p^3 + p^5 + p^6 + 2p^7 + \cdots = \ldots 21102010.1.$$

Notice that the expansion got shorter as we divided by 3; to get more digits here would require starting with more digits before. Each time we divide by p we lose some precision.

Provided that we treat the whole process formally, it is easy to check that this always works, and that the resulting series reflects the properties of the rational number $x = a/b$ as regards the prime number p (we will get into the habit of saying "locally at p" or even "near p," to emphasize the analogy). So the upshot is that for each prime p, we can write any (positive, for now) rational number a/b in the form

$$x = \frac{a}{b} = \sum_{n \geq n_0} a_n p^n,$$

and, for example, we have $n_0 \geq 0$ if and only if $p \nmid b$, and $n_0 > 0$ if and only if $p \nmid b$ and $p | a$ (assuming a/b is in lowest terms). In fact, the number n_0 (which is something like the order of a zero or pole) reflects the "multiplicity" of p in a/b; it is characterized by the equation

$$x = p^{n_0} \frac{a_1}{b_1} \qquad \text{with} \quad p \nmid a_1 b_1.$$

It remains to see how to get the negative rational numbers, but since our power series in p can be multiplied, it is enough to get an expansion for -1. Keeping in mind that we are working formally, and with a little imagination, that is not too hard to do. We find, for any p, that

$$-1 = (p - 1) + (p - 1)p + (p - 1)p^2 + (p - 1)p^3 + \cdots,$$

since, if we add 1, we get

$$\underbrace{1 + (p - 1)} + (p - 1)p + (p - 1)p^2 + (p - 1)p^3 + \cdots$$

$$= \underbrace{p + (p - 1)p} + (p - 1)p^2 + (p - 1)p^3 + \cdots$$

$$= \underbrace{p^2 + (p - 1)p^2} + (p - 1)p^3 + \cdots$$

$$= \cdots$$

$$= 0.$$

For example, if $p = 7$ and we use the notation "in base p" this looks like

$$-1 = \ldots 66666.$$

Now just multiply the expression of a positive number by -1 to get the expression of its negative. Or, if the idea of multiplying by an infinitely long -1 seems like a pain (it is!), then we can find the negative directly:

Problem 4 Consider a p-adic number

$$y = a_0 + a_1 p + a_2 p^2 + a_3 p^3 + \cdots .$$

What is $-y$? (This means: what is its p-adic expansion?)

For each rational number, we'll end up with a series that has finitely many negative powers of p. We call this a "finite-tailed Laurent series in p." The term "finite-tailed" refers, of course, to the fact that the expansion is finite *to the left*, i.e. there are finitely many negative powers of p. It is usually infinite to the right. (Alas, if we choose to write digits in base p, then this reverses: finite to the right, infinite to the left.)

The conclusion is that, at least in a formal sense, every rational number x can be written as a "finite-tailed Laurent series in powers of p"

$$x = a_{n_0} p^{n_0} + a_{n_0+1} p^{n_0+1} + \ldots$$

We will call this the *p-adic expansion* of x; remember that if x is a positive integer, it is just its expansion "in base p."

Let's not bother to check, at this point, that this process is injective, i.e., that two different rational numbers will have different p-adic expansions. (This will come out more naturally later on.) The upshot is that we now have a new way to represent rational numbers.

It turns out that the set of *all* finite-tailed Laurent series in powers of p (i.e., of all p-adic expansions) is a field (see Problem 5), just as $\mathbb{C}((X-\alpha))$ is a field. We will denote this field by $\mathbb{Q}_p$, and call it the *field of p-adic numbers*. As before, we can describe much of what we have done by saying that the function

$$x \mapsto p\text{-adic expansion of } x$$

gives an inclusion of fields

$$\mathbb{Q} \hookrightarrow \mathbb{Q}_p.$$

(We have not yet shown that $\mathbb{Q}_p$ is strictly bigger than $\mathbb{Q}$, but we will soon.)

The definition of a p-adic number as a formal object (a finite-tailed Laurent expansion in powers of p) is of course rather unsatisfactory according to the tastes of today. We will remedy this in Chapter 3, where we will show how to construct the field $\mathbb{Q}_p$ as an analogue of the field of real numbers. For now, note only that whatever the "real" definition is, it must allow our series to converge, so that powers p^n must get *smaller* as n grows. This is pretty strange, so let's give ourselves time to get used to the idea. The problems in this section are intended to help the reader feel a little more comfortable with p-adic expansions.

Problem 5 Show that $\mathbb{Q}_p$ is indeed a field. (At this point this is a bit annoying, so you can also take it on trust and wait for the proof in Chapter 3.)

You will have to begin by making explicit what the operations are, and this is a bit tricky because of "carrying." For example, the coefficient of a given power of p in the sum of two expansions depends on the coefficients of *all* the lower powers in the summands; however, this is still a finite rule. Then show that the map $\mathbb{Q} \longrightarrow \mathbb{Q}_p$ given by sending each rational number to its expansion is a homomorphism.

By analogy with the real numbers, it's natural to guess that every rational number will have a periodic (or eventually periodic) p-adic expansion, and that conversely any such expansion represents a rational number. That is harder to prove than one might expect, so let's break it up into several problems.

Problem 6 Suppose y has a p-adic expansion that is eventually periodic. Show that y is a rational number.

Problem 7 Suppose $y \in \mathbb{Z}$. Show that the p-adic expansion of y must end either in an infinite string of 0s or an infinite string of $(p-1)$s.

Problem 8 Suppose $y = \frac{a}{b} \in \mathbb{Q}$, $p \nmid b$, and $-1 < y < 0$. Show that the p-adic expansion of y is purely periodic.

Problem 9 Show that the p-adic expansion of any rational number is eventually periodic. (This is tricky; there's no shame in looking at the solution in Appendix B.)

Notice that it follows that $\mathbb{Q}_p$ is bigger than $\mathbb{Q}$, since not all p-adic expansions are periodic. For example,

$$\sum_{n=0}^{\infty} p^{n^2} = 1 + p + p^4 + p^9 + p^{16} + \dots$$

is not periodic, so it is an element of $\mathbb{Q}_p$ that is not in $\mathbb{Q}$.

Problem 10 (Some abstract algebra required!) Another point at which our analogy seems to break down is the fact that rational functions $f(X) \in \mathbb{C}(X)$ are really *functions*: one can really compute their value at a complex number α. This problem explains a highfalutin' way of interpreting rational numbers as functions too.

  i) First of all, show that we can identify the set of complex numbers α with the set of maximal ideals in $\mathbb{C}[X]$ via the correspondence $\alpha \leftrightarrow (X - \alpha)$.

  ii) Fix a complex number α. Show that the map $f \mapsto f(\alpha)$ defines a homomorphism from the ring $\mathbb{C}[X]$ to $\mathbb{C}$, whose kernel is exactly the ideal $(X - \alpha)$.

  iii) Now let $f(X)$ be a rational function. Show that the map $f \mapsto f(\alpha)$ still makes sense provided the denominator of f is not divisible by $X - \alpha$. If the denominator is divisible by $(X - \alpha)^n$ but not by $(X - \alpha)^{n+1}$, we say that f has a pole of order n at α.

  iv) Now take $x = a/b \in \mathbb{Q}$, and choose a prime $p \in \mathbb{Z}$. If p does not divide b, define the *value of* x *at* p to be $a/b \pmod{p}$, which means $ab' \pmod{p}$, where b' is an integer satisfying $bb' \equiv 1 \pmod{p}$. We think of this value as an element of $\mathbb{F}_p$, the field with p elements. If p does divide b, we say that x has a pole at p. Explain how to define the order of the pole. This interprets the elements of

$\mathbb{Q}$ as a sort of "function" on the primes $p \in \mathbb{Z}$. It is a bit weird, because this "function" doesn't have a "range:" the value at each p belongs to a different field $\mathbb{F}_p$.

v) Discuss whether this way of thinking of rational numbers as functions is reasonable. Does it make the analogy any tighter?

1.2   How to Compute

Computing with p-adic expansions by hand is mostly easy: just do the usual thing, but in base p. This section gives some examples of this, then goes on to explain how to do these computations with a computer.

Let's take $p = 11$ for our examples. Then our digits must be the numbers from zero to ten. We'll use the same convention as in the previous section, but we'll add an extra digit x to stand for ten. So our 11-adic digits are

$$0, 1, 2, \ldots, 9, \mathrm{x}.$$

For example,

$$100 = 1 + 9 \cdot 11 \qquad \text{and} \qquad 230 = 10 + 9 \cdot 11 + 1 \cdot 11^2$$

which we will write as

$$100 = 91 \qquad \text{and} \qquad 230 = 19\mathrm{x}.$$

Notice that we are following the usual convention of writing the digits with the higher powers of p first, which means that our p-adic expansions will go on forever to the left, not to the right.

Now let's add our two numbers:

$$\begin{array}{r} 91 \\ 19\mathrm{x} \\ \hline 280 \end{array}$$

Because $1 + \mathrm{x} = 10$, so we write zero and carry a 1, then $1 + 9 + 9 = 18$, and so on. (Do it!)

If we are dealing with an infinite expansion, we can only do part of the computation, of course, but sometimes we can see where it will go. For example, when we add $\ldots \mathrm{xxxxxxxxxx} + 1$ we can see that we will get 0 even though there are infinitely many carries.

Sticking to $p = 11$, try these:

Problem 11 I claim that $\frac{1}{2} = \ldots 555556$. Multiply that by 2 and check that the answer is 1.

Problem 12 Similarly, check that $\frac{1}{7} = \ldots 7947947947948 = \overline{7948}$, where the bar marks a repeating block.

Problem 13 When you solved Problem 4 you obtained a recipe for finding the expansion of $-y$ from that of y: take the p-complement of the last digit and the $(p-1)$-complement of all the others. Since $123 = 102$, this gives $-123 = \ldots$ xxxxx9x9. Check it twice: first by adding

$$102 + \ldots \text{xxxxx9x9}$$

to get zero, then by multiplying

$$\ldots \text{xxxxxxxxxxxxxxxxx} \times 102.$$

Problem 14 According to Problem 8, the following numbers should have purely periodic 11-adic expansions. Check that they do.

i) $-2/3$ iii) $-4/9$

ii) $-3/5$ iv) $-11/12$

It's easy to come up with many more examples. Instead, let's do a trickier one, returning to the example from the previous section.

Problem 15 Set $p = 3$, and consider the rational number $24/17$. We have $24 = 220$ and $17 = 122$.

i) We said $\frac{24}{17} = \ldots 2110201010$. We can find that by repeated division by 3:

$$\frac{24}{17} = 3 \times \frac{8}{17} + 0$$
$$\frac{8}{17} = 3 \times \frac{-3}{17} + 1$$
$$\frac{-3}{17} = 3 \times \frac{-1}{17} + 0$$
$$\ldots$$

Carry this out until you know that the digits will repeat.

ii) Multiply the result by 122 and check that the answer is 220.

So we can add and multiply, and since we can take negatives we can also subtract. Division is harder, mostly because the school algorithm for long division starts with the highest-order digits, and in our infinite expansions there might not be a top digit at all. Still, there are algorithms to do this. But the truth is that no one does arithmetic by hand any more. So let's learn to do this on a computer.

Sage and GP handle the p-adics in very similar ways (not surprising, since Sage includes GP).

A p-adic expansion is an infinite object and computers are finite, so we have to work with finite chunks of the expansion. Just as we do with real numbers, we need to choose a certain number of digits to keep, which we think of as the "p-adic precision" in which we are working. In GP, this can be done by adding +0(p^k) at the end of a number.[4] So if we write 117+0(5^20), we will get 2 + 3*5 + 4*5^2 + 0(5^20), which is the (finite) base-5 representation of 117. If we write 117/2+0(5^20), we get

```
1 + 4*5 + 4*5^2 + 2*5^3 + 2*5^4 + 2*5^5 + 2*5^6 + 2*5^7
+ 2*5^8 + 2*5^9 + 2*5^10 + 2*5^11 + 2*5^12 + 2*5^13 + 2*5^14
+ 2*5^15 + 2*5^16 + 2*5^17 + 2*5^18 + 2*5^19 + 0(5^20)
```

(except that we have added line breaks). It's easy to see that this is $\overline{2}441$; of course, if the period is long and we haven't chosen high enough precision, it will be hard to know when the repetition starts.

If we want to see the number written "in base p," we can make GP do it like this:

```
gp > a=1/42+0(5^20)
%1 = 3 + 2*5^3 + 4*5^4 + 4*5^5 + 2*5^6 + 2*5^9 + 4*5^10
      + 4*5^11 + 2*5^12 + 2*5^15 + 4*5^16 + 4*5^17
      + 2*5^18 + 0(5^20)
gp > digits(lift(a),a.p)
%2 = [2, 4, 4, 2, 0, 0, 2, 4, 4, 2, 0, 0, 2, 4, 4, 2, 0, 0, 3]
gp > concat("...",concat([Str(x)|x<-digits(lift(a),a.p)]))
%3 = "...2442002442002442003"
```

In other words, digits gives the sequence of digits as a vector, and the magical invocation involving concat writes them as a string.

In Sage, the right way to do it is to first specify in what context you are working: K=Qp(5) tells Sage that K is the p-adic numbers[5] with $p = 5$. Then you can find p-adic expansions by asking Sage to create a number in K, like this: a=K(1/42). So the commands

```
K=Qp(5)
a=K(1/42)
print(a)
```

produce the output

```
3 + 2*5^3 + 4*5^4 + 4*5^5 + 2*5^6 + 2*5^9 + 4*5^10 + 4*5^11
    + 2*5^12 + 2*5^15 + 4*5^16 + 4*5^17 + 2*5^18 + 0(5^20)
```

If you prefer to see the digits, you can vary the options when you create the field K:

---

[4]Computers are literal-minded. If you write +0(125), GP will think you mean $p = 125$ and $k = 1$. It will happily proceed until something blows up because "p" is not actually prime.

[5]The default precision is 5^{20}. There are lots of options one can add to the command Qp, for example to specify the precision and the way p-adic numbers are displayed; check the Sage manual if you want details.

```
K=Qp(5,print_mode="digits")
a=K(1/42)
print(a)
```

produces

...02442002442002442003

You can even create both K and another entity, say Kd, with the "digits" option. Then Kd(a) gives you the digits version of a.

Once we can enter numbers, it's smooth sailing: you can add, subtract, multiply, divide, and more. Whatever operation you ask it to do, GP will either perform the operation or tell you that it can't.

Here are some examples in GP. The same commands work in Sage, but remember that in Sage you may need to use **print** to get output.

```
gp > a=6+O(5^20)
%1 = 1 + 5 + O(5^20)

gp > b=17+O(5^20)
%2 = 2 + 3*5 + O(5^20)

gp > c=a/b
%3 = 3 + 3*5 + 4*5^2 + 2*5^3 + 4*5^5 + 3*5^6 + 5^7 + 2*5^8
    + 5^9 + 2*5^11 + 4*5^12 + 5^14 + 3*5^15 + 2*5^16 + 3*5^17
    + 4*5^18 + 2*5^19 + O(5^20)

gp > d=b/a
%4 = 2 + 5 + 4*5^2 + 4*5^4 + 4*5^6 + 4*5^8 + 4*5^10 + 4*5^12
    + 4*5^14 + 4*5^16 + 4*5^18 + O(5^20)

gp > c*d
%5 = 1 + O(5^20)
```

How about putting 5 in the denominator?

```
gp > a/5
%6 = 5^-1 + 1 + O(5^19)

gp > a/5*b
%7 = 2*5^-1 + 4*5 + O(5^19)
```

Did you see what happened there? Here's the right way:

```
gp > a/(5*b)
%8 = 3*5^-1 + 3 + 4*5 + 2*5^2 + 4*5^4 + 3*5^5 + 5^6 + 2*5^7
    + 5^8 + 2*5^10 + 4*5^11 + 5^13 + 3*5^14 + 2*5^15
    + 3*5^16 + 4*5^17 + 2*5^18 + O(5^19)
```

Notice that dividing by 5 reduces the 5-adic precision. This shouldn't be surprising.

Now let's try something exotic:

```
gp > sqrt(a)
%9 = 1 + 3*5 + 4*5^3 + 2*5^4 + 5^5 + 2*5^6 + 3*5^7 + 5^8
    + 3*5^9 + 3*5^10 + 3*5^12 + 3*5^13 + 2*5^14 + 3*5^15
    + 2*5^16 + 2*5^17 + 2*5^18 + 4*5^19 + O(5^20)

gp > sqrt(2*a)
  ***    at top-level: sqrt(2*a)
  ***                   ^---------
  *** sqrt: not an n-th power residue
            in Qp_sqrt: 2 + 2*5 + O(5^20).

gp > log(a)
%10 = 5 + 2*5^2 + 4*5^3 + 2*5^4 + 5^6 + 4*5^7 + 2*5^8 + 3*5^9
    + 5^10 + 2*5^11 + 2*5^12 + 3*5^14 + 3*5^15 + 4*5^16
    + 4*5^17 + 5^18 + 2*5^19 + O(5^20)
```

So apparently even square roots and logs can (sometimes) work. We needn't worry right now about what the error message when we tried to compute $\sqrt{2a}$ means. We'll find out soon.

1.3 Solving Congruences Modulo p^n

The "p-adic numbers" we have just constructed are closely related to the problem of solving congruences modulo powers of p. We will look at some examples of this.

Let's start with the easiest possible case, an equation which has solutions in $\mathbb{Q}$, such as

$$X^2 = 25.$$

We want to consider it modulo p^n for every n, i.e., to solve the congruences

$$X^2 \equiv 25 \pmod{p^n}.$$

Now, of course, our equation has solutions already in the integers: $X = \pm 5$. This automatically gives solutions of the congruence for every n; just put $X \equiv \pm 5 \pmod{p^n}$ for every n.

Problem 16 Check that these are the only solutions, up to congruence, of $X^2 \equiv 25$ (mod p^n), at least when $p \neq 2, 5$. What happens in these special cases?

Let's try to understand these solutions a little better from the p-adic point of view. To make our life easier, we take $p = 3$ once again. We begin by rewriting our solutions using residue class representatives between 0 and $3^n - 1$

for the solutions modulo 3^n. The first solution, $X = 5$, gives:

$$X \equiv 2 \pmod{3}$$
$$X \equiv 5 = 2 + 3 \pmod{9}$$
$$X \equiv 5 = 2 + 3 \pmod{27}$$

etc.

which doesn't change any more, and therefore just gives the 3-adic expansion of this solution:

$$X = 5 = 2 + 1 \times 3 = 12.$$

For $X = -5$, the results are a little more interesting; let's give the representatives mod 3^n as integers and also in base 3:

$$X \equiv -5 \equiv 1 = 1 \pmod{3}$$
$$X \equiv -5 \equiv 4 = 1 + 3 = 11 \pmod{9}$$
$$X \equiv -5 \equiv 22 = 1 + 3 + 2 \times 3^2 = 211 \pmod{27}$$
$$X \equiv -5 \equiv 76 = 1 + 3 + 2 \times 3^2 + 2 \times 3^3 = 2211 \pmod{81}$$

etc.

Again, continuing this gives the 3-adic expansion of the solution, which is a bit more interesting because it is infinite:

$$X = -5 = 1 + 1 \times 3 + 2 \times 3^2 + 2 \times 3^3 + 2 \times 3^4 + \cdots = \ldots 22211 = \overline{2}11.$$

But of course we already knew that this is the 3-adic expansion of -5.

Notice that the two systems of solutions are "coherent," in the sense that when we look at, say, $X = 76$ (which is a solution modulo 3^4) and reduce it modulo 3^3, we get $X = 22$ (which is the corresponding solution modulo 3^3). Let's give this a formal definition:

Definition 1.3.1 *Let p be a prime. We say a sequence of integers α_n such that $0 \leq \alpha_n \leq p^n - 1$ is coherent if, for every $n \geq 1$, we have*

$$\alpha_{n+1} \equiv \alpha_n \pmod{p^n}.$$

If we need to emphasize the choice of prime p, we will say the sequence is p-adically coherent.

We can picture our two coherent sequences of solutions as branches in a tree (see Figure 1.1). Of course this is all rather painfully obvious in the case we are considering, since the sequences of solutions are coherent simply because they "are" solutions in $\mathbb{Z}$ (76 is congruent to 22 just because both are congruent to -5). The only real bit of information we have obtained is the connection between expressing the roots as a coherent sequence and obtaining their p-adic expansions.

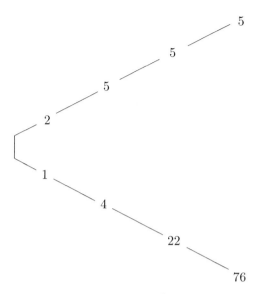

Figure 1.1: Solutions of $X^2 \equiv 25 \pmod{3^n}$

Problem 17 Interpret the definition of a coherent sequence in terms of expansions in base p.

Problem 18 Before we go on to something more interesting, do a couple of similar examples (in the sense that there are integer solutions) on your own, say with $X^2 = 49$ and $p = 5$, and $X^3 = 27$ and $p = 2$.

Problem 19 Things already get slightly more interesting if we take $p = 2$ and the equation $X^2 = 81$. In this case, the "tree" of solutions modulo 2^n is much more complex: there are two infinite branches that correspond to the solutions $X = \pm 9$, but there are also lots of finite branches (solutions modulo 2^n that do not "lift" to solutions modulo 2^{n+1}). We will later consider what is special about this situation.

Things become much more interesting if we follow the same process with an equation that does *not* have rational roots. For example, take the system of congruences

$$X^2 \equiv 2 \pmod{7^n}, \qquad n = 1, 2, 3, \dots$$

For $n = 1$, the solutions are $X \equiv 3 \pmod 7$ and $X \equiv 4 \equiv -3 \pmod 7$. To find the solutions for $n = 2$, note that their reductions modulo 7 must be solutions for $n = 1$. Hence we set $X = 3 + 7k$ or $X = 4 + 7k$ and solve for k:

$$(3 + 7k)^2 \equiv 2 \pmod{49}$$
$$9 + 42k \equiv 2 \pmod{49}$$

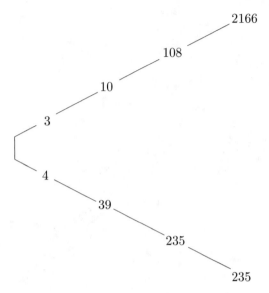

Figure 1.2: Solutions of $X^2 \equiv 2 \,(\mathrm{mod}\ 7^n)$

(notice that the term involving $(7k)^2$ is congruent to zero)

$$7 + 42k \equiv 0 \quad (\mathrm{mod}\ 49)$$
$$1 + 6k \equiv 0 \quad (\mathrm{mod}\ 7)$$
$$k \equiv 1 \quad (\mathrm{mod}\ 7)$$

which, since $X = 3 + 7k$, gives the solution $X \equiv 10 \,(\mathrm{mod}\ 49)$. Using $X = 4 + 7k$ gives the other solution $X \equiv 39 \equiv -10 \,(\mathrm{mod}\ 49)$.

Problem 20 Prove that for each n there can be at most two solutions. (All you need is $p \neq 2$.)

Problem 21 Show that the process above can be continued indefinitely, that is, that given a solution α_n of the congruence $X^2 \equiv 2 \,(\mathrm{mod}\ 7^n)$, there always exists a unique solution α_{n+1} of $X^2 \equiv 2 \,(\mathrm{mod}\ 7^{n+1})$ satisfying $\alpha_{n+1} \equiv \alpha_n \,(\mathrm{mod}\ 7^n)$. Find a few more terms in each of the sequences of solutions above.

Again, the solutions can be represented as branches in a tree (see Figure 1.3). This time, however, we can't predict *a priori* what the numbers that appear will be; instead, all we can do is convince ourselves that the process will continue as long as we want it to. The fact that one can continue finding roots indefinitely shows that there are two coherent sequences of solutions:

$$x_1 = (3,\ 10,\ 108,\ 2166,\ \ldots)$$

and

$$x_2 = (4, 39, 235, 235 \ldots) = (-3, -10, -108, -2166 \ldots) = -x_1.$$

Just as before, we can expand each number in each sequence 7-adically. The fact that the sequence is coherent means that the expansion of each root is the truncation of the expansion of the following root, so that, for example,

$$3 = 3 = 3$$
$$10 = 3 + 1 \times 7 = 13$$
$$108 = 3 + 1 \times 7 + 2 \times 7^2 = 213.$$

This gives us two 7-adic numbers:

$$x_1 = 3 + 1 \times 7 + 2 \times 7^2 + 6 \times 7^3 + \cdots = \ldots 6213$$

and

$$x_2 = 4 + 5 \times 7 + 4 \times 7^2 + 0 \times 7^3 + \cdots = \ldots 0454 = -x_1.$$

It probably bears repeating: we are not claiming that we can predict the pattern here; indeed we already know that it cannot be periodic. All we know is that we can *continue* the pattern for as long as necessary, if we have enough time and patience. It's just like finding the decimal expansion of the square root of two: we can get as close as we like, and we can *prove* that, though we can't predict what the expansion will actually be like.

In any case, we do get two 7-adic numbers, and they are indeed roots of the equation $X^2 = 2$ in $\mathbb{Q}_7$, in the usual sense:

Problem 22 Show that the 7-adic number x_1 obtained as above satisfies $x_1^2 = 2$ in $\mathbb{Q}_7$. This shows once again that the field $\mathbb{Q}_7$ is strictly bigger than $\mathbb{Q}$.

The tie between solving sequences of congruences modulo higher and higher powers of p and solving the corresponding equation in $\mathbb{Q}_p$ is quite close, as the problems below try to emphasize. We will return to it when we study Hensel's Lemma. It is also one of the more important reasons for using p-adic methods in number theory.

Problem 23 Check that $X^2 = 2$ has no solutions in the field $\mathbb{Q}_5$. (Begin by expressing the putative solution as a 5-adic expansion. Show that it must be of the form $a_0 + a_1 5 + a_2 5^2 + \ldots$, and conclude that a_0 must satisfy a congruence modulo 5. Finally, check that the congruence you obtained has no solutions modulo 5.) Notice that this shows (in a very roundabout way) that 2 has no square root in $\mathbb{Q}$, since any square root in $\mathbb{Q}$ would be a square root in any of the $\mathbb{Q}_p$ (remember that there is an inclusion $\mathbb{Q} \hookrightarrow \mathbb{Q}_p$), hence in particular in $\mathbb{Q}_5$.

Problem 24 Check that $X^2 + 1 = 0$ has a solution in $\mathbb{Q}_5$, but not in $\mathbb{Q}_7$.

Problem 25 Show that a p-adic number

$$x = a_0 + a_1 p + a_2 p^2 + a_3 p^3 + \cdots$$

is a solution in $\mathbb{Q}_p$ of the equation $X^2 = m$ if and only if the sequence

$$a_0, a_0 + a_1 p, a_0 + a_1 p + a_2 p^2, \ldots$$

is a coherent sequence of solutions of the congruences $X^2 \equiv m \,(\mathrm{mod}\ p^n)$. (Hint: compute x^2 up to a certain power of p, and compare it with m to read off a congruence modulo that power of p.)

We have already mentioned that there is some analogy between p-adic numbers and real numbers. The next problem gives an example of this. Over $\mathbb{R}$, there is a simple condition that determines whether the equation $X^2 = m$ has a solution (just check the sign of m). In $\mathbb{Q}_p$, the condition is also simple:

Problem 26 Let m be any integer, and suppose that the congruence $X^2 \equiv m$ (mod p) has a solution; show that if $p \neq 2$ and $p \nmid m$ it is always possible to "extend" this solution to a full coherent sequence of solutions of $X^2 \equiv m \,(\mathrm{mod}\ p^n)$. Use this to find a necessary and sufficient condition for the equation $X^2 = m$ to have a root in $\mathbb{Q}_p$ for $p \neq 2$. What is special about $p = 2$?

Notice that this explains what happened when we tried to compute $\sqrt{2a}$ with GP above: our a is a square mod 5, but $2a$ is not.

Problem 27 Show that for every p, there is a polynomial equation that has solutions in $\mathbb{Q}_p$ but not in $\mathbb{Q}$. (Hint: the basic work has all been done; when $p \neq 2$ an equation of the form $X^2 - m$ will work if you choose a good m. For $p = 2$ you need a different kind of equation.)

Problem 28 In the same spirit as the previous problem, show that $\mathbb{Q}_p$ is never algebraically closed; more precisely, for each p one can find an algebraic equation with rational coefficients that has no roots in $\mathbb{Q}_p$.

1.4   Other Examples

Working with p-adic numbers is useful in all sorts of contexts. We round off this chapter by giving two rather whimsical examples.

Consider the equation $X = 1 + 3X$. This is of course easy to solve, but let's try something strange and look at it as a fixed-point problem, i.e., as the problem of finding a solution of $f(x) = x$ for some function $f(x)$. Such problems are often solved by iteration, plugging in an arbitrary initial value, then computing $f(x)$ over and over in the hope that we will get closer and closer to a fixed point. To try this in our case, we take $x_0 = 1$ and iterate, so that $x_{n+1} = 1 + 3x_n$. Here's what we get:

$$x_0 = 1$$
$$x_1 = 1 + 3x_0 = 1 + 3$$
$$x_2 = 1 + 3x_1 = 1 + 3 + 3^2$$
$$\cdots$$
$$x_n = 1 + 3 + 3^2 + \cdots + 3^n$$

In $\mathbb{R}$, this is a divergent sequence, and we were all taught in calculus classes never to have any dealings with them. On the other hand, it is the sequence of partial sums of a geometric series, and we all know that

$$1 + a + a^2 + a^3 + \cdots = \frac{1}{1-a}.$$

(Well, we know it for $|a| < 1$, but what the heck...) Plugging in blindly gives $x = 1/(1-3) = -1/2$, which is (surprise!) the correct answer.

This dubious playing around with divergent sequences is clearly illegal in calculus class, but it works. Here's one way to understand why. While the sequence is certainly divergent in $\mathbb{R}$, there is nothing to keep us from looking at the sequence in $\mathbb{Q}_3$ (the elements in the sequence are in $\mathbb{Q}$, which is contained in both $\mathbb{R}$ and $\mathbb{Q}_3$). Now, in $\mathbb{Q}_3$, the sequence is obviously convergent, to the 3-adic number

$$1 + 3 + 3^2 + \cdots + 3^n + \cdots = \ldots 33331.$$

One then easily checks (by the same argument used over $\mathbb{R}$!) that this is equal to $-1/2$.

Of course it is silly to solve a linear equation in such a roundabout way, but the remarkable fact here is that an argument that was either dubious or outright illegal at first sight turns out to work perfectly well in the p-adic context. The series we used is divergent only if we insist in thinking of it as a series of real numbers. Once we put it in the "right" context, it becomes quite nice. In fact, we will see in the next chapter that there is an absolute value in $\mathbb{Q}_3$, and that with respect to the notion of size determined by that absolute value our series is convergent.

The point, then, is that introducing the p-adic fields broadens our world in such a way as to allow arguments that were previously impossible. This toy example points the way to many analogous situations where considering the p-adic numbers simplifies matters tremendously.

Problem 29 Show that, for any prime p, the formula

$$1 + p + p^2 + p^3 + \cdots = \frac{1}{1-p}$$

is true in $\mathbb{Q}_p$.

The next example is perhaps even more interesting. It shows that sometimes introducing p-adic ideas allows a more conceptual proof of a fact that seems obscure (and hard to prove) otherwise. This example is a bit more advanced, and we will take for granted things that we will prove only later, but the reader should be able to follow it. We will work with $p = 2$, that is, in the field $\mathbb{Q}_2$ of 2-adic numbers.

Consider the usual MacLaurin series for the logarithm of $1 + X$:

$$\log(1 + X) = X - \frac{X^2}{2} + \frac{X^3}{3} - \frac{X^4}{4} + \cdots$$

Since powers of 2 are "small" in $\mathbb{Q}_2$, it turns out that we can plug in $X = -2$ to compute the logarithm of -1:

$$\log(-1) = \log(1 - 2) = -\left(2 + \frac{2^2}{2} + \frac{2^3}{3} + \frac{2^4}{4} + \cdots\right).$$

(This is of course wildly divergent in $\mathbb{R}$, but it turns out to be convergent in $\mathbb{Q}_2$; this is not completely obvious because of the denominators, but it does work—see ahead.) Now, if the series converges, it must converge to zero, by the usual properties of the logarithm:

$$2\log(-1) = \log((-1)^2) = \log(1) = 0.$$

So the series

$$2 + \frac{2^2}{2} + \frac{2^3}{3} + \frac{2^4}{4} + \cdots + \frac{2^n}{n}$$

must be equal to 0. Remember that what this means is that the terms in the 2-adic expansion "disappear to the right," that is, that the partial sums, written in base 2, end with longer and longer stretches of zeros. Here's the upshot:

Fact 1.4.1 *For each integer $M > 0$ there exists an n such that the partial sum*

$$2 + \frac{2^2}{2} + \frac{2^3}{3} + \frac{2^4}{4} + \cdots + \frac{2^n}{n}$$

is divisible by 2^M.

Problem 30 Can you give a direct proof of this fact?

What this example points out is that using p-adic methods, and in particular the methods of the calculus in the p-adic context, we can often prove facts about divisibility by powers of p which are otherwise quite hard to understand. The proofs are often, as in this case, "cleaner" than any direct proof would be, and therefore easier to understand. We will see more examples of this before we are done.

2 Foundations

The goal of this chapter is to begin to lay a solid foundation for the theory we described informally in Chapter 1. The main idea will be to introduce a different absolute value function on the field of rational numbers. This will give us a different way to measure distances, hence a different calculus, one in which the formal series of the first chapter actually converge. Once we have that, we will use it (in Chapter 3) to construct the p-adic numbers.

To get the p-adic numbers, we need to start with the field $\mathbb{Q}$ of rational numbers. However, rather than deal exclusively with $\mathbb{Q}$, we will devote this chapter to studying absolute values on fields in general. Of course, the main example we will have in mind will be $\mathbb{Q}$, but the general theory is easy enough that it would be a waste to specialize to rational numbers too soon. (Later, when the generality would cost us some effort, we will speedily go back to the special case of the rationals.)

So, for this chapter, $\mathbb{k}$ will be an arbitrary field, and we will be interested in constructing an abstract theory of absolute values on $\mathbb{k}$. We will do this by starting from the basic properties of the absolute values we already know and love, and then looking for other functions with similar properties.

One thing to notice from the start is that we will want to think of our new absolute values as giving alternative ways to measure the "size" of things. This can feel rather strange at first, so it's wise to keep many concrete examples in mind as we go.

2.1 Absolute Values on a Field

Let $\mathbb{k}$ be a field and let $\mathbb{R}_+ = \{x \in \mathbb{R} : x \geq 0\}$ be the set of all non-negative real numbers. We begin by defining an absolute value on $\mathbb{k}$ and exploring the possibilities implicit in the definition. The definition just tries to capture what seem to be the most important properties of the everyday absolute value.

Definition 2.1.1 *An* absolute value *on $\mathbb{k}$ is a function*

$$| \ | : \mathbb{k} \longrightarrow \mathbb{R}_+$$

that satisfies the following conditions:

   i) $|x| = 0$ if and only if $x = 0$;

© Springer Nature Switzerland AG 2020

F. Q. Gouvêa, *p-adic Numbers*, Universitext, https://doi.org/10.1007/978-3-030-47295-5_2

ii) $|xy| = |x|\,|y|$ *for all* $x, y \in \Bbbk$;

iii) $|x + y| \leq |x| + |y|$ *for all* $x, y \in \Bbbk$.

We will say an absolute value on $\Bbbk$ *is* non-archimedean *if it satisfies the additional condition:*

iv) $|x + y| \leq \max\{|x|, |y|\}$ *for all* $x, y \in \Bbbk$;

otherwise, we will say that the absolute value is archimedean.

Note that condition *(iv)* implies condition *(iii)*, since $\max\{|x|, |y|\}$ is certainly nor larger than the sum $|x| + |y|$. We will later discuss in more detail why non-archimedean absolute values are important, and where their name comes from; for now, let's just mention that they are quite common.

The most obvious example of an absolute value is, of course, our model: take $\Bbbk = \mathbb{Q}$, and take the usual absolute value $|\ |$ defined by

$$
|x| = \begin{cases} x & \text{if } x \geq 0 \\ -x & \text{if } x < 0 \end{cases}
$$

A more sophisticated way of describing this absolute value is to say that it is actually the absolute value on the field $\mathbb{R}$ of real numbers, applied to $\mathbb{Q}$ via the inclusion $\mathbb{Q} \hookrightarrow \mathbb{R}$. It is easy to see that this absolute value is *archimedean*. (Take $x = y = 1$ to see that condition *(iv)* does not hold.) For reasons that we will discuss later, this absolute value is usually called the *infinite absolute value* on $\mathbb{Q}$, or the *absolute value at infinity*, and is written as $|\ |_\infty$.

At the other extreme is the most boring example: the one we get by setting $|x| = 1$ if $x \neq 0$ and $|0| = 0$. This works for any field $\Bbbk$, and defines a non-archimedean absolute value. It is known, for obvious reasons, as the *trivial absolute value*. It will often have to be excluded in the theorems to follow.

There are many simple properties that one can deduce quickly from the conditions above. We will try to develop them systematically in the next section. For now, let's try to be as concrete as we can. First of all, it's worth pointing out that for the special case of *finite* fields, the whole theory is trivial:

Problem 31 Let $\Bbbk$ be a finite field. Show that the only absolute value on $\Bbbk$ is the trivial absolute value.

We now go on to introduce the example that we will focus on for most of this book. Take $\Bbbk = \mathbb{Q}$, and choose any prime $p \in \mathbb{Z}$. Any integer $n \in \mathbb{Z}$ can be written as $n = p^v n'$, with $p \nmid n'$, and this representation is unique. Since v is determined by p and n, it makes sense to define a function v_p by setting $v_p(n) = v$, so that $v_p(n)$ is just the multiplicity of p as a divisor of n. Formally:

Definition 2.1.2 *Fix a prime number $p \in \mathbb{Z}$. The p-adic valuation on $\mathbb{Z}$ is the function*

$$v_p : \mathbb{Z} - \{0\} \longrightarrow \mathbb{R}$$

defined as follows: for each integer $n \in \mathbb{Z}$, $n \neq 0$, let $v_p(n)$ be the unique positive integer satisfying

$$n = p^{v_p(n)}\, n' \quad \text{with} \quad p \nmid n'.$$

We extend v_p to the field of rational numbers as follows: if $x = a/b \in \mathbb{Q}^\times$, then

$$v_p(x) = v_p(a) - v_p(b).$$

It is often convenient to set $v_p(0) = +\infty$, with the usual conventions on how to handle this symbol. The reasoning here is that we can certainly divide 0 by p, and the answer is 0, which we can divide by p, and the answer is 0, which we can divide by p...

Problem 32 Check that for any $x \in \mathbb{Q}$, the value of $v_p(x)$ does not depend on its representation as a quotient of two integers. In other words, if $a/b = c/d$, then $v_p(a) - v_p(b) = v_p(c) - v_p(d)$.

It is in fact easy to see that the p-adic valuation of any $x \in \mathbb{Q}^\times$ is determined by the formula

$$x = p^{v_p(x)} \cdot \frac{a}{b} \qquad p \nmid ab.$$

Problem 33 Compute a few examples, to get a feel for the thing. For example, determine $v_5(400)$, $v_7(902)$, $v_2(621)$, $v_3(123/48)$, $v_5(180/3)$.
  To check your work, go to Sage or GP and enter valuation(n,p).

The basic properties of the p-adic valuation v_p are the following:

Lemma 2.1.3 *For all x and $y \in \mathbb{Q}$, we have*

i) $v_p(xy) = v_p(x) + v_p(y)$ *and*

ii) $v_p(x + y) \geq \min\{v_p(x), v_p(y)\}$,

with the obvious conventions with respect to $v_p(0) = +\infty$.

Problem 34 Prove Lemma 2.1.3. (Hint: the first property is easy to see by writing out factorizations of x and y; the second comes from the fact that common powers of p can be factored out from a sum.)

Now here comes the really tricky thing: if we compare the two properties in this lemma with conditions (ii) and (iv) in the definition of absolute values, we see that they are very similar, except that the product in the first has been turned into a sum (as when taking a logarithm) and that the inequality in the second has been reversed. We can "unreverse" the inequality by changing the sign, and then turn the sum into a product by putting it into an exponent. This suggests the following, which is the crucial definition:

Definition 2.1.4 *For any nonzero $x \in \mathbb{Q}$, we define the p-adic absolute value of x by*

$$|x|_p = p^{-v_p(x)}.$$

We extend this to all of $\mathbb{Q}$ by defining $|0|_p = 0$.

Notice that the definition of $|0|_p$ matches our convention that $v_p(0) = +\infty$ if we interpret $p^{-\infty}$ in the only reasonable way. To see that our definition really does give an absolute value, we need to check that our requirements have been satisfied.

Proposition 2.1.5 *The function $|\;|_p$ is a non-archimedean absolute value on $\mathbb{Q}$.*

PROOF: Everything follows at once from Lemma 2.1.3.   □

To get a general impression about what the p-adic absolute value is doing, notice that when a number n is very divisible by our prime p the valuation $v_p(n)$ will be large, and then the absolute value $|n|_p$ will be small. (Look at that minus sign in the exponent!) So the p-adic absolute value gives, in a strange sort of way, a measure of how divisible by p a number is.

Problem 35 More practice: take $\Bbbk = \mathbb{Q}$, $p = 7$, and let $|\;| = |\;|_7$ be the 7-adic absolute value. Compute $|35|$, $|56/12|$, $|177553|$, $|3/686|$.

The connection between a non-archimedean absolute value and a function such as in Lemma 2.1.3 (called a *valuation*, or sometimes an *additive valuation*) is quite general. In fact, one can develop the theory taking either object (valuation or absolute value) as the primitive one. In this book, we will stick to absolute values, because they are closer to our intuition, but it is often convenient to go the other way.

Problem 36 (Some abstract algebra required) Let A be an integral domain, and let K be its field of fractions. Let $v : A - \{0\} \longrightarrow \mathbb{R}$ be a function satisfying the conditions of Lemma 2.1.3, i.e., a valuation on A. Extend v to K by setting $v(a/b) = v(a) - v(b)$. Show that the function $|\;|_v : K \longrightarrow \mathbb{R}_+$ defined by

$$|x|_v = e^{-v(x)} \qquad \text{for } x \neq 0$$

and $|0| = 0$ is a non-archimedean absolute value on K. Conversely, show that if $|\;|$ is a non-archimedean absolute value, then $-\log|\;|$ is a valuation.

Problem 37 Let $v : \Bbbk^\times \longrightarrow \mathbb{R}$ be a valuation. Show that the image of v is an additive subgroup of $\mathbb{R}$. This is sometimes called the *value group* of the valuation v. What is the value group of the p-adic valuation?

Though the p-adic absolute value is certainly the most interesting one from the point of view of this book, it's worth pointing out that there are other interesting absolute values on other fields. Before we go on to look at them, however, here are two problems to force another look at the p-adic absolute value.

Problem 38 Show that $|p^n|_p \to 0$ when $n \to \infty$, so that high powers of p are *small* with respect to the p-adic absolute value.

Problem 39 Show that for any $c \in \mathbb{R}$, $c > 1$, the equation $|x| = c^{-v_p(x)}$ defines a non-archimedean absolute value on $\mathbb{Q}$. Make a conjecture about the relation between this absolute value and the p-adic absolute value $|\ |_p$. Make a conjecture about why we chose $c = p$ for the p-adic absolute value.

Our final example is intended to show that the theory we are developing is indeed quite general, and in fact can be applied, almost without change, in all sorts of contexts. The example we want to consider also serves to confirm Hensel's intuition on the similarity between $\mathbb{Q}$ and fields of rational functions. So let F be any field (for example, a finite field, or $\mathbb{C}$), let $\mathsf{F}[t]$ be the ring of polynomials with coefficients in F, and let $\mathsf{F}(t)$ be the field of rational functions over F, which is the field of fractions of the form $f(t)/g(t)$ where $f(t)$ and $g(t)$ belong to $\mathsf{F}[t]$ (and $g(t) \neq 0$, of course). We will define several valuations (and therefore several absolute values) on $\mathsf{F}(t)$. The first is very specific to this situation, since it depends on the notion of degree of a polynomial; by contrast, the others are closely analogous to the p-adic absolute value.

First, for any polynomial $f(t) \in \mathsf{F}[t]$, we set $v_\infty(f) = -\deg(f(t))$, and extend this to rational functions as before, by setting $v_\infty(0) = +\infty$ and

$$v_\infty\left(\frac{f(t)}{g(t)}\right) = v_\infty(f(t)) - v_\infty(g(t)) = \deg(g(t)) - \deg(f(t)).$$

It is easy to check that this is a valuation:

Problem 40 Check that for any $f(t)$, $g(t) \in \mathsf{F}(t)$ we have $v_\infty(f(t)g(t)) = v_\infty(f(t)) + v_\infty(g(t))$ and also $v_\infty(f(t) + g(t)) \geq \min\{v_\infty(f(t)), v_\infty(g(t))\}$. (Is it enough to check for polynomials? Why?)

This gives us a non-archimedean absolute value just as before:

$$|f(t)|_\infty = e^{-v_\infty(f)} = e^{\deg(f)}$$

for any $f(t) \in \mathsf{F}(t)$. (As we hinted in Problem 39, any real number greater than one will do for the base of the exponential; if F is a finite field, a nicer choice might be the number of elements in F.)

Problem 41 When is a rational function "small" with respect to $|\ |_\infty$? Is a polynomial ever small?

One can get other valuations on $F(t)$ by imitating the definition of the p-adic valuation, since $F[t]$ is a unique factorization domain. Just choose an irreducible polynomial $p(t)$ and proceed as before: define a valuation by counting the multiplicity of $p(t)$ as a factor.

Problem 42 Do it! For an irreducible polynomial $p(t) \in F[t]$, define the $p(t)$-adic valuation and absolute value on $F(t)$.

Problem 43 Since F is a subfield of $F(t)$, any absolute value on $F(t)$ also gives an absolute value on F. For the examples we have just constructed, describe the absolute value on F obtained in this way.

Problem 44 Suppose $F = \mathbb{C}$. What are the irreducible polynomials in this case? Are we getting close to realizing Hensel's analogy?

Problem 45 All of the absolute values we have constructed on $F(t)$ are non-archimedean. Try to construct an archimedean absolute value on some $F(t)$. (First of all, this may or may not be possible, depending on F. If you're very sneaky, it can be done for $F = \mathbb{Q}$. Can it be done in such a way that the induced absolute value on F is the trivial one?)

Problem 46 The field $F(t)$ contains the subring of polynomials $F[t]$, but it also contains the subring $F[1/t]$ of "polynomials in $1/t$." In fact, every element of $F(t)$ can be written as a quotient of elements in $F[1/t]$, so this subring serves just as well as $F[t]$ as a starting point. Very well, in $F[1/t]$ the "polynomial" $1/t$ is clearly irreducible, so we can construct, as in Problem 42, a $1/t$-adic valuation v_1. Check that v_1 is the same as the v_∞ constructed above. This means that all of the valuations we have constructed on $F(t)$ are of the "$p(t)$-adic" type.

Problem 47 Let $\mathbb{k} = \mathbb{Q}(i)$ be the field obtained by adjoining $i = \sqrt{-1}$ to the rational numbers, so that any element of $\mathbb{k}$ can be written as $a + bi$ with $a, b \in \mathbb{Q}$. The "integers" in $\mathbb{k}$ are the elements of $\mathbb{Z}[i] = \{a + bi : a, b \in \mathbb{Z}\}$. It is not too hard to check that this is a unique factorization domain, so that its properties are much like those of the usual integers.[1] The primes of $\mathbb{Z}[i]$ are of three kinds:

  i) $1 + i$ is prime,

  ii) if $p \in \mathbb{Z}$ is a prime number and $p \equiv 3 \pmod 4$, then p is a prime in $\mathbb{Z}[i]$,

  iii) for each prime $p \in \mathbb{Z}$ which is congruent to 1 modulo 4, there are two primes $x + yi$ and $x - yi$ in $\mathbb{Z}[i]$ satisfying $(x + yi)(x - yi) = x^2 + y^2 = p$.

In each case, we can use the prime $\pi \in \mathbb{Z}[i]$ to construct a π-adic valuation v_π and (from it) a π-adic absolute value $|\ |_\pi$ on $\mathbb{k}$ as before:

$$|\alpha|_\pi = c^{-v_\pi(\alpha)}$$

(can you come up with a "good" choice for the constant c?). Check that this works, and explore the resulting situation. For example, since $\mathbb{Q}$ is contained in $\mathbb{k}$, this induces an absolute value on $\mathbb{Q}$; describe the induced absolute value. In particular, for a fixed π, can you compute $v_\pi(p)$ as p ranges through the primes in $\mathbb{Z}$?

---

[1] See most introductory texts in algebra or number theory, or just take it for granted.

There is an extensive theory of how valuations extend (or not) from sub-fields to larger fields, and this theory turns out to be closely connected to algebraic number theory. (In fact, many texts on algebraic numbers develop the theory in terms of valuations and absolute values rather than in terms of ring theory; the best example is probably [34].) Some aspects of this subject are discussed in Chapter 6.

Keeping in mind this set of examples, and of course especially the p-adic absolute value, let's go on to look at absolute values in general in a more careful way.

2.2 Basic Properties

In this section, $\Bbbk$ will be an arbitrary field, and $|\ |$ will be a (usually non-trivial) absolute value on $\Bbbk$, which may or may not be archimedean. The first few things to prove are some "obvious" facts, which we had better make sure work in a general setting.

Lemma 2.2.1 *For any absolute value $|\ |$ on any field $\Bbbk$, we have:*

i) $|1| = 1$.

ii) *If $x \in \Bbbk$ and $x^n = 1$, then $|x| = 1$.*

iii) $|-1| = 1$.

iv) *For any $x \in \Bbbk$, $|-x| = |x|$.*

v) *If $\Bbbk$ is a finite field, then $|\ |$ is trivial.*

PROOF: The crucial fact to remember is that $|x|$ is a positive real number. Then, to prove the first statement, all one needs to note is that

$$|1| = |1^2| = |1|^2,$$

since the only non-zero positive real number α for which $\alpha^2 = \alpha$ is $\alpha = 1$. The remaining statements follow in a similar fashion. □

Problem 48 Prove the remaining statements in the lemma.

Our first serious theorem will give a necessary and sufficient condition for an absolute value to be non-archimedean. Let's begin with some easier conditions.

Lemma 2.2.2 *Let $\Bbbk$ be a field and let $|\ |$ be an absolute value on $\Bbbk$. The following are equivalent.*

i) *For all $x, y \in \Bbbk$, $|x + y| \leq \max\{|x|, |y|\}$.*

ii) For all $z \in \Bbbk$, $|z + 1| \leq \max\{|z|, 1\}$.

PROOF: If (i) is true, we get (ii) by putting $x = z$, $y = 1$, so it's clear that (i) implies (ii).

Now assume (ii). If $y = 0$ then (i) holds automatically, so we can assume $y \neq 0$. Let $z = x/y$. Then we have

$$\left| \frac{x}{y} + 1 \right| \leq \max\left\{ \frac{|x|}{|y|}, 1 \right\}.$$

Multiplying both sides by $|y|$ now gives (i).   □

Here is a useful minor variant of this result that we will need in Chapter 6:

Lemma 2.2.3 *Let $\Bbbk$ be a field and let $|\ | : \Bbbk \longrightarrow \mathbb{R}_+$ satisfy*

i) $|x| = 0$ if and only if $x = 0$,

ii) $|xy| = |x||y|$ for all $x, y \in \Bbbk$, and

iii) $|x| \leq 1 \Longrightarrow |x - 1| \leq 1$.

Then $|\ |$ is a non-archimedean absolute value on K.

PROOF: Given the first two conditions and Lemma 2.2.2, what we need to show is that for any $x \in \Bbbk$ we have

$$|x + 1| \leq \max\{|x|, 1\}.$$

Notice first that $x + 1 = -(-x - 1)$, so that we also have

$$|x| = |-x| \leq 1 \Longrightarrow |x + 1| = |-x - 1| \leq 1.$$

In other words, condition (iii) implies that

$$|x| \leq 1 \Longrightarrow |x + 1| \leq 1$$

as well. Now consider cases:

- If $|x| \leq 1$, then $\max\{|x|, 1\} = 1$ and we get $|x + 1| \leq 1 = \max\{|x|, 1\}$.

- If $|x| > 1$, then $|1/x| < 1$, and (iii) implies $|1 + 1/x| \leq 1$. So we have

$$\left| \frac{x + 1}{x} \right| = \left| 1 + \frac{1}{x} \right| \leq 1,$$

which says $|x + 1| \leq |x| = \max\{|x|, 1\}$.   □

For a more interesting criterion, begin by noticing (or remembering) that for any field $\Bbbk$ we have a map $\mathbb{Z} \longrightarrow \Bbbk$ defined by

$$
n \mapsto \begin{cases} \underbrace{1+1+\cdots+1}_{n} & \text{if } n > 0 \\ 0 & \text{if } n = 0 \\ -\underbrace{(1+1+\cdots+1)}_{-n} & \text{if } n < 0 \end{cases}
$$

If $\mathbb{Q} \subset \Bbbk$, this is just the usual inclusion of $\mathbb{Z}$ into $\mathbb{Q}$; if $\Bbbk$ is a finite field, the image is a subfield (it is clearly a subring with no zero-divisors, but finite domains are fields) of $\Bbbk$, which will have a prime number of elements.

Theorem 2.2.4 *Let $A \subset \Bbbk$ be the image of $\mathbb{Z}$ in $\Bbbk$. An absolute value $|\ |$ on $\Bbbk$ is non-archimedean if and only if $|a| \le 1$ for all $a \in A$. In particular, an absolute value on $\mathbb{Q}$ is non-archimedean if and only if $|n| \le 1$ for every $n \in \mathbb{Z}$.*

PROOF: One part is easy: we have $|\pm 1| = 1$ always; hence, if $|\ |$ is non-archimedean, we get that

$$
|a \pm 1| \le \max\{|a|, 1\}.
$$

By induction, if follows that $|a| \le 1$ for every $a \in A$.

The converse requires some hocus-pocus: suppose that $|a| \le 1$ for all $a \in A$. By Lemma 2.2.2, we need to prove that for any $x \in \Bbbk$ we have

$$
|x + 1| \le \max\{|x|, 1\}.
$$

Let m be any positive integer. Then we have

$$
|x + 1|^m = \left| \sum_{k=0}^{m} \binom{m}{k} x^k \right| \le \sum_{k=0}^{m} \left| \binom{m}{k} \right| |x^k|.
$$

Now, since $\binom{m}{k}$ is an integer, we have $\left| \binom{m}{k} \right| \le 1$, so we can continue with

$$
|x + 1|^m \le \sum_{k=0}^{m} \left| \binom{m}{k} \right| |x^k| \le \sum_{k=0}^{m} |x^k| = \sum_{k=0}^{m} |x|^k.
$$

Now notice that the largest value of $|x|^k$ for $k = 0, 1, 2, \ldots m$ is equal to $|x|^m$ if $|x| > 1$ and is equal to 1 otherwise, because $x^0 = 1$. So

$$
|x + 1|^m \le (m + 1) \max\{1, |x|^m\}.
$$

Taking the m-th root on both sides gives

$$|x + 1| \leq \sqrt[m]{m + 1} \, \max\{1, |x|\}.$$

This strange inequality holds for *every* positive integer m, no matter how large, and we know (from calculus) that

$$\lim_{m \to \infty} \sqrt[m]{m + 1} = 1.$$

Therefore, if we let $m \to \infty$ we get

$$|x + 1| \leq \max\{|x|, 1\},$$

which is what we wanted to prove. $\square$

Problem 49 Show that if $\sup\{|n| : n \in \mathbb{Z}\} = C < +\infty$, then $|\ |$ is non-archimedean, and $C = 1$.

This helps explain the difference between archimedean and non-archimedean absolute values. It allows us to restate things in the following way. An absolute value is *archimedean* if the image of $\mathbb{Z}$ in $\Bbbk$ is unbounded. It follows that an absolute value is archimedean if it has the following property:

> *Archimedean Property: Given x, $y \in \Bbbk$, $x \neq 0$, there exists a positive integer n such that $|nx| > |y|$.*

It is easy to see that the Archimedean Property is equivalent to the assertion that there are "arbitrarily big" integers (translation: that there are integers whose absolute values are arbitrarily big). This property holds for the usual absolute value on $\mathbb{Q}$ and in the real numbers. (In a slightly different form, this observation does go back to Archimedes.)

2.3   Topology

The whole point of an absolute value is that it provides us with a notion of "size." In other words, once we have an absolute value, we can use it to measure distances between numbers, that is, to put a *metric* on our field. Having the metric, we can define open and closed sets, and in general investigate what is called the *topology* of our field.[2]

The first step is measuring distances, in the obvious way:

Definition 2.3.1 *Let $\Bbbk$ be a field and $|\ |$ an absolute value on $\Bbbk$. We define the distance $d(x, y)$ between two elements x, $y \in \Bbbk$ by*

$$d(x, y) = |x - y|.$$

The function $d(x, y)$ is called the metric *induced by the absolute value.*

---

[2] The reader who has never met topology or metric spaces before should not feel spooked; all we are doing is repeating the usual constructions of the calculus, but using our unusual absolute values.

The definition of $d(x, y)$ parallels, of course, the usual way we define the distance between two real numbers. The first point we need to make is that a great many of the notions that we can define using the usual distance on $\mathbb{R}$ work just as well for any old distance.

Problem 50 Show that $d(x, y)$ has the following properties.

i) For any x, $y \in \mathbb{k}$, $d(x, y) \geq 0$, and $d(x, y) = 0$ if and only if $x = y$.

ii) For any x, $y \in \mathbb{k}$, $d(x, y) = d(y, x)$.

iii) For any x, y, $z \in \mathbb{k}$, $d(x, z) \leq d(x, y) + d(y, z)$.

These are the general defining properties for a metric; the last inequality is called the *triangle inequality*, since it expresses the usual fact that the sum of the lengths of two legs of a triangle is bigger than the length of the other side. ("A line is the shortest path between two points.") A set on which a metric is defined is called a *metric space*, so we can read the statement of this last problem as saying that any field with an absolute value can be made into a metric space by defining $d(x, y) = |x - y|$. For more on metric spaces in general, check a book on real analysis or an introductory text on general topology.[3]

If we have a metric, we can talk about continuity. We won't need it very much right now, but let's at least record the definition.

Definition 2.3.2 *Let $\mathbb{k}$ and $\mathbb{F}$ both be fields with absolute values, and let $f : \mathbb{k} \longrightarrow \mathbb{F}$ be a function. We say f is* continuous *at $x_0 \in \mathbb{k}$ if given any $\varepsilon > 0$ we can find $\delta > 0$ (possibly depending on both x_0 and ε) so that*

$$d(x, x_0) < \delta \implies d(f(x), f(x_0)) < \varepsilon.$$

We say f is uniformly continuous *on $\mathbb{k}$ if δ does not depend on x_0, i.e., if given any $\varepsilon > 0$ we can find $\delta > 0$ so that for any $x, y \in \mathbb{k}$ we have*

$$d(x, y) < \delta \implies d(f(x), f(y)) < \varepsilon.$$

If the function is defined only on a subset of $\mathbb{k}$, both definitions still make sense when restricted to that subset.

Problem 51 The point of this problem is to check that the metric $d(x, y)$ (or, equivalently, the absolute value it is derived from) relates well to the operations in the field $\mathbb{k}$, in the sense that the field operations are continuous functions.

i) Fix x_0, $y_0 \in \mathbb{k}$. Show that for any $\varepsilon > 0$ there exists a $\delta > 0$ such that, whenever $d(x, x_0) < \delta$ and $d(y, y_0) < \delta$, we have $d(x + y, x_0 + y_0) < \varepsilon$. In other words, addition is a continuous function.

---

[3]I learned metric spaces from [55, Ch. 2], which is notoriously terse. An alternative is [59]. A good introduction to topology is [47].

ii) Fix x_0, $y_0 \in \Bbbk$. Show that for any $\varepsilon > 0$ there exists a $\delta > 0$ such that, whenever $d(x, x_0) < \delta$ and $d(y, y_0) < \delta$, we have $d(xy, x_0 y_0) < \varepsilon$. In other words, multiplication is a continuous function.

iii) Fix $x_0 \in \Bbbk$, $x_0 \neq 0$. Show that for any $\varepsilon > 0$ there exists a $\delta > 0$ such that, whenever $d(x, x_0) < \delta$, we have $x \neq 0$ and $d(1/x, 1/x_0) < \varepsilon$. In other words, taking inverses is a continuous function.

iv) Show that if $|x - y| < \varepsilon$ then $\big||x| - |y|\big| < \varepsilon$ as well. This means that if we give $\mathbb{R}$ its usual topology, the function $\Bbbk \longrightarrow \mathbb{R}$ that sends x to $|x|$ is uniformly continuous.

This shows that the metric $d(x, y)$ makes $\Bbbk$ a *topological field*.

The upshot is that general fields with absolute values behave very much like the real numbers. For us, the important differences happen when the absolute value is non-archimedean.

The fact that an absolute value is non-archimedean can also be expressed in terms of the metric:

Lemma 2.3.3 *Let* $|\ |$ *be an absolute value on a field* $\Bbbk$, *and define a metric by* $d(x, y) = |x - y|$. *Then* $|\ |$ *is non-archimedean if and only if for any* x, y, $z \in \Bbbk$, *we have*

$$d(x, y) \leq \max\{d(x, z), d(z, y)\}.$$

PROOF: To go one way, apply the non-archimedean property to the equation

$$(x - y) = (x - z) + (z - y).$$

For the converse, take $y = -y_1$ and $z = 0$ in the inequality satisfied by $d(\cdot, \cdot)$. $\qquad\square$

Problem 52 Give the details of the proof of the lemma. Prove also that the inequality $d(x, y) \leq \max\{d(x, z), d(z, y)\}$ implies the triangle inequality from Problem 50.

This inequality is known as the "ultrametric inequality," and a metric for which it is true is sometimes called an "ultrametric." A space with an ultrametric is called an "ultrametric space." Such spaces have rather curious properties, and we will spend the rest of this section exploring them.[4] The main point in what follows is that, once we have a way to measure distances, we can do geometry. Since our way to measure distances is rather strange, the geometry is also rather strange. As we explore it, we will focus on fields with a non-archimedean valuation, but almost all our results are in fact true in any ultrametric space.

---

[4]Ultrametric spaces sound like the sort of thing only a mathematician would dream up. Surprisingly, they have recently turned up in physics (in the theory of "spin glasses"). This may be one more example of the "unreasonable effectiveness of mathematics in the physical sciences"—see [66].

Proposition 2.3.4 *Let* $\Bbbk$ *be a field and let* $|\ |$ *be a non-archimedean absolute value on* $\Bbbk$. *If* $x, y \in \Bbbk$ *and* $|x| \neq |y|$, *then*

$$|x + y| = \max\{|x|, |y|\}.$$

PROOF: Exchanging x and y if necessary, we may suppose that $|x| > |y|$. Then we know that

$$|x + y| \leq |x| = \max\{|x|, |y|\}.$$

On the other hand, $x = (x + y) - y$, so that

$$|x| \leq \max\{|x + y|, |y|\}.$$

Since we know that $|x| > |y|$, this inequality can hold only if

$$\max\{|x + y|, |y|\} = |x + y|.$$

This gives the reverse inequality $|x| \leq |x + y|$, and from it (using our first inequality) we can conclude that $|x| = |x + y|$.   □

Alain Robert refers to this as the *strongest wins* principle in [53] (see page 429, for example). It has an interesting corollary that captures in a memorable statement a property that ends up having a big role later on:

Corollary 2.3.5 *In an ultrametric space, all "triangles" are isosceles.*

PROOF: Let x, y and z be three elements of our space (the vertices of our "triangle"). The lengths of the sides of the "triangle" are the three distances $d(x, y) = |x - y|$, $d(y, z) = |y - z|$, and $d(x, z) = |x - z|$. Now, of course,

$$(x - y) + (y - z) = (x - z),$$

so that we can invoke the proposition to show that if $|x - y| \neq |y - z|$, then $|x - z|$ is equal to the bigger of the two. In either case, two of the "sides" are equal.   □

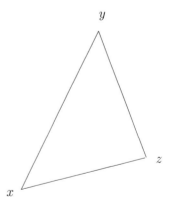

Figure 2.1: All isosceles!

This is a rather unintuitive result (and it will have an enormous impact on the topology on our field). Thus, rather than simply barging on, it may be worth a brief look at the case of the p-adic absolute value to try to understand what is behind the truth of the proposition. As before, we put $|x| = p^{-v_p(x)}$. Since we're looking for insight, not for proof, we will only look at the case where x, $y \in \mathbb{Z}$. Say that $v_p(x) = n$ and $v_p(y) = m$, so that

$$x = p^n x' \qquad y = p^m y' \qquad p \nmid x'y'.$$

Translating into the absolute values, we get

$$|x| = p^{-n} \qquad \text{and} \qquad |y| = p^{-m}.$$

We will have $|x| > |y|$ when $n < m$; say $m = n + \varepsilon$, with $\varepsilon > 0$. Then

$$x + y = p^n x' + p^{n+\varepsilon} y' = p^n (x' + p^\varepsilon y').$$

Now, since $p \nmid x'$, we have $p \nmid (x' + p^\varepsilon y')$, and therefore $v_p(x + y) = n$, which means $|x + y| = p^{-n} = |x|$, as the proposition states. In this situation $|y|$ is smaller and $|x| = |x + y|$.

On the other hand, suppose that $|x| = |y|$, that is, $n = m$. Then we get

$$x + y = p^n (x' + y')$$

with $p \nmid x'$ and $p \nmid y'$, and it is perfectly possible that $p|(x' + y')$. If so, the most we can say is that $v_p(x + y) \geq n = \min\{v_p(x), v_p(y)\}$, which translates to

$$|x + y| \leq \max\{|x|, |y|\} = |x| = |y|.$$

So when $|x|$ and $|y|$ are equal it is possible for $|x + y|$ to be smaller (or not). So in either case two of the three absolute values $|x|$, $|y|$ and $|x+y|$ are equal.

Problem 53 Give $\mathbb{Q}$ the 5-adic topology, and consider the triangle whose vertices are $x = 2/15$, $y = 1/5$, $z = 7/15$; what are the lengths of the three sides?

In metric spaces, more important than triangles are the "balls" or "disks." These also turn out to be pretty strange in the case of an ultrametric.

Definition 2.3.6 *Let $\Bbbk$ be a field with an absolute value $|\ |$. Let $a \in \Bbbk$ be an element and $r \in \mathbb{R}_+$ be a real number. The* open ball *of radius r and center a is the set*

$$B(a, r) = \{x \in \Bbbk : d(x, a) < r\} = \{x \in \Bbbk : |x - a| < r\}.$$

The closed ball *of radius r and center a is the set*

$$\overline{B}(a, r) = \{x \in \Bbbk : d(x, a) \leq r\} = \{x \in \Bbbk : |x - a| \leq r\}.$$

These are standard definitions in any metric space. The open balls are the prototypes of the open sets, and the closed balls of the closed sets.[5]

---

[5]Here are the definitions: a set U is open if any element in U is the center of a (usually small) an open ball that is contained in U; a set S is closed if its complement is an open set. A point x is a boundary point of a set S if any open ball with center x contains both points that are in S and points that are not in S. S is closed exactly when it contains all of its boundary points.

Problem 54 Show that open balls are always open sets, and that closed balls are always closed sets. (This is true for any absolute value, archimedean or not.)

For non-archimedean absolute values, we get some surprising properties:

Proposition 2.3.7 *Let $\Bbbk$ be a field with a non-archimedean absolute value.*

i) *If $b \in B(a,r)$, then $B(a,r) = B(b,r)$; in other words, every point that is contained in an open ball is a center of that ball.*

ii) *If $b \in \overline{B}(a,r)$, then $\overline{B}(a,r) = \overline{B}(b,r)$; in other words, every point that is contained in a closed ball is a center of that ball.*

iii) *The set $B(a,r)$ is both open and closed. $B(a,r)$ has empty boundary.*

iv) *If $r \neq 0$, the set $\overline{B}(a,r)$ is both open and closed and has empty boundary.*

v) *If $a, b \in \Bbbk$ and $r, s \in \mathbb{R}_+^\times$, we have $B(a,r) \cap B(b,s) \neq \varnothing$ if and only if $B(a,r) \subset B(b,s)$ or $B(a,r) \supset B(b,s)$; in other words, any two open balls are either disjoint or contained in one another.*

vi) *If $a, b \in \Bbbk$ and $r, s \in \mathbb{R}_+^\times$, we have $\overline{B}(a,r) \cap \overline{B}(b,s) \neq \varnothing$ if and only if $\overline{B}(a,r) \subset \overline{B}(b,s)$ or $\overline{B}(a,r) \supset \overline{B}(b,s)$; in other words, any two closed balls are either disjoint or contained in one another.*

PROOF: Most of this is easy. The weird parts all depend on the fact that "all triangles are isosceles;" drawing pictures may help understand what is going on.

i) By the definition, $b \in B(a,r)$ if and only if $|b - a| < r$. Now, taking any x for which $|x - a| < r$, the non-archimedean property tells us that

$$|x - b| \leq \max\{|x - a|, |b - a|\} < r,$$

so that $x \in B(b,r)$; this shows that $B(a,r) \subset B(b,r)$. Switching a and b, we get the opposite inclusion, so that the two balls are equal.

ii) Replace $<$ with $\leq$ in the proof of *(i)*.

iii) The open ball $B(a,r)$ is always an open set in any metric space, as we showed above. But in an ultrametric space it follows at once from *(i)*, since the open ball of radius r around any point in the open ball is the same as $B(a,r)$.

What we need to show is that in our non-archimedean case, it is also closed. This is equivalent to saying its complement

$$C = \{x \in \Bbbk : d(x,a) \geq r\}$$

is open. Choose any $y \in C$, so that $|y - a| \geq r$, and let $s < r$. We claim the open ball $B(y,s)$ is contained in C. We have $|z - y| < s < r \leq |y - a|$, so by "all triangles are isosceles" we get

$$|z - a| = \max\{|z - y|, |y - a|\} = |y - a| \geq r,$$

These are allowed...   but not this!

Figure 2.2: Balls for non-archimedean absolute values

so $z \in C$. So there is an open ball around every $y \in C$ that is entirely contained in C, which says C is an open set. Therefore its complement $B(a, r)$ is closed.

The final claim is a general fact: any set that is both closed and open has empty boundary. This is easy to see: if a point is on the boundary of $B(a, r)$, then any ball around it contains points in both $B(a, r)$ and C. That means it is also on the boundary of the complement C. Since both $B(a, r)$ and C are closed, we conclude that any boundary point must belong to $B(a, r) \cap C = \varnothing$.

iv) This is a lot like (iii).

v) We can assume that $r \leq s$ (otherwise switch them around). If the intersection is not empty, there exists a $c \in B(a, r) \cap B(b, s)$. Then we know, from ($i$), that $B(a, r) = B(c, r)$ and $B(b, s) = B(c, s)$. Hence

$$B(a, r) = B(c, r) \subset B(c, s) = B(b, s),$$

as claimed.

vi) Identical to the preceding, using (ii).   □

Problem 55 Supply the missing portions of the proof (parts (iv) and (vi)). Why is the condition $r \neq 0$ necessary for closed, but not for open, balls?

In an ultrametric space, open balls are closed sets and our closed balls are open sets! This suggests that the names "open ball" and "closed ball" are not quite appropriate, but it's hard to think of alternatives (but see [53], which goes for "stripped" and "dressed" balls). Any alternative would conflict with standard language anyway, so it's best to leave the terms alone. Nevertheless, the reader should keep in mind that the bar in the notation $\overline{B}(a, r)$ does not mean closure.

The geometry of the balls in an ultrametric space seems very strange at first sight; getting a good feeling for it may be the most important initial step toward understanding the p-adic absolute value. The next problems are intended to help with that goal.

Problem 56 Describe the closed ball of radius 1 around the point $x = 0$ in $\mathbb{Q}$ with respect to the p-adic absolute value. Describe the open ball of radius 1 around $x = 3$; which integers belong to this ball?

Problem 57 Let $\Bbbk = \mathbb{Q}$ and $|\,| = |\,|_p$. Show that the closed ball $\overline{B}(0,1)$ can be written as a disjoint union of open balls, as follows:

$$\overline{B}(0,1) = B(0,1) \cup B(1,1) \cup B(2,1) \cup \cdots \cup B(p-1,1)$$

(both the equality and the disjointness need to be checked). This gives another proof that the closed unit ball is open, since unions of open sets are always open.

Problem 58 Take the 5-adic absolute value on $\mathbb{Q}$. Show that $B(1,1) = B(1,1/2) = \overline{B}(1,1/5)$. What is going on here?

Problem 59 Under the hypotheses of the proposition, show that for $a \in \Bbbk$ and $r \in \mathbb{R}_+$, $r \neq 0$, the "sphere" $\{x \in \Bbbk : |x - a| = r\}$ is both an open and a closed set. (Notice that the "sphere" is *not* the boundary of the open ball. In fact, the boundary of the open ball is empty.)

Sets that are both open and closed are rather rare in the usual calculus, but are very common when we are dealing with non-archimedean absolute values. (As we've just seen!) So we give them a name.

Definition 2.3.8 *Let $\Bbbk$ be a field with an absolute value $|\,|$ (or, more generally, any metric space). We say a set $S \subset \Bbbk$ is* clopen *if it is both an open and a closed set.*

The fact that there are so many clopen sets around makes the topology of fields with non-archimedean valuations rather strange. For example, recall that a set S is called *disconnected* if one can find two open sets U_1 and U_2 such that

- $S = (S \cap U_1) \cup (S \cap U_2)$,

- $(S \cap U_1) \cap (S \cap U_2) = \varnothing$, and

- neither $S \cap U_1$ nor $S \cap U_2$ is empty.

The idea, of course, is that such an S is made up of two "pieces" (namely, the intersections with each of the open sets). Sets which *cannot* be divided in this way are called *connected*.

Problem 60 Show that a set S is disconnected if and only if we can write it as a union $S = A \cup B$ of two nonempty sets satisfying the condition

$$\overline{A} \cap B = A \cap \overline{B} = \varnothing,$$

where, for a set X, $\overline{X}$ means the *closure* of X, that is, the union of X and all of its boundary points.

Problem 61 What are the connected sets in $\mathbb{R}$? (Hint: they appear all the time in elementary calculus.)

Problem 62 Show that in a field with a (non-trivial) non-archimedean absolute value every closed ball with radius $r > 0$ is disconnected. Is the same true for open balls?

If we take a point $x \in \Bbbk$, we define the *connected component of x* to be the union of all the connected sets that contain x. Since the union of two non-disjoint connected sets is connected, this is a connected set, so we can describe it as the largest connected set containing x. For example, if $\Bbbk = \mathbb{R}$ is the real numbers, then the connected component of any point $x \in \mathbb{R}$ is all of $\mathbb{R}$ (simply because $\mathbb{R}$ is connected). Things are quite different in the non-archimedean case:

Proposition 2.3.9 *In a field $\Bbbk$ with a non-archimedean absolute value, the connected component of any point $x \in \Bbbk$ is the set $\{x\}$ consisting of only that point.*

Problem 63 Prove the proposition by showing that if a set contains two distinct points then it is disconnected. In the language of general topology, this says that $\Bbbk$ is a *totally disconnected* topological space.

What this says is that there are really no interesting connected sets in $\Bbbk$: only the sets with only one element are connected. On the other hand, provided the absolute value on $\Bbbk$ is non-trivial, the set $\{x\}$ is *not* open (if every set $\{x\}$ were open, the topology on $\Bbbk$ would be *discrete*, i.e., every set would be open, which only happens with the trivial absolute value). One consequence is worth noticing:

Corollary 2.3.10 *If $\Bbbk$ is a field with a non-archimedean absolute value and $\mathbb{R}$ is given its usual absolute value then there are no non-constant continuous functions $\mathbb{R} \longrightarrow \Bbbk$.*

PROOF: We know $\mathbb{R}$ is connected and the image of a connected set under a continuous function is a connected set. Since $\Bbbk$ is totally disconnected, the image can only be a single point, which means f is constant.   □

If we allow functions defined on subsets of $\mathbb{R}$, then the conclusion is that f is *locally constant*, i.e., constant on every interval.

Problem 64 Take the *usual* absolute value on $\mathbb{Q}$, which of course is archimedean. Are there any clopen sets in $\mathbb{Q}$ with respect to this absolute value? Is $\mathbb{Q}$ totally disconnected with respect to this absolute value?

The same questions make sense in the real numbers, of course. Are there any clopen sets in $\mathbb{R}$?

Problem 65 Take the p-adic absolute value on $\mathbb{Q}$. Show that with respect to this absolute value every open ball is the disjoint union of open balls. (So that open balls are disconnected in a rather dramatic way.) Do you think this is true for any field with a non-archimedean absolute value? If not, can you come up with a counter-example?

2.4   Algebra

So far, we have mostly concentrated on the geometry we obtain from an absolute value on a field $\Bbbk$. In this section, we take a more algebraic point-of-view, and look for connections between (non-archimedean) absolute values and the algebraic structure[6] of the underlying field. These connections turn out to be quite strong. In fact, they point to a tight relationship between geometric and algebraic properties of such fields. (This section necessarily requires a little more background in abstract algebra than the preceding ones, but shouldn't be very hard to manage.) The main message is that the close connection between the p-adic absolute value and the prime number p is actually typical of non-archimedean valued fields.

To begin with, every non-archimedean absolute value is attached to a subring of the field $\Bbbk$, and this subring has some rather nice properties:

Proposition 2.4.1 *Let $\Bbbk$ be a field, and let $|\ |$ be a non-archimedean valuation on $\Bbbk$. The set*

$$\mathcal{O} = \overline{B}(0,1) = \{x \in \Bbbk : |x| \le 1\}$$

is a subring of $\Bbbk$. Its subset

$$\mathfrak{P} = B(0,1) = \{x \in \Bbbk : |x| < 1\}$$

is an ideal of $\mathcal{O}$. Furthermore, $\mathfrak{P}$ is a maximal ideal in $\mathcal{O}$, and every element of the complement $\mathcal{O} - \mathfrak{P}$ is invertible in $\mathcal{O}$.

Problem 66 Prove the proposition. It is all a matter of using the definitions directly, and remembering that the absolute value is non-archimedean. Notice that the statement about the complement of $\mathfrak{P}$ implies at once that $\mathfrak{P}$ is a maximal ideal.

Rings that contain a unique maximal ideal whose complement consists of invertible elements are called *local rings*. The proposition, then, shows us how to attach to any non-archimedean absolute value on $\Bbbk$ a subring of $\Bbbk$ which is a local ring. Let's give it a name:

Definition 2.4.2 *Let $\Bbbk$ be a field and $|\ |$ be a non-archimedean absolute value on $\Bbbk$. The subring*

$$\mathcal{O} = \overline{B}(0,1) = \{x \in \Bbbk : |x| \le 1\} \subset \Bbbk$$

is called the valuation ring *of $|\ |$. The ideal*

$$\mathfrak{P} = B(0,1) = \{x \in \Bbbk : |x| < 1\} \subset \mathcal{O}$$

is called the valuation ideal *of $|\ |$. The quotient*

$$\kappa = \mathcal{O}/\mathfrak{P}$$

is called the residue field *of $|\ |$.*

---

[6]in the sense of abstract algebra

(For the residue field, remember that the quotient of a commutative ring by a maximal ideal is always a field.)

It is natural to expect that many of the properties of the absolute value are connected to algebraic properties of its associated valuation ring. In fact, one can develop the theory by concentrating on this side of things (so that finding an absolute value on a field gets translated into finding a subring with certain properties). Exactly what properties characterize the rings that arise in this way is a question that will be touched upon in one of the problems for this section.

Since we'll be mostly interested in the p-adic absolute values, let's record what we get in that case:

Proposition 2.4.3 *Let* $\Bbbk = \mathbb{Q}$ *and let* $|\ | = |\ |_p$ *be the p-adic absolute value. Then:*

i) the associated valuation ring is $\mathcal{O} = \mathbb{Z}_{(p)} = \{a/b \in \mathbb{Q} : p \nmid b\}$;

ii) the valuation ideal is $\mathfrak{P} = p\mathbb{Z}_{(p)} = \{a/b \in \mathbb{Q} : p \nmid b \text{ and } p|a\}$;

iii) the residue field is $\kappa = \mathbb{F}_p$ *(the field with p elements).*

PROOF: All we need is to remember the definitions. We have

$$\left|\frac{a}{b}\right| = p^{-v} \qquad \text{when} \qquad \frac{a}{b} = p^v \frac{a_1}{b_1} \quad \text{with } p \nmid a_1 b_1.$$

So we get that $a/b \in \mathcal{O}$ if and only if $v \geq 0$. If a/b is in lowest terms, this just means $p \nmid b$, as claimed. Similarly, $a/b \in \mathfrak{P}$ happens when $v > 0$, hence when $p \nmid b$ and $p|a$. The last statement is an easy exercise in quotient rings.□

The notation $\mathbb{Z}_{(p)}$ comes from commutative algebra: it is the standard notation for the localization of the commutative ring $\mathbb{Z}$ at the prime ideal (p). See, for example, [33, §5.9].

Problem 67 Prove the last statement in the proposition. (The jazzy proof begins with the inclusion $\mathbb{Z} \hookrightarrow \mathbb{Z}_{(p)}$, and checks that it induces a map on quotient rings.)

Problem 68 Compute the valuation ring, valuation ideal, and the residue field for the non-archimedean valuations on $F(t)$ introduced above.

One could go further in exploring these connections between absolute values and algebraic structure, but we will stop here, at least for now. As we go along, we will develop a clearer feeling for how the connection works by finding out more and more about the specific case of the p-adic absolute value. The following problems use a little more background from abstract algebra.

Problem 69 Consider $\mathbb{Q}$ with a p-adic absolute value, and let $a \in \mathbb{Z}$. Describe the open ball $B(a, 1)$ with center a and radius 1 in terms of the algebraic structure. Use your description to interpret algebraically the fact (Problem 57) that the closed ball $\bar{B}(0, 1)$ is the disjoint union of finitely many open balls of radius 1.

Problem 70 In the case of the p-adic absolute value, the valuation ideal is a *principal* ideal, that is, it is the set of multiples of an element of $\mathcal{O}$ (to wit, the element p). Is this always the case for the examples we have considered? Make a conjecture as to whether it will always be the case for any non-archimedean absolute value. (Hint: if so, it shouldn't be too hard to prove in general...)

Problem 71 Let k be a field, and let $|\ |$ be an absolute value on k. Define a valuation v on k by
$$v(x) = -\log|x|$$
for $x \neq 0$ and $v(0) = +\infty$. Check that if $|\ |$ is non-archimedean then this is indeed a valuation (i.e., it has the properties listed in Lemma 2.1.3).

i) If $|\ |$ is the p-adic absolute value, how does v relate to the p-adic valuation v_p? What is the image of v in this case?

ii) Show that the valuation ideal of $|\ |$ is a principal ideal if and only if the image of v is a *discrete* additive subgroup of $\mathbb{R}$. (We showed above that the image is a subgroup; the point here is the discreteness, which means that each element of the subgroup is contained in an open interval that does not contain any other elements of the subgroup.)

iii) Show that if the image of v is a discrete subgroup of $\mathbb{R}$ then the valuation ring $\mathcal{O}$ is a principal ideal domain whose only prime ideals are 0 and $\mathfrak{P}$. (For example, check that this happens for the p-adic absolute values.)

3 The p-adic Numbers

Having built our foundation, we can now apply the general theory to the specific case of the field $\mathbb{Q}$ of rational numbers. Extending our scope to include all fields of algebraic numbers (i.e., finite extensions of $\mathbb{Q}$), or even to include what the experts call "global fields" in general, would not be very hard. Nevertheless, we have preferred to stick, at first, to the most concrete example available. In a later chapter, we will consider some aspects of the problem of extending valuations from $\mathbb{Q}$ to larger fields. More details about the theory of valuations on global fields can be found in several of the references.

3.1 Absolute Values on $\mathbb{Q}$

We have already found a few examples of absolute values on the field $\mathbb{Q}$ of rational numbers. The next step will be to show that these are essentially all the possible absolute values; for that we will need to introduce a refined notion of what it means for two absolute values to be "the same." Up to that notion of equivalence, we will be able to show that the absolute values we have are the complete list of possible absolute values on $\mathbb{Q}$. Finally, we will prove the *product formula* as an initial example of how all the absolute values work together in the arithmetic of $\mathbb{Q}$.

We begin by recording what has been achieved so far, namely that we have constructed the following absolute values on the field $\mathbb{Q}$:

- the trivial absolute value;

- the "usual" absolute value $|\ |_\infty$, which we have called the "absolute value at infinity," and which is associated to the real numbers;

- for each prime p, the p-adic absolute value $|\ |_p$.

Notice that, except for the trivial absolute value (which we will tend to ignore), we have written all of these in the form $|\ |_p$, where p is either a prime or ∞. It turns out to be convenient to think of the symbol ∞ as some sort of prime number in $\mathbb{Z}$, and refer to it as "the infinite prime," and to the corresponding absolute value as the "∞-adic" absolute value. This will allow us to say things like "$|\ |_p$ for all primes $p \le \infty$." Though there are

F. Q. Gouvêa, *p-adic Numbers*, Universitext, https://doi.org/10.1007/978-3-030-47295-5_3

some reasons[1] for doing this, at this point we will use it only as a notational convenience.

To be able to state our main theorem in this section, we must first make a good definition of when two absolute values are "the same." The main idea here is that we use absolute values on a field $\Bbbk$ to introduce a topology (open and closed sets, connectedness, etc.) on $\Bbbk$. So it is reasonable to define:

Definition 3.1.1 *Two absolute values $|\ |_1$ and $|\ |_2$ on a field $\Bbbk$ are called equivalent if they define the same topology on $\Bbbk$, that is, if every set that is open with respect to one is also open with respect to the other.*

This is easier to say than to check, so we had better find a more accessible criterion. Notice first that the fact that a sequence converges can be expressed in terms of open sets, and so equivalent absolute values will have the same convergent sequences.

Lemma 3.1.2 *Let $\Bbbk$ be a field with an absolute value $|\ |$. The following are equivalent:*

i) $\displaystyle\lim_{n\to\infty} x_n = a$.

ii) Any open set containing a also contains all but finitely many of the x_n.

(This lemma serves also as an excuse to remind the reader what it means for a sequence to converge.)

PROOF: Assume (*ii*). Since an open ball $B(a, \varepsilon)$ centered at a is an open set, all but finitely many x_n will be in the open ball, and so there is an N such that $n \geq N$ implies $a \in B(a, \varepsilon)$. Therefore for any ε an N such that $n \geq N$ implies $|x - a| < \varepsilon$, i.e., $x_n \to a$.

Conversely, suppose $x_n \to a$ and let U be an open set containing a. Since U is open there exists an r such that $B(a, r) \subset U$. Therefore there is an N such that $|x - a| < r$ for all $n \geq N$. Hence for all but finitely many n we have $x_n \in B(a, r) \subset U$. $\qquad\qquad\Box$

This gives a definition of convergent sequence (in a valued field) in terms of open sets. Hence if two absolute values define the same open sets they will also define the same convergent sequences. This is the first of several equivalent ways to characterize equivalent of absolute values.

---

[1]The reasons hinge on the close connection between primes and absolute values that we are about to establish. If all the other absolute values correspond to primes, then so should the usual absolute value. As to why it should be called the *infinite* prime, that is far less clear. In fact, John H. Conway has been heard to argue quite vigorously that the "usual" absolute value should be attached to the "prime" -1, and this does seem to make more sense. (Think of the ± 1 that appears in prime factorizations.) Unfortunately, number theorists are too used to talking of "primes at infinity" for this to change easily, and we have preferred to go along with convention.

Proposition 3.1.3 *Let $|\ |_1$ and $|\ |_2$ be absolute values on a field $\mathbb{k}$. The following statements are equivalent.*

i) *$|\ |_1$ and $|\ |_2$ are equivalent absolute values.*

ii) *For any sequence (x_n) in $\mathbb{k}$ we have $x_n \to a$ with respect to $|\ |_1$ if and only if $x_n \to a$ with respect to $|\ |_2$.*

iii) *For any $x \in \mathbb{k}$ we have $|x|_1 < 1$ if and only if $|x|_2 < 1$.*

iv) *There exists a positive real number α such that for every $x \in \mathbb{k}$ we have*

$$|x|_1 = |x|_2^\alpha.$$

PROOF: We follow the usual method of proving a circle of implications.

- First, suppose (i), i.e., that $|\ |_1$ and $|\ |_2$ are equivalent. By the lemma, any sequence that converges with respect to one absolute value must also converge in the other, which is (ii).

- Suppose (ii). Given any $x \in \mathbb{k}$, it is easy to see that $\lim_{n \to \infty} x^n = 0$ with respect to the topology induced by an absolute value $|\ |$ if and only if $|x| < 1$. This gives (iii).

- We leave it to the reader to prove that (iii) implies (iv), not because it is easy, but because it is the hardest part of the theorem, and the convoluted argument that one ends up resorting to can only be appreciated after one has become convinced that easier methods don't work. The next problem includes some hints and a complete solution is in Appendix B.

- Finally, if we assume (iv), we get that

$$|x - a|_1 < r \iff |x - a|_2^\alpha < r \iff |x - a|_2 < r^{1/\alpha},$$

so that any open ball with respect to $|\ |_1$ is also an open ball (albeit of different radius) with respect to $|\ |_2$. This is enough to show that the topologies defined by the two absolute values are identical.

□

Problem 72 Prove the missing step, i.e., that $(iii) \Longrightarrow (iv)$. The first hurdle is finding the number α. For that, just choose any appropriate x_0 and choose α to be the unique real number that will make $|x_0|_1 = |x_0|_2^\alpha$. The proof will be done if you can show that the same equation will hold for every $x \in \mathbb{k}$; it is here that you have to find a way to use condition (iii). (This is quite hard, but worth a try. The argument suggested in Appendix B is quite sophisticated, and it will be hard to understand why it is needed unless some effort has been expended to try to do it in an easier way.)

Problem 73 Let $|\ |_1$ and $|\ |_2$ be two absolute values on a field $\Bbbk$. If every open ball with respect to one of these is also an open ball with respect to the other, show that the induced topologies are identical, i.e., that every set that is open with respect to one is open with respect to the other.

Problem 74 Show that we can add the following condition to the list in the proposition:

v) for any $x \in \Bbbk$, we have $|x|_1 \le 1$ if and only if $|x|_2 \le 1$.

Problem 75 Suppose that $|\ |$ is an absolute value that is equivalent to the trivial absolute value. Must it *be* the trivial absolute value? Do we need to change the definition of "nontrivial"?

Problem 76 Show that if p and q are two different primes, the p-adic and the q-adic absolute values are not equivalent. Do the same when p is a prime and $q = \infty$.

Problem 77 Show that in general a non-archimedean absolute value cannot be equivalent to an archimedean absolute value.

As an example, recall that we considered, in Problem 39, an absolute value defined by

$$|x| = c^{-v_p(x)},$$

where $c > 1$ was a real number. Now we can check that this is equivalent to the p-adic absolute value—just choose α so that $c^\alpha = p$. We will see later that the choice $c = p$ is dictated by "global" considerations (namely, the product formula).

The main theorem in this section says that we have already found all the absolute values on $\mathbb{Q}$.

Theorem 3.1.4 (Ostrowski) *Every non-trivial absolute value on $\mathbb{Q}$ is equivalent to one of the absolute values $|\ |_p$, where either p is a prime number or $p = \infty$.*

PROOF: Let $|\ |$ be a non-trivial absolute value on $\mathbb{Q}$. We will consider the possible cases.

a) Suppose, first, that $|\ |$ is *archimedean*. We want to show that it is equivalent to the "usual" (∞-adic) absolute value.

Let n_0 be the least positive integer for which $|n_0| > 1$ (there has to be one, because otherwise $|\ |$ would be non-archimedean). Now of course we can find a positive real number α so that

$$|n_0| = n_0^\alpha.$$

(Finding a formula for α is an easy exercise on logarithms.) We claim that this α will do, that is, that it will realize the equivalence between $|\ |$ and $|\ |_\infty$. This means that we want to prove that for every $x \in \mathbb{Q}$ we have $|x| = |x|_\infty^\alpha$.

Given the known properties of absolute values, this will follow if we know it for positive integers, that is, if we show that $|n| = n^\alpha$ for any positive integer n. (Check this!)

We know that the equality holds for $n = n_0$. To prove it in general, we use a little trick. Take an arbitrary integer n, and write "in base n_0," i.e., in the form

$$n = a_0 + a_1 n_0 + a_2 n_0^2 + \cdots + a_k n_0^k,$$

with $0 \le a_i \le n_0 - 1$ and $a_k \neq 0$. Notice that k is determined by the inequality $n_0^k \le n < n_0^{k+1}$, which says that

$$k = \left\lfloor \frac{\log n}{\log n_0} \right\rfloor,$$

where $\lfloor x \rfloor$ denotes the "floor" of x, that is, the largest integer that is less than or equal to x. Now take absolute values. We get

$$|n| = |a_0 + a_1 n_0 + a_2 n_0^2 + \cdots + a_k n_0^k|$$
$$\le |a_0| + |a_1|n_0^\alpha + |a_2|n_0^{2\alpha} + \cdots + |a_k|n_0^{k\alpha}.$$

Since we chose n_0 to be the *smallest* integer whose absolute value was greater than 1, we know that $|a_i| \le 1$, so we get

$$|n| \le 1 + n_0^\alpha + n_0^{2\alpha} + \cdots + n_0^{k\alpha}$$

$$= n_0^{k\alpha}\left(1 + n_0^{-\alpha} + n_0^{-2\alpha} + \cdots + n_0^{-k\alpha}\right)$$

$$= n_0^{k\alpha} \sum_{i=0}^{k} n_0^{-i\alpha}$$

$$\le n_0^{k\alpha} \sum_{i=0}^{\infty} n_0^{-i\alpha}$$

$$= n_0^{k\alpha} \frac{n_0^\alpha}{n_0^\alpha - 1}.$$

If we set $C = n_0^\alpha/(n_0^\alpha - 1)$ (which is, the reader will note, a positive number), we can read this as saying that

$$|n| \le C n_0^{k\alpha} \le C n^\alpha.$$

Now we use a dirty trick. This formula applies for every n (since the one we chose was arbitrary); applying it to an integer of the form n^N we get

$$|n^N| \le C n^{N\alpha}$$

(the crucial point is that the number C does not depend on n—check its definition above!). Taking N-th roots, we get

$$|n| \leq \sqrt[N]{C}\, n^\alpha.$$

Since any N will do, we can let $N \to \infty$, which makes $\sqrt[N]{C} \to 1$, and so gives an inequality: $|n| \leq n^\alpha$. This is half of what we want.

Now we need to show the inequality in the opposite direction. For that, we go back to the expression in base n_0

$$n = a_0 + a_1 n_0 + a_2 n_0^2 + \cdots + a_k n_0^k.$$

Since $n_0^{k+1} > n \geq n_0^k$, we get

$$n_0^{(k+1)\alpha} = |n_0^{k+1}| = |n + n_0^{k+1} - n| \leq |n| + |n_0^{k+1} - n|,$$

so that

$$|n| \geq n_0^{(k+1)\alpha} - |n_0^{k+1} - n| \geq n_0^{(k+1)\alpha} - (n_0^{k+1} - n)^\alpha,$$

where we have made use of the inequality proved in the previous paragraph. Now since $n \geq n_0^k$, it follows that

$$|n| \geq n_0^{(k+1)\alpha} - (n_0^{k+1} - n_0^k)^\alpha$$

$$= n_0^{(k+1)\alpha}\left(1 - \left(1 - \frac{1}{n_0}\right)^\alpha\right)$$

$$= C' n_0^{(k+1)\alpha}$$

$$> C' n^\alpha,$$

and once again $C' = 1 - (1 - 1/n_0)^\alpha$ does not depend on n and is positive. Using precisely the same trick as before, we get the reverse inequality $|n| \geq n^\alpha$, and hence $|n| = n^\alpha$. This proves that $|\ |$ is equivalent to the "usual" absolute value $|\ |_\infty$, as claimed.

b) Now suppose $|\ |$ is *non-archimedean*. Then, as we have shown, we have $|n| \leq 1$ for every integer n. Since $|\ |$ is non-trivial, there must exist a smallest integer n_0 such that $|n_0| < 1$.

The first thing to see is that n_0 must be a prime number. To see why, suppose that $n_0 = a \cdot b$ with a and b both smaller than n_0. Then, by our choice for n_0, we would have $|a| = |b| = 1$ and $|ab| = |n_0| < 1$, which cannot be. Thus, n_0 is prime, so let's call it by a prime-like name; set $p = n_0$. Now, of course, we want to show that $|\ |$ is equivalent to the p-adic absolute value, where p is this particular prime.

The next step is to show that if $n \in \mathbb{Z}$ is not divisible by p, then $|n| = 1$. This is not too hard. If we divide n by p we will have a remainder, so that we can write

$$n = rp + s$$

with $0 < s < p$. By the minimality of p (see the preceding paragraph), we have $|s| = 1$. We also have $|rp| < 1$, because $|r| \leq 1$ (because $|\ |$ is non-archimedean) and $|p| < 1$ (by construction). Since $|\ |$ is non-archimedean (and therefore "all triangles are isosceles"), it follows that $|n| = 1$.

Finally, given any $n \in \mathbb{Z}$, write it as $n = p^v n'$ with $p \nmid n'$. Then

$$|n| = |p|^v |n'| = |p|^v = c^{-v},$$

where $c = |p|^{-1} > 1$, so that $|\ |$ is equivalent to the p-adic absolute value, as claimed.   □

Problem 78 There's one fishy thing about the first part of the proof: once we have the conclusion we know $n_0 = 2$, but while we're proving we have to consider the possibility that n_0 is large. So we might have $n < n_0$, which would make the k in the expansion in base n_0 equal to zero. In other words, if $n < n_0$ its expansion in base n_0 is just n. Do we need to modify the proof to account for this case?

This theorem is the main reason for thinking of the "usual" absolute value $|\ |_\infty$ (or of the inclusion $\mathbb{Q} \hookrightarrow \mathbb{R}$ from which it comes) as some sort of "prime" of $\mathbb{Q}$. The point is that then it is true that every absolute value of $\mathbb{Q}$ "comes from" a (finite or infinite) prime.

There are lots of contexts in arithmetic where it is useful to work with "all of the primes," that is, to use information obtained from all of the absolute values of $\mathbb{Q}$. In terms of general "feeling," the real absolute value records information related to *sign*, while the other absolute values record information related to the various primes. Here is the most fundamental example of this:

Proposition 3.1.5 (Product Formula) *For any $x \in \mathbb{Q}^\times$, we have*

$$\prod_{p \leq \infty} |x|_p = 1,$$

where $p \leq \infty$ means that we take the product over all of the primes of $\mathbb{Q}$, including the "prime at infinity."

PROOF: It is easy to see that we only need to prove the formula when x is a positive integer, and that the general case will then follow. So let x be a positive integer, which we can factor as $x = p_1^{a_1} p_2^{a_2} \cdots p_k^{a_k}$. Then we have

$$\begin{cases} |x|_q = 1 & \text{if } q \neq p_i \\ |x|_{p_i} = p_i^{-a_i} & \text{for } i = 1, 2, \ldots, k \\ |x|_\infty = p_1^{a_1} p_2^{a_2} \cdots p_k^{a_k} \end{cases}$$

The result then follows.   □

This formula establishes a close relation between the absolute values of $\mathbb{Q}$; for example, it says that if we know all but one of the absolute values of a number $x \in \mathbb{Q}$, then we can determine the missing one. This turns out to be surprisingly important in many applications (for example, the theory of heights on algebraic varieties).

A similar result is true for finite extensions of $\mathbb{Q}$, except that in that case we must use *several* "infinite primes" (one for each different inclusion into $\mathbb{R}$ or $\mathbb{C}$). Of course, we also need an extension of Ostrowski's theorem for this to make sense, and a correct notion of a "prime" in such a field. It is because of these technicalities that we have chosen to deal only with the theory over $\mathbb{Q}$. See, for example, [20] for the general case.

3.2   Completions

We are now ready to construct, for each prime number p, the p-adic field $\mathbb{Q}_p$. The main point will be to pursue the idea that all of the absolute values on $\mathbb{Q}$ are "equally important," and hence should be treated equally. We first need to recall three important concepts from basic topology (we only state them in the context of fields with absolute values, but they are really general concepts for metric spaces).

Definition 3.2.1 *Let $\Bbbk$ be a field and let $|\ |$ be an absolute value on $\Bbbk$.*

   i) A sequence of elements $x_n \in \Bbbk$ is called a Cauchy sequence *if for every $\varepsilon > 0$ one can find a bound M such that we have $|x_n - x_m| < \varepsilon$ whenever $m, n \geq M$.*

   ii) The field $\Bbbk$ is called complete *with respect to $|\ |$ if every Cauchy sequence of elements of $\Bbbk$ has a limit in $\Bbbk$.*

   iii) A subset $S \subset \Bbbk$ is called dense *in $\Bbbk$ if every open ball around every element of $\Bbbk$ contains an element of S; in symbols, if for every $x \in \Bbbk$ and every $\varepsilon > 0$ we have*

$$B(x, \varepsilon) \cap S \neq \varnothing.$$

The reader has probably met these concepts in a course on real analysis, since one of the big things about the field $\mathbb{R}$ of real numbers is that it is a *complete* field, i.e., that every Cauchy sequence converges.[2] In intuitive terms, a Cauchy sequence is a sequence that "ought to" have a limit, because

---

[2]In most real analysis classes, the completeness of $\mathbb{R}$ is given by another condition, known as the "sup axiom," which is expressed in terms of the order relations in $\mathbb{R}$. In $\mathbb{R}$, the sup axiom is equivalent to completeness in our sense, but in general ordered fields the relationship between the two is complicated. In any case, our p-adic fields will not have an order, so the notion of a "least upper bound" makes no sense for them.

its terms get crowded into smaller and smaller balls (think of choosing a sequence of smaller and smaller values for ε). In other words, a field is complete if sequences that ought to converge do converge.

Problem 79 [Fancy] Show that completeness is equivalent to the following statement: suppose we find a decreasing sequence $r_n \in \mathbb{R}$, $r_n \to 0$, and a sequence $a_n \in \mathbb{k}$ such that

$$\overline{B}(a_1, r_1) \supset \overline{B}(a_2, r_2) \supset \cdots \supset \overline{B}(a_n, r_n) \supset \overline{B}(a_{n+1}, r_{n+1}) \supset \cdots$$

Then the balls $\overline{B}(a_n, r_n)$ have nontrivial intersection.

Most valued fields are not complete, but it turns out that one can always construct a completion, i.e., a bigger valued field that is complete and that contains the original field as a dense subset.

Problem 80 Show that $\mathbb{Q}$ is not complete with respect to the usual absolute value $|\ |_\infty$. (This was done in real analysis, too; one way is to construct a Cauchy sequence whose limit, if it existed, would have to be the square root of 2. Since 2 has no square root in $\mathbb{Q}$, there can be no limit.)

The following problem is intended to deal with a very common misunderstanding (which the reader also probably met in her course on real analysis). It is especially important to get this straight now, because things will get confusing for non-archimedean absolute values.

Problem 81 Show that the condition

$$\lim_{n \to \infty} |x_{n+1} - x_n| = 0$$

is *not the same* as the Cauchy condition, by showing that there exists a sequence of real numbers that satisfies this condition but is not a Cauchy sequence. In informal terms, the Cauchy condition is *stronger* than the assertion that successive terms of the sequence get closer and closer together. (Hint: one example of such a sequence was met in Calculus, in the portion on series.)

Our reason for recalling these notions is that, as our theory now stands, the archimedean absolute value $|\ |_\infty$ is different from all the rest, because there exists an inclusion $\mathbb{Q} \hookrightarrow \mathbb{R}$ of $\mathbb{Q}$ into a field $\mathbb{R}$ (yes, we do mean the real numbers) which is a completion:

- the absolute value $|\ |_\infty$ extends to $\mathbb{R}$,

- $\mathbb{R}$ is *complete* with respect to the metric given by this absolute value, and

- $\mathbb{Q}$ is *dense* in $\mathbb{R}$ (with respect to the metric given by $|\ |_\infty$).

This is all probably well-known to the reader (see the standard references for proofs). We summarize this list of properties by saying that $\mathbb{R}$ is the *completion* of $\mathbb{Q}$ with respect to the absolute value $|\ |_\infty$. The point is that

$\mathbb{R}$ is the smallest field containing $\mathbb{Q}$ which is complete with respect to this absolute value. We can see this because any such field would have to include the limit of any Cauchy sequence of elements of $\mathbb{Q}$, and, since $\mathbb{Q}$ is dense in $\mathbb{R}$, any element of $\mathbb{R}$ is a limit of such a sequence.

Problem 82 Can you prove the assertions of the preceding paragraph?

Our main goal in this section is to restore the parity between the absolute values on $\mathbb{Q}$, by constructing, for each of the other absolute values, a completion analogous to $\mathbb{R}$. That is, we want to show that for each prime p there exists some field to which we can extend the p-adic absolute value, which is then complete with respect to the extended absolute value, and in which $\mathbb{Q}$ is dense. The existence of such a field is a general theorem about metric spaces, which the reader may have met in another context; if so,[3] she may prefer to skip directly to the next chapter (or at least to Theorem 3.2.14). This section is for those who wish to see the full construction of such a completion.

One remark is important: as in the case of the construction of the real numbers, the method of constructing our completion is less important than the properties of the resulting field. In other words, the construction itself is important *only* because it establishes the existence of a completion. It will not be of any further use after that, so that skipping this section is a real possibility.

Problem 83 Should we bother trying to construct a completion of $\mathbb{Q}$ with respect to the trivial absolute value?

While our process for constructing a completion is valid in general, we will focus only on constructing completions of $\mathbb{Q}$. For the rest of this section, we let $|\ | = |\ |_p$ be the p-adic absolute value on $\mathbb{Q}$, for some prime p. The first useful thing to note is that Cauchy sequences can be characterized much more simply when the absolute value is non-archimedean.

Lemma 3.2.2 *A sequence (x_n) in a field $\Bbbk$ with a non-archimedean absolute value $|\ |$ is a Cauchy sequence if and only if we have*

$$\lim_{n \to \infty} |x_{n+1} - x_n| = 0.$$

PROOF: If $m = n + r > n$, we get

$$|x_m - x_n| = |x_{n+r} - x_{n+r-1} + x_{n+r-1} - x_{n+r-2} + \cdots + x_{n+1} - x_n|$$

$$\leq \max\{|x_{n+r} - x_{n+r-1}|, |x_{n+r-1} - x_{n+r-2}|, \ldots, |x_{n+1} - x_n|\},$$

because the absolute value is non-archimedean. The result then follows at once. $\qquad \square$

---

[3]or if she is willing to grant the existence of a completion

This makes analysis much simpler when the field is non-archimedean, as we will see later. We should insist, once again, that this lemma is false for archimedean absolute values, as Problem 81 shows.

The next step is to show that $\mathbb{Q}$ is not complete with respect to the p-adic absolute values, so that the completion process is really going to accomplish something.

Lemma 3.2.3 *The field $\mathbb{Q}$ of rational numbers is not complete with respect to any of its nontrivial absolute values.*

PROOF: Given Ostrowski's Theorem (3.1.4), we need to check this for $|\ |_p$ for $p \leq \infty$. That $\mathbb{Q}$ is not complete for $|\ |_\infty$ is well known (and left as a problem above), so we look at the p-adic absolute values.

If we take $|\ | = |\ |_p$ for some prime p, we need to construct a Cauchy sequence in $\mathbb{Q}$ which does not have a limit in $\mathbb{Q}$. This was essentially the content of section 3 of Chapter 1. To construct the necessary Cauchy sequence, we need only find a coherent sequence of solutions modulo p^n of an equation that has no solution in $\mathbb{Q}$. We work this out in the case $p \neq 2$, and leave the case $p = 2$ to the reader.

Thus, suppose $p \neq 2$ is a prime. Choose an integer $a \in \mathbb{Z}$ such that

- a is not a square in $\mathbb{Q}$;

- p does not divide a;

- a is a quadratic residue modulo p, i.e., the congruence $X^2 \equiv a \pmod{p}$ has a solution.

For example, we might take any square in $\mathbb{Z}$ and add a multiple of p to get a suitable a.

Now we can construct a Cauchy sequence (with respect to $|\ |_p$) in the following way:

- choose x_0 to be any solution of $x_0^2 \equiv a \pmod{p}$;

- choose x_1 so that $x_1 \equiv x_0 \pmod{p}$ and $x_1^2 \equiv a \pmod{p^2}$ (the existence of x_1 was proved in one of the problems in Chapter 1, and is easy to see in any case);

- in general, choose x_n so that

$$x_n \equiv x_{n-1} \pmod{p^n} \quad \text{and} \quad x_n^2 \equiv a \pmod{p^{n+1}}.$$

It was in fact checked in Problem 26 that such sequences do exist whenever a is a quadratic residue mod p (it is here that we need to know that $p \neq 2$).

The next step is to check that we really have a Cauchy sequence. It is clear from the construction that we have

$$|x_{n+1} - x_n| = |\lambda p^{n+1}| \leq p^{-(n+1)} \to 0,$$

which shows, together with Lemma 3.2.2, that the sequence of the x_n is indeed a Cauchy sequence. On the other hand, we also know that

$$|x_n^2 - a| = |\mu p^{n+1}| \le p^{-(n+1)} \to 0,$$

so that the limit, if it existed, would have to be a square root of a. Since a is not a square in $\mathbb{Q}$, there can be no limit in $\mathbb{Q}$, which shows $\mathbb{Q}$ is not complete with respect to $|\ |_p$. $\qquad\square$

Problem 84 Finish the proof, by showing that $\mathbb{Q}$ is also not complete with respect to the 2-adic absolute value. (Hint: the easiest way is probably to use cube roots instead of square roots...)

Since $\mathbb{Q}$ is not complete, we need to construct a completion. There are several ways to do so. We will follow the path of least resistance. What we want to do is to "add to $\mathbb{Q}$ the limits of all the Cauchy sequences." Since at first no such limits exist, one cannot literally do that. What we do instead is to use a standard bit of mathematical skullduggery, replacing the limit we do not have with the sequence we do have (so that in the end the sequence will be sort of like a limit of itself!). To do that, we begin with the set of all Cauchy sequences as the basic object, then use the algebraic operations on $\mathbb{Q}$ to handle the resulting object. (The construction uses some notions from abstract algebra; these can be avoided, but doing so would make our life much harder.)

Definition 3.2.4 *Let* $|\ | = |\ |_p$ *be a non-archimedean absolute value on* $\mathbb{Q}$. *We denote by* $\mathcal{C}$, *or* $\mathcal{C}_p(\mathbb{Q})$ *if we want to emphasize* p *and* $\mathbb{Q}$, *the set of all Cauchy sequences of elements of* $\mathbb{Q}$:

$$\mathcal{C} = \mathcal{C}_p(\mathbb{Q}) = \{(x_n) : (x_n) \text{ is a Cauchy sequence with respect to } |\ |_p\}.$$

The first thing to check is that $\mathcal{C}$ has a natural ring structure, using the "obvious" definitions for the sum and product of two sequences.

Proposition 3.2.5 *Defining*

$$(x_n) + (y_n) = (x_n + y_n)$$

$$(x_n) \cdot (y_n) = (x_n y_n)$$

makes $\mathcal{C}$ *a commutative ring with unity.*

PROOF: Easy; the only thing that really needs checking is that the sequences on the right-hand side are Cauchy. $\qquad\square$

Problem 85 Check that the sum and product of two Cauchy sequences, as defined above, are also Cauchy sequences.

Problem 86 What is the zero element of the ring $\mathcal{C}$? What is the unit element? Can you decide which elements are invertible?

Problem 87 Suppose (x_n) is a Cauchy sequence and (y_n) is a sequence such that $\lim\limits_{n\to\infty} |x_n - y_n| = 0$. Show that y_n is a Cauchy sequence. Show also that if $x_n \to a$ then $y_n \to a$ as well.

The ring $\mathcal{C}$ is not a field (as is clear from the previous exercise, since not all non-zero elements are invertible). In fact, it contains "zero divisors," i.e., non-zero elements whose product is zero.

Problem 88 Find two non-zero Cauchy sequences (say, with respect to the p-adic absolute value, but it doesn't really matter) whose product is the zero sequence.

We should check at once that this huge ring does contain the field of rational numbers, since, after all, the point of the whole exercise is to construct something which extends $\mathbb{Q}$. For that, all we need to do is to notice that if $x \in \mathbb{Q}$ is any number, the sequence

$$x, \; x, \; x, \; x, \ldots$$

is certainly Cauchy; we will call it the *constant sequence associated to x* and denote it by $\tilde{x}$. Then we have

Lemma 3.2.6 *The map $x \mapsto \tilde{x}$ is an injective ring homomorphism from $\mathbb{Q}$ into $\mathcal{C}$.*

PROOF: This is clear from the definitions. □

The main problem with $\mathcal{C}$ is that it does not yet capture the idea of "adding the limits of all Cauchy sequences," because different Cauchy sequences whose terms get close to each other "ought" to have the same limit, but they are different objects in $\mathcal{C}$. This sort of situation calls for identifying two sequences which "ought" to have the same limit, which means we must pass to a quotient[4] of $\mathcal{C}$.

It is here that the algebraic structure helps us, because it makes it easy to describe when it is that two sequences "ought" to have the same limit: this should happen when their terms get close to each other, i.e., when the difference of the sequences tends to zero. So we begin by looking at the set of sequences that tend to zero.

---

[4]The operation of passing to a quotient to identify objects is one of those absolutely basic ideas that one meets over and over in mathematics. In every case, one has to introduce an equivalence relation of some sort, then identify equivalent elements. In our situation, we will take advantage of the machinery of abstract algebra to do this, since $\mathcal{C}$ is a commutative ring.

Definition 3.2.7 *We define* $\mathcal{N} \subset \mathcal{C}$ *to be the ideal*

$$\mathcal{N} = \{(x_n) : x_n \to 0\} = \{(x_n) : \lim_{n \to \infty} |x_n|_p = 0\}$$

of sequences that tend to zero with respect to the absolute value $|\ |_p$.

Problem 89 Check that $\mathcal{N}$ is in fact an ideal of $\mathcal{C}$. (This is really already known from way back when in Calculus class.)

Lemma 3.2.8 $\mathcal{N}$ *is a* maximal *ideal of* $\mathcal{C}$.

PROOF: Let $(x_n) \in \mathcal{C}$ be a Cauchy sequence that does not tend to zero (i.e., does not belong to $\mathcal{N}$), and let I be the ideal generated by (x_n) and $\mathcal{N}$. What we want to show is that I must be all of $\mathcal{C}$. We will do that by showing that the unit element $\tilde{1}$ (i.e., the constant sequence corresponding to 1) is in I. This is enough, because any ideal that contains the unit element must be the whole ring.

Now, since (x_n) does not tend to zero and is a Cauchy sequence, it must "eventually" be away from zero, that is, there must exist a number $c > 0$ and an integer N such that $|x_n| \geq c > 0$ whenever $n \geq$ N. (If this is not clear, the reader should find a proof!) Now in particular this means that $x_n \neq 0$ for $n \geq$ N, so that we may define a new sequence (y_n) by setting $y_n = 0$ if $n <$ N and $y_n = 1/x_n$ if $n \geq$ N.

The first thing to check is that (y_n) is a Cauchy sequence. But that is clear because if $n \geq$ N we have

$$|y_{n+1} - y_n| = \left| \frac{1}{x_{n+1}} - \frac{1}{x_n} \right| = \frac{|x_{n+1} - x_n|}{|x_n x_{n+1}|} \leq \frac{|x_{n+1} - x_n|}{c^2} \longrightarrow 0,$$

which shows $(y_n) \in \mathcal{C}$ because $|\ |$ is non-archimedean. (One can modify the argument slightly so that it works also if $|\ |$ is archimedean, but this is easier.)
Now notice that

$$x_n y_n = \begin{cases} 0 & \text{if } n < N \\ 1 & \text{if } n \geq N \end{cases}$$

This means that the product sequence $(x_n)(y_n)$ consists of a finite number of 0s followed by an infinite string of 1s. In particular, if we subtract it from the constant sequence $\tilde{1}$, we get a sequence that tends to zero (in fact, which goes to zero and then stays there). In other words

$$\tilde{1} - (x_n)(y_n) \in \mathcal{N}.$$

But this says that $\tilde{1}$ can be written as a multiple of (x_n) plus an element of $\mathcal{N}$, and hence belongs to I, as we had claimed. Hence $I = \mathcal{C}$ and we have proved that $\mathcal{N}$ is maximal.   □

Problem 90 To make sure that you understand the proof, check that it works just as well for any field $\Bbbk$ with an absolute value $|\ |$. (The only catch is to supply a version of the check that the "almost inverse" sequence is Cauchy that does not depend on $|\ |$ being non-archimedean. But this is easy: the use of Lemma 3.2.2 is really a red herring.)

We want to identify sequences that differ by elements of $\mathcal{N}$, on the grounds that they ought to have the same limit. This is done in the standard way, by taking the quotient of the ring $\mathcal{C}$ by the ideal $\mathcal{N}$. To make things even nicer, taking a quotient of a commutative ring by a maximal ideal gives a field.

Definition 3.2.9 *We define the field of p-adic numbers to be the quotient of the ring $\mathcal{C}$ by its maximal ideal $\mathcal{N}$:*

$$\mathbb{Q}_p = \mathcal{C}/\mathcal{N}.$$

Notice that two different constant sequences never differ by an element of $\mathcal{N}$ (their difference is just another constant sequence, and the constant is not zero). Hence, we still have an inclusion

$$\mathbb{Q} \hookrightarrow \mathbb{Q}_p$$

by sending $x \in \mathbb{Q}$ to the equivalence class of the constant sequence $\tilde{x}$.

Very well: we now have a field, and an inclusion of $\mathbb{Q}$ into the field. It remains to check that it has the stated properties of the completion. The first is that the absolute value $|\ |_p$ extends to $\mathbb{Q}_p$. This follows easily from the following lemma.

Lemma 3.2.10 *Let $(x_n) \in \mathcal{C}$, $(x_n) \notin \mathcal{N}$. The sequence of real numbers $|x_n|_p$ is eventually stationary, that is, there exists an integer N such that $|x_n|_p = |x_m|_p$ whenever $m, n \geq N$.*

PROOF: Since (x_n) is a Cauchy sequence which does not tend to zero, we can (as in the previous lemma) find c and N_1 such that

$$n \geq N_1 \implies |x_n| \geq c > 0.$$

On the other hand, there also exists an integer N_2 for which

$$n, m \geq N_2 \implies |x_n - x_m| < c.$$

We want both conditions to be true at once, so set $N = \max\{N_1, N_2\}$. Then we have

$$n, m \geq N \implies |x_n - x_m| < \min\{|x_n|, |x_m|\},$$

which gives $|x_n| = |x_m|$ by the non-archimedean property ("all triangles are isosceles"). $\qquad\square$

This means that the following definition makes sense:

Definition 3.2.11 *If $\lambda \in \mathbb{Q}_p$ is an element of $\mathbb{Q}_p$, and (x_n) is any Cauchy sequence representing λ, we define*

$$|\lambda|_p = \lim_{n \to \infty} |x_n|_p.$$

(Recall that we have defined $\mathbb{Q}_p$ as a quotient, so that elements of $\mathbb{Q}_p$ are equivalence classes of Cauchy sequences.) There are several things to check here, but they are all quite easy to verify, so we leave them to the reader.

Problem 91 Let $\lambda \in \mathbb{Q}_p$. Explain why the limit defining $|\ |_p$ exists.

Problem 92 Let $\lambda \in \mathbb{Q}_p$. Show that $|\lambda|_p$, as defined above, does not depend on the choice of the sequence (x_n) representing λ. In other words, show that if we replace (x_n) by an equivalent sequence (y_n) (which means, recall, that the difference $(x_n - y_n)$ is a sequence that tends to zero), then

$$\lim_{n \to \infty} |x_n|_p = \lim_{n \to \infty} |y_n|_p.$$

(One can either do this directly, or note that the definition of $|\ |_p$ defines a function on $\mathcal{C}$ which maps $\mathcal{N}$ to zero, and hence descends to the quotient.)

Problem 93 Let $\lambda \in \mathbb{Q}_p$. Show that $|\lambda|_p = 0$ if and only if $\lambda = 0$. (You will need to remember what it means for an element to equal zero in the quotient.)

Problem 94 Show that the function $|\ |_p : \mathbb{Q}_p \longrightarrow \mathbb{R}_+$ is a non-archimedean absolute value.

Problem 95 Let $x \in \mathbb{Q}$, and let $\tilde{x}$ be the constant sequence which is the image of x in $\mathbb{Q}_p$. Check that the two definitions of $|\ |_p$ are consistent, that is, that $|\tilde{x}|_p = |x|_p$. (Yes, this is essentially obvious.)

These problems, taken together, show that we have indeed defined an absolute value on $\mathbb{Q}_p$ which extends the p-adic absolute value on $\mathbb{Q}$. There is one more important fact which should be recorded, which is that the image of the absolute value function is the same for both fields.

Problem 96 Show that the image of $\mathbb{Q}$ under $|\ |_p$ is equal to the image of $\mathbb{Q}_p$ under $|\ |_p$. In other words, for any $\lambda \in \mathbb{Q}_p$ which is different from zero, there exists an $n \in \mathbb{Z}$ such that $|\lambda|_p = p^{-n}$.

To check that we have indeed obtained the completion, we must now check the remaining two requirements: that $\mathbb{Q}$ is dense in $\mathbb{Q}_p$, and that $\mathbb{Q}_p$ is complete. The first is easy:

Proposition 3.2.12 *The image of $\mathbb{Q}$ under the inclusion $\mathbb{Q} \hookrightarrow \mathbb{Q}_p$ is a dense subset of $\mathbb{Q}_p$.*

PROOF: We need to show that any open ball around an element $\lambda \in \mathbb{Q}_p$ contains an element of (the image of) $\mathbb{Q}$, i.e., a constant sequence. So fix a radius $\varepsilon > 0$. We will show that there is a constant sequence belonging to the open ball $B(\lambda, \varepsilon)$.

First of all, let (x_n) be a Cauchy sequence representing λ, and let $\varepsilon' > 0$ be a number slightly smaller than ε. By the Cauchy property, there exists a number N such that $|x_n - x_m|_p < \varepsilon'$ whenever $n, m \geq N$. Let $y = x_N$ and consider the constant sequence $\tilde{y}$. We claim that

$$\tilde{y} \in B(\lambda, \varepsilon),$$

i.e., that $|\lambda - \tilde{y}|_p < \varepsilon$. To see this, recall that $\lambda - \tilde{y}$ is represented by the sequence $(x_n - y)$, and that we have defined

$$|(x_n - y)|_p = \lim_{n \to \infty} |x_n - y|_p.$$

But for any $n \geq N$ we have

$$|x_n - y|_p = |x_n - x_N|_p < \varepsilon'$$

so that, in the limit, we get

$$\lim_{n \to \infty} |x_n - y|_p \leq \varepsilon' < \varepsilon,$$

so that (y) does indeed belong to $B(\lambda, \varepsilon)$, and we are done.   □

Problem 97 Why does $<$ become $\leq$ in the limit? Do we really need the business of decreasing ε slightly to ε'?

It remains to show that $\mathbb{Q}_p$ is complete, i.e., that every Cauchy sequence in $\mathbb{Q}_p$ converges to an element of $\mathbb{Q}_p$. This seems almost obvious, until one realizes that a Cauchy sequence of elements of $\mathbb{Q}_p$ amounts to a sequence of Cauchy sequences and that the limit will have to be an equivalence class of Cauchy sequences. This seems to make everything very confusing. It is not so hard if one keeps one's wits about one, but that is easier said than done! The key is to use the fact that $\mathbb{Q}$ is dense in $\mathbb{Q}_p$.

Theorem 3.2.13 $\mathbb{Q}_p$ *is complete with respect to* $|\ |_p$.

PROOF: Let $\lambda_1, \lambda_2, \ldots, \lambda_n, \ldots$ be a Cauchy sequence of elements of $\mathbb{Q}_p$, so that each λ_i is a Cauchy sequence $(x_k^{(i)})$ of elements of $\mathbb{Q}$, taken up to equivalence.

Since (the image of) $\mathbb{Q}$ is dense in $\mathbb{Q}_p$, we can find, for each i, a number $y_i \in \mathbb{Q}$ such that the constant sequence $\tilde{y}_i \in \mathbb{Q}_p$ is as close to λ_i as we like. Taking, say,

$$|\lambda_i - \tilde{y}_i|_p < \frac{1}{i},$$

we can make sure that

$$\lim_{n\to\infty} |\lambda_n - \tilde{y}_n|_p = 0.$$

Now, using Exercise 87, we can conclude that the sequence $(\tilde{y}_n)$ (a sequence of constant sequences in $\mathbb{Q}_p$) is Cauchy. Therefore, since the absolute value of a constant sequence is the absolute value of the constant, the sequence (y_n) (a sequence of rational numbers) is Cauchy, so defines an element of $\mathbb{Q}_p$. Let λ be the element of $\mathbb{Q}_p$ corresponding to (y_n).

The sequence λ is, of course, the limit we are looking for. Let's prove it in two steps.

Let $\varepsilon > 0$. Since $\lambda = (y_n)$ is Cauchy, there exists an N such that $n, m \geq N$ implies $|y_m - y_m| < \frac{1}{2}\varepsilon$. Consider the sequence of constant sequences $(\tilde{y}_n)$. The difference $\lambda - \tilde{y}_n$ is represented by $(y_m - y_n)$, where n is fixed and m varies. So if $m \geq N$ we have

$$|\lambda - \tilde{y}_n|_p = \lim_{m\to\infty} |y_m - y_n|_p \leq \frac{1}{2}\varepsilon < \varepsilon.$$

Therefore the sequence $\lambda - (\tilde{y}_n)$ converges to 0 in $\mathbb{Q}_p$. In other words, the sequence of constant sequences $(\tilde{y}_n)$ converges to the Cauchy sequence $\lambda = (y_n)$ (as, indeed, we would have guessed!).

Now put it all together. We know that $|\lambda_n - \tilde{y}_n|$ converges to zero, and we know that $(\tilde{y}_n)$ converges to λ. Therefore (using Exercise 87 again), (λ_n) converges to λ. Since (λ_n) was an arbitrary Cauchy sequence in $\mathbb{Q}_p$, we have proved that any Cauchy sequence in $\mathbb{Q}_p$ has a limit.   □

That was an annoying proof, but we are now done. Putting it all together, we have proved the following theorem:

Theorem 3.2.14 *For each prime $p \in \mathbb{Z}$ there exists a field $\mathbb{Q}_p$ with a non-archimedean absolute value $|\ |_p$, such that:*

   i) there exists an inclusion $\mathbb{Q} \hookrightarrow \mathbb{Q}_p$, and the absolute value induced by $|\ |_p$ on $\mathbb{Q}$ via this inclusion is the p-adic absolute value;

   ii) the image of $\mathbb{Q}$ under this inclusion is dense in $\mathbb{Q}_p$ (with respect to the absolute value $|\ |_p$); and

   iii) $\mathbb{Q}_p$ is complete with respect to the absolute value $|\ |_p$.

The field $\mathbb{Q}_p$ satisfying (i), (ii) and (iii) is unique up to unique isomorphism preserving the absolute values.

PROOF: We've done it all except the uniqueness statement. To get that, suppose we have another such field K. Then we have two fields $\mathbb{Q}_p$ and K and inclusions $\mathbb{Q} \hookrightarrow \mathbb{Q}_p$ and $\mathbb{Q} \hookrightarrow K$, both of which preserve absolute values. If we have a Cauchy sequence (x_n) with $x_n \in \mathbb{Q}$, we can look at its image in $\mathbb{Q}_p$ and in K. Both will be Cauchy sequences (since the absolute value doesn't change) and so both will converge.

Now take $\lambda \in \mathbb{Q}_p$. Since $\mathbb{Q}$ is dense in $\mathbb{Q}_p$, there is a Cauchy sequence (x_n) with $x_n \in \mathbb{Q}$ whose limit is λ. Since the $x_n \in \mathbb{Q}$ we can take their images in K, and since the absolute value is preserved they will form Cauchy sequence as well. Since K is complete this sequence converges. Call the limit $f(\lambda)$. This defines a function $\mathbb{Q}_p \longrightarrow K$. Of course, f is the identity on $\mathbb{Q}$.

It is now easy to check that f is an isomorphism that preserves the absolute values, and its uniqueness is clear because it must induce the identity on the dense subset $\mathbb{Q}$. □

Problem 98 Fill in the gaps in the uniqueness proof:

i) Since the inclusion preserves the operations on $\mathbb{Q}$, and these operations are continuous, show that the extended map is a homomorphism of fields (and hence is injective).

ii) Perform precisely the same construction in reverse to get a map in the opposite direction, and show that the resulting map is the inverse of the first. (Hint: the composition is a continuous map which restricts to the identity on $\mathbb{Q}$!)

iii) Check that the isomorphism thus constructed preserves absolute values. (Hint: is the absolute value function itself continuous?)

The strong uniqueness statement is important because it says we can now forget the construction of $\mathbb{Q}_p$, and work only with the properties specified in the theorem. This is precisely what we will do.

Problem 99 Why is it important that something be "unique up to unique isomorphism"? Can you give an example of some mathematical object that is unique up to isomorphism, but not up to unique isomorphism?

Notice that this also shows that there is no nontrivial continuous automorphism $\mathbb{Q}_p \longrightarrow \mathbb{Q}_p$. In fact, we can even drop the assumption of continuity: the only field isomorphism $\mathbb{Q}_p \longrightarrow \mathbb{Q}_p$ is the identity. See [53, p. 53] for a proof. The same is true for $\mathbb{R}$.

A final remark: when we move from $\mathbb{Q}$ to $\mathbb{R}$, there is a very natural picture of what is happening: we visualize $\mathbb{R}$ as a line, and the rational numbers sit on that line but do not fill out all its points. We can think of the completion process as "filling in" the missing points. There is no such simple picture for the move from $\mathbb{Q}$ to $\mathbb{Q}_p$, because the topology we get from the p-adic absolute value is much stranger, as we already know. In particular, $\mathbb{Q}_p$ is totally disconnected, so it looks nothing like the real line.

4 Exploring $\mathbb{Q}_p$

The goal of this chapter is to explore the field $\mathbb{Q}_p$ which we have just constructed. The basic idea is to get away from the explicit construction we gave above and to come up with other ways to represent and understand the elements of $\mathbb{Q}_p$. In particular, we will show that every element of $\mathbb{Q}_p$ can be represented uniquely by a series in ascending powers of p, as in Chapter 1.

4.1 What We Already Know

At the end of the previous chapter, we showed the field $\mathbb{Q}_p$ is entirely determined by its properties:

- There is an absolute value $|\ |_p$ on $\mathbb{Q}_p$, and $\mathbb{Q}_p$ is complete with respect to this absolute value.

- There is an inclusion $\mathbb{Q} \hookrightarrow \mathbb{Q}_p$ whose image is dense in $\mathbb{Q}_p$, and the restriction of the absolute value $|\ |_p$ to (the image of) $\mathbb{Q}$ coincides with the p-adic absolute value.

- The image of both $\mathbb{Q}$ and $\mathbb{Q}_p$ under $|\ |_p$ is the same; specifically, the two sets

$$\{x \in \mathbb{R}_+ \mid x = |\lambda|_p \text{ for some } \lambda \in \mathbb{Q}\}$$

and

$$\{x \in \mathbb{R}_+ \mid x = |\lambda|_p \text{ for some } \lambda \in \mathbb{Q}_p\}$$

are both equal to the set $\{p^n \mid n \in \mathbb{Z}\} \cup \{0\}$ of powers of p, together with 0.

As the wording suggests, we will from now on *identify* $\mathbb{Q}$ with its image under the inclusion in $\mathbb{Q}_p$, that is, we will think of $\mathbb{Q}$ as a subfield of $\mathbb{Q}_p$. The last property turns out to be very useful, so we will re-state it as a lemma.

Lemma 4.1.1 *For each $x \in \mathbb{Q}_p$, $x \neq 0$, there exists an integer $n \in \mathbb{Z}$ such that $|x|_p = p^{-n}$. Conversely, for each $n \in \mathbb{Z}$ we can find $x \in \mathbb{Q}_p$ such that $|x|_p = p^n$.*

Another way of saying this is in terms of the p-adic valuation v_p. Remember that for $x \in \mathbb{Q}$ we had $|x|_p = p^{-v_p(x)}$; so what the lemma says is:

© Springer Nature Switzerland AG 2020

F. Q. Gouvêa, *p-adic Numbers*, Universitext, https://doi.org/10.1007/978-3-030-47295-5_4

Lemma 4.1.2 *For each* $x \in \mathbb{Q}_p$, $x \neq 0$, *there exists an integer* $v_p(x)$ *such that* $|x|_p = p^{-v_p(x)}$. *In other words, the p-adic valuation* v_p *extends to* $\mathbb{Q}_p$.

As before, we extend v_p to all of $\mathbb{Q}_p$ by setting $v_p(0) = +\infty$. Later in this section (when we have a good way to describe elements of $\mathbb{Q}_p$) we will be able to describe v_p in another way.

4.2   *p*-adic Integers

Now we begin to explore the structure of $\mathbb{Q}_p$. Since $\mathbb{Q}_p$ is a field with a non-archimedean valuation, everything we did in Chapter 2 applies. In particular, we can consider the corresponding valuation ring, which has a name of its own:

Definition 4.2.1 *The* ring of *p*-adic integers *is the valuation ring*

$$\mathbb{Z}_p = \{x \in \mathbb{Q}_p : |x|_p \leq 1\}.$$

Of course, $\mathbb{Z}_p$ is also the closed unit ball with center 0, so we already know a few things about it. Since $\mathbb{Z}_p$ is a closed set, every convergent sequence of elements of $\mathbb{Z}_p$ has a limit in $\mathbb{Z}_p$. Since $\mathbb{Q}_p$ is complete, all Cauchy sequences converge. So $\mathbb{Z}_p$ is a complete metric space. $\mathbb{Z}_p$ is also an open set, because every ball is. Here is a much more precise description:

Proposition 4.2.2 *The ring* $\mathbb{Z}_p$ *of p-adic integers is a local ring whose maximal ideal is the principal ideal* $p\mathbb{Z}_p = \{x \in \mathbb{Q}_p : |x|_p < 1\}$. *Furthermore,*

i) $\mathbb{Q} \cap \mathbb{Z}_p = \mathbb{Z}_{(p)} = \left\{ \dfrac{a}{b} \in \mathbb{Q} : p \nmid b \right\}$.

ii) *The inclusion* $\mathbb{Z} \hookrightarrow \mathbb{Z}_p$ *has dense image. Specifically, given* $x \in \mathbb{Z}_p$ *and* $n \geq 1$, *there exists an* $\alpha \in \mathbb{Z}$, $0 \leq \alpha \leq p^n - 1$, *such that* $|x - \alpha|_p \leq p^{-n}$. *The integer* α *with these properties is unique.*

iii) *For any* $x \in \mathbb{Z}_p$, *there exists a Cauchy sequence* (α_n) *converging to* x, *of the following type:*

- $\alpha_n \in \mathbb{Z}$ *satisfies* $0 \leq \alpha_n \leq p^n - 1$
- *for every* $n \geq 2$ *we have* $\alpha_n \equiv \alpha_{n-1} \pmod{p^{n-1}}$.

The sequence (α_n) *with these properties is unique.*

PROOF: Most of this follows directly from things we have already checked. To begin with, $\mathbb{Z}_p$ is a valuation ring, hence (Prop. 2.4.3) it is a local ring, i.e., there is a unique maximal ideal and every element of $\mathbb{Z}_p$ not in the maximal ideal is invertible in $\mathbb{Z}_p$. To see that the valuation ideal is indeed generated by p, we use Lemma 4.1.1: if $|x|_p < 1$, then in fact $|x|_p \leq \frac{1}{p}$; since $|p|_p = 1/p$, this implies $\left|\frac{x}{p}\right|_p \leq 1$, so $x \in p\mathbb{Z}_p$. This shows that the valuation ideal is *contained* in $p\mathbb{Z}_p$, but that is enough, since the valuation ideal is a *maximal* ideal, and $p\mathbb{Z}_p \neq \mathbb{Z}_p$. Now to the other statements:

(i) is clear, because we already know that $\mathbb{Z}_{(p)}$ is the valuation ring in $\mathbb{Q}$ corresponding to the p-adic valuation.

To check (ii), choose $x \in \mathbb{Z}_p$ and $n \geq 1$. Since $\mathbb{Q}$ is dense in $\mathbb{Q}_p$, one can certainly find $a/b \in \mathbb{Q}$ which is as close as we like to x; choose it so that

$$\left|x - \frac{a}{b}\right|_p \leq p^{-n} < 1.$$

The point is to show that we can in fact choose an *integer*. But notice that for a/b as above, we will have

$$\left|\frac{a}{b}\right|_p \leq \max\left\{|x|_p, \left|x - \frac{a}{b}\right|_p\right\} \leq 1,$$

which says that $a/b \in \mathbb{Z}_{(p)}$, that is, $p \nmid b$. Now recall that, from the elementary theory of congruences, if $p \nmid b$ there exists an integer $b' \in \mathbb{Z}$ such that $bb' \equiv 1$ (mod p^n), unique mod p^n. This implies (the reader will check) that

$$\left|\frac{a}{b} - ab'\right|_p \leq p^{-n},$$

and of course $ab' \in \mathbb{Z}$. Finally, we need to check that we can find an integer between zero and $p^n - 1$, but this is clear from the connection between congruences modulo powers of p and the p-adic absolute value: two integers are p^{-n}-close if and only if they are congruent mod p^n. Choosing α to be the unique integer such that

$$0 \leq \alpha \leq p^n - 1 \qquad \text{and} \qquad \alpha \equiv ab' \pmod{p^n}$$

gives $|x - \alpha|_p \leq p^{-n}$ (check it!), which is what we want.

Finally, (iii) follows directly from (ii); just use (ii) for a sequence of integers $n = 1, 2, \ldots$. For the uniqueness, notice that at each step of the construction in (ii) our choices were unique mod p^n. □

This proposition says several important things (and implies a bunch of others—see the next few corollaries). For example, it shows that every element of $\mathbb{Z}_p$ is the limit of a sequence of integers, so that

Corollary 4.2.3 $\mathbb{Z}$ *is dense in* $\mathbb{Z}_p$.

This would, in fact, be another way to begin the whole story, by creating $\mathbb{Z}_p$ as the completion of $\mathbb{Z}$ with respect to the p-adic absolute value. Notice, too, that the sequence in (iii) is exactly one of our "coherent sequences" from Chapter 1, so that things are coming together rather nicely.

Here are some more consequences.

Corollary 4.2.4 $\mathbb{Q}_p = \mathbb{Z}_p[1/p]$, *that is, for every* $x \in \mathbb{Q}_p$ *there exists an* $n \geq 0$ *such that* $p^n x \in \mathbb{Z}_p$. *The map* $\mathbb{Q}_p \longrightarrow \mathbb{Q}_p$ *given by* $x \mapsto px$ *is a homeomorphism. (This means that it is a continuous map with a continuous inverse, so it preserves the topology of* $\mathbb{Q}_p$.) *The sets* $p^n \mathbb{Z}_p$, $n \in \mathbb{Z}$ *form a fundamental system of neighborhoods of* $0 \in \mathbb{Q}_p$ *which covers all of* $\mathbb{Q}_p$.

PROOF: If $x \in \mathbb{Q}_p$, we can compute its valuation $v_p(x)$. If $v_p(x) \geq 0$, then x is already an element of $\mathbb{Z}_p$, by the definition of a valuation ring. Otherwise, $v_p(x)$ is negative, and we have

$$v_p(p^{-v_p(x)}x) = -v_p(x) + v_p(x) = 0,$$

which means that $p^{-v_p(x)}x \in \mathbb{Z}_p$, as claimed. That multiplication by p is a homeomorphism is immediate from the fact that the field operations are continuous functions. The remaining statements will be checked in the next problem. □

Problem 100 Prove the corollary. Recall that a *neighborhood* of a point x is a set containing an open ball around x, and that a *fundamental system of neighborhoods* is a bunch of neighborhoods with the property that any other neighborhood contains one of them. Finally, a collection of sets *covers* a set X if the union of all the sets in the collection contains (or is) the set X.

It may be useful to remember that a map is continuous exactly when the inverse image of any open set is an open set (this is often easier to work with than the ε-δ definition).

Problem 101 (Just to keep us awake.) Describe a fundamental system of neighborhoods of 0 in $\mathbb{R}$ which also covers $\mathbb{R}$.

Recall that we pointed out that the p-adic valuation v_p can be extended to $\mathbb{Q}_p$, because for any $x \in \mathbb{Q}_p$ there exists an integer $v_p(x)$ such that $|x|_p = p^{-v_p(x)}$. The last corollary allows us to understand this a little better:

Problem 102 Show that we can give the following more natural description of $v_p(x)$: by the corollary, x belongs to some $p^n \mathbb{Z}_p$; let n_0 be the *largest* n for which this is true; then $v_p(x) = n_0$. (Be careful: n_0 may very well be negative.)

Hence, for example, $v_p(x) = 0$ if $x \in \mathbb{Z}_p$ but $x \notin p\mathbb{Z}_p$, so that $n_0 = 0$. This agrees, of course, with the original definition, since $v_p(x) = 0$ means $|x| = 1$.

One of the main points of these results is that the topology (neighborhoods, open sets,...) of $\mathbb{Q}_p$ is closely connected to its algebraic structure (multiplication by p, subrings). For example, it is very useful to burn into one's brain that for x, $y \in \mathbb{Q}_p$ we have

$$|x - y|_p \leq p^{-n} \qquad \text{if and only if} \qquad x - y \in p^n \mathbb{Z}_p.$$

In particular,
$$\overline{B}(0, p^{-n}) = p^n \mathbb{Z}_p.$$
The next few results forge ahead in this direction.

Corollary 4.2.5 *For any $n \geq 1$, the sequence*

$$0 \longrightarrow \mathbb{Z}_p \xrightarrow{p^n} \mathbb{Z}_p \longrightarrow \mathbb{Z}/p^n\mathbb{Z} \longrightarrow 0,$$

where the map $\mathbb{Z}_p \longrightarrow \mathbb{Z}_p$ is given by $x \mapsto p^n x$, is exact, and the maps are continuous (where we give $\mathbb{Z}/p^n\mathbb{Z}$ the discrete topology). In particular,

$$\mathbb{Z}_p / p^n \mathbb{Z}_p \cong \mathbb{Z}/p^n\mathbb{Z}.$$

Recall that a sequence $A \xrightarrow{f} B \xrightarrow{g} C$ is *exact* if image(f) = ker(g). A five-term sequence as above is exact when it is exact at each stage, so that the claims above are:

- the map $\mathbb{Z}_p \longrightarrow \mathbb{Z}_p$ given by multiplication by p^n is injective (its kernel is the image of zero, which is zero)

- there is a map $\mathbb{Z}_p \longrightarrow \mathbb{Z}/p^n\mathbb{Z}$ which is surjective

- the kernel of this map is precisely the image of $\mathbb{Z}_p$ under the first map, which of course is $p^n \mathbb{Z}_p$.

Recall, too, that the discrete topology is the one where *all* sets are open.

Problem 103 Check that the corollary is true.

The sets $a + p^n \mathbb{Z}_p$, with $a \in \mathbb{Q}$ and $n \in \mathbb{Z}$ are closed balls in $\mathbb{Q}_p$ (with center a and radius p^{-n}), hence are clopen sets. Since $\mathbb{Q}$ is dense in $\mathbb{Q}_p$, they cover all of $\mathbb{Q}_p$. As we have already shown for general ultrametric spaces (Prop. 2.3.9), $\mathbb{Q}_p$ is totally disconnected (the connected component of any point is the set consisting of only that point). Furthermore, given any two points we can always find balls around them that do not intersect (which is a useful thing to know about a topology: points can be separated). In big words:

Corollary 4.2.6 $\mathbb{Q}_p$ *is a totally disconnected Hausdorff topological space.*

A more interesting topological property is compactness, which plays a big role in classical analysis. A subset X of a topological space is called *compact* if it has the following property:

- any collection of open sets which covers X has a *finite* subcollection which also covers X.

This is a rather unintuitive definition, but it turns out to be quite important. For example, the compact sets in $\mathbb{R}$ are precisely the closed and bounded sets, which play a big role in real analysis.

Problem 104 Read up on compactness in any introductory book on general topology. In particular, prove, or find out how to prove, or read the proof of, the following:

i) The image of a compact set by a continuous map is a compact set.

ii) In $\mathbb{R}$, any closed and bounded set, and in particular a closed interval, is compact. (In a general metric space, any compact set is closed and bounded, but the converse is not always true.)

iii) In a metric space, a set X is compact if and only if every sequence (x_n) with $x_n \in X$ has a convergent subsequence.

iv) In a metric space, a set X will be compact if and only if it is complete (every Cauchy sequence in X converges to a point in X) and totally bounded (for every $\varepsilon > 0$, there exists a *finite* covering of X by balls of radius ε).

Problem 105 A space is called *locally compact* when every point has a neighborhood which is a compact set. Show that $\mathbb{R}$ is locally compact. (This property is very important in classical analysis.)

Problem 106 If $\Bbbk$ is a field with an absolute value, show that $\Bbbk$ is locally compact if and only if there exists a neighborhood of zero that is compact. (Hint: if a set X is compact, the set $\{a + x : x \in X\}$ is the image of X under a continuous map, hence is also compact.)

Corollary 4.2.7 $\mathbb{Z}_p$ *is compact, and* $\mathbb{Q}_p$ *is locally compact.*

PROOF: Since $\mathbb{Z}_p$ is a neighborhood of zero, proving that it is compact is enough to prove that $\mathbb{Q}_p$ is locally compact, so that the second statement follows from the first.

To prove the first statement, remember that we already know that $\mathbb{Z}_p$ is complete (because it is a closed set in a complete field), so that (using one of the statements above) what we need to prove is that it is totally bounded, that is, that for any $\varepsilon > 0$ one can cover $\mathbb{Z}_p$ with finitely many balls of radius ε. It is enough to check this for every $\varepsilon = p^{-n}$, $n \geq 0$. But remember that

$$\mathbb{Z}_p / p^n \mathbb{Z}_p \cong \mathbb{Z} / p^n \mathbb{Z},$$

and that the cosets of $p^n \mathbb{Z}_p$ in $\mathbb{Z}_p$ are also balls in the p-adic topology. This means that as a ranges through $0, 1, \ldots, p^n - 1$ (or any other set of coset representatives), the p^n balls

$$a + p^n \mathbb{Z}_p = \{a + p^n x : x \in \mathbb{Z}_p\} = \{y \in \mathbb{Z}_p : |y - a| \leq p^{-n}\} = \overline{B}(a, p^{-n})$$

cover $\mathbb{Z}_p$, and we are done.   □

Problem 107 Why is it enough to check for this special family of values for ε?

One should notice that the crucial element in our proof of the compactness is the finiteness of the quotients. In fact, one can check that knowing that one quotient is finite will do the trick.

Problem 108 Let $\Bbbk$ be a field, $|\ |$ a non-archimedean absolute value on $\Bbbk$, $\mathcal{O} \subset \Bbbk$ the valuation ring, and $\mathfrak{P}$ the valuation ideal. Suppose that $\Bbbk$ is complete and that $\mathfrak{P}$ is principal. Show that $\Bbbk$ is locally compact if and only if the residue field $\mathcal{O}/\mathfrak{P}$ is finite. Do we really need the completeness of $\Bbbk$? Do we really need to know that $\mathfrak{P}$ is principal?

The *p-adic units* are the invertible elements of $\mathbb{Z}_p$. We will denote the set of all such elements by $\mathbb{Z}_p^\times$. Since $x \in \mathbb{Z}_p$ means $|x|_p \le 1$ and $x^{-1} \in \mathbb{Z}_p$ means $|x^{-1}|_p = |x|_p^{-1} \le 1$, we see that

$$\mathbb{Z}_p^\times = \{x \in \mathbb{Q}_p : |x|_p = 1\}.$$

It is also easy to see that

$$\mathbb{Z}_p^\times \cap \mathbb{Q} = \left\{ \frac{a}{b} \in \mathbb{Q} : p \nmid ab \right\}.$$

Like the invertible elements of every ring, the p-adic units form a group. In our case, this group contains quite a few elements (notice that $\mathbb{Z}_p^\times \cap \mathbb{Q}$ is already quite large). We will later study its structure a little more closely. For now, notice that it is a closed subset of $\mathbb{Z}_p$, and therefore is compact.

Problem 109 What are the invertible elements of $\mathbb{Z}$? Of $\mathsf{F}[t]$? Of $\mathbb{C}[[t]]$? ($\mathbb{C}[[t]]$ is the ring of power series in one variable with coefficients in $\mathbb{C}$)

Problem 110 Let $a, b \in \mathbb{Z}$ be integers with b not divisible by p. Define sets

$$A^+(a, b) = \{n \in \mathbb{Z} \mid n \equiv a \pmod{b} \text{ and } n > 0\},$$

and

$$A^-(a, b) = \{n \in \mathbb{Z} \mid n \equiv a \pmod{b} \text{ and } n < 0\}.$$

Show that both sets are dense in $\mathbb{Z}_p$.

4.3   The Elements of $\mathbb{Q}_p$

The elements of $\mathbb{Q}_p$ are, at this point, hard to grab hold of, because we only know $\mathbb{Q}_p$ via its basic properties. To counteract this, we will now give two different descriptions of the elements of $\mathbb{Q}_p$, both of which we have already met in Chapter 1: as "coherent sequences," and as "p-adic expansions."

The description in terms of coherent sequences, which we will give first, is interesting for theoretical reasons, while the description in terms of expansions will give us the most "concrete" version of $\mathbb{Q}_p$. The first description will be stated in rather sophisticated terms, and the reader may want to skim through it rather than check all the details.

We begin from item (iii) in Proposition 4.2.2: given $x \in \mathbb{Z}_p$, we can find a rather special kind of Cauchy sequence converging to x. This sequence has the property of being "coherent," which we met in Chapter 1:

- $\alpha_n \in \mathbb{Z}$, $0 \leq \alpha_n \leq p^n - 1$

- $\alpha_{n+1} \equiv \alpha_n \pmod{p^n}$

and in addition converges to x because $|x - \alpha_n|_p \leq p^{-n}$. Finally, we checked that this sequence is unique.

On the other hand, suppose we have such a sequence (α_n). The coherence property clearly makes it a Cauchy sequence, because $|\alpha_{n+1} - \alpha_n|_p \leq p^{-n}$. Hence, it must converge to some element, which will be in $\mathbb{Z}_p$ because the α_n are in $\mathbb{Z}$.

Problem 111 Check that a limit of a Cauchy sequence of integers must be an element of $\mathbb{Z}_p$ (rather than merely of $\mathbb{Q}_p$).

This means that we can *identify* the elements of $\mathbb{Z}_p$ with such sequences. We will summarize this in the next proposition, but in a rather sophisticated language. To set it up, let's write φ_n for the projection on the quotient

$$\varphi_n : \mathbb{Z}_p \longrightarrow \mathbb{Z}/p^n\mathbb{Z}.$$

As an element of $\mathbb{Z}/p^n\mathbb{Z}$, we then have $\varphi_n(x) \equiv \alpha_n \pmod{p^n}$ (just because the set of integers between 0 and $p^n - 1$ gives representatives for the cosets, and the α_n are chosen as the representatives corresponding to x). We also set

$$A_n = \mathbb{Z}/p^n\mathbb{Z}$$

and think of it as a topological ring with a discrete topology.[1] We have an obvious map $\psi_n : A_n \longrightarrow A_{n-1}$, which sends $(a \mod p^n)$ to $(a \mod p^{n-1})$. We want to consider the product of all these rings, that is, the ring of sequences (α_n) such that $\alpha_n \in A_n$. (The operations are defined in the obvious way, term by term.) There is a standard way to put a topology on this ring (it is called the *product topology*). This topology is rather tricky to describe, and we do not really need to know much about it. We just point out that the product ring will be compact with this topology.

With all that set up, we can state:

---

[1] This is mumbo-jumbo. All the other rings will have a topology because they have absolute values. The ring A_n, on the other hand, doesn't come with such a "built-in" topology, so we just give it the simplest one of all: the one where all sets are open, which corresponds to the trivial absolute value. The point is that this makes the important maps (the projection from $\mathbb{Z}_p$ to A_n and the homomorphisms $A_n \longrightarrow A_{n-1}$) be continuous.

Proposition 4.3.1 *The projection maps φ_n together give an inclusion*

$$\varphi : \mathbb{Z}_p \hookrightarrow \prod_{n \geq 1} A_n$$

which identifies $\mathbb{Z}_p$, as a topological ring, with the closed subring of $\prod A_n$ consisting of the coherent sequences, i.e., those sequences (α_n) for which we have $\psi_n(\alpha_n) = \alpha_{n-1}$ for every $n > 1$.

PROOF: If all the concepts are understood, this is just a re-statement of known facts. See the next problem. □

Problem 112 Prove the proposition. Notice that we could use this to give another construction of $\mathbb{Z}_p$ with a more algebraic flavor (and a bit more subtle to handle). For example, the fact that closed subsets of a compact set are necessarily compact would provide the proof that $\mathbb{Z}_p$ is compact in this version of the theory.

It is often useful to describe how several functions are related by drawing what is called a "commutative diagram." One says a diagram of homomorphisms is *commutative* if the homomorphisms obtained by "following different routes around the diagram" always coincide. For example, the diagram

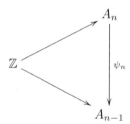

is commutative, because reducing modulo p^n and then reducing modulo p^{n-1} is the same as reducing modulo p^{n-1} by itself, so that one can follow either path from $\mathbb{Z}$ to A_{n-1} and get the same map. Using the language of commutative diagrams, one can describe a very important property of $\mathbb{Z}_p$:

Problem 113 Show that $\mathbb{Z}_p$ has the following property, which is an instance of what are usually called *universal properties*: given any ring R plus homomorphisms $R \longrightarrow A_n$ (one for each $n \geq 1$) such that all the triangles

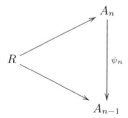

are commutative, there exists a unique homomorphism $R \longrightarrow \mathbb{Z}_p$ from which all the maps to the A_n are obtained (i.e., all the obvious triangles commute). In highfalutin' terms, this says that $\mathbb{Z}_p$ is the *inverse limit* of the A_n.

One can begin the theory from this point, and deduce all the rest from the universal property; this is the approach in [60]. For ordinary mortals, however, this may all be a little too abstract, so we go on to obtain a canonical way to represent the elements of $\mathbb{Q}_p$ as "power series in p." This will finally return us to the picture we sketched in Chapter 1.

We begin with a p-adic integer $x \in \mathbb{Z}_p$. As we have just shown, there exists a coherent sequence of integers α_n converging to x such that:

- $\alpha_n \equiv x \pmod{p^n}$

- $\alpha_{n+1} \equiv \alpha_n \pmod{p^n}$

- $0 \le \alpha_n \le p^n - 1$.

To understand the α_n a little better, we write them in base p. As we saw in Chapter 1, for integers written in base p the process of reducing modulo p^n is very simple: just strip off all but the last n digits.[2] This means that the coherence condition

$$\alpha_{n+1} \equiv \alpha_n \pmod{p^n}$$

simply says that the last n digits of both numbers are the same. Going up the sequence, what we get is

$$
\begin{array}{ll}
\alpha_1 = b_0 & 0 \le b_0 \le p - 1 \\
\alpha_2 = b_0 + b_1 p & 0 \le b_1 \le p - 1 \\
\alpha_3 = b_0 + b_1 p + b_2 p^2 & 0 \le b_2 \le p - 1 \\
\alpha_4 = b_0 + b_1 p + b_2 p^2 + b_3 p^3 & 0 \le b_3 \le p - 1
\end{array}
$$

and so on. But these are just a sequence of partial sums of a series. So we get a series expansion

$$x = b_0 + b_1 p + b_2 p^2 + \cdots + b_n p^n + \cdots$$

Of course, to be able to really write that equals sign with a clear conscience, we must check that the series on the right does converge to x. But that is easy:

Lemma 4.3.2 *Given any $x \in \mathbb{Z}_p$, the series*

$$b_0 + b_1 p + b_2 p^2 + \cdots + b_n p^n + \cdots$$

obtained as above converges to x.

---

[2] Just as in base 10: to get your number modulo 10, keep the last digit only; to get it modulo 100, keep the last two, and so on.

PROOF: Remember that a series converges to x if and only if the sequence of its partial sums converges to x. But the partial sums of our series are exactly the a_n, which we already know converge to x (we picked them that way). □

To sum up, this gives

Corollary 4.3.3 *Every $x \in \mathbb{Z}_p$ can be written in the form*

$$x = b_0 + b_1 p + b_2 p^2 + \cdots + b_n p^n + \cdots$$

with $0 \leq b_i \leq p - 1$, and this representation is unique.

PROOF: We have checked all but the uniqueness. To see that, notice that we already know the a_n are unique, and this implies that the b_n are too (because they are just the digits[3] in base p). □

Now, we need to get all of $\mathbb{Q}_p$. But remember that any element of $\mathbb{Q}_p$ can be written in the form $p^m y$ with $y \in \mathbb{Z}_p$ and $m \in \mathbb{Z}$ (the interesting case for us is when m is negative, of course). If we express y as a power series in p, then multiply by p^m, we just get a power series in p where some of the powers may be negative. So:

Corollary 4.3.4 *Every $x \in \mathbb{Q}_p$ can be written in the form*

$$x = b_{-m} p^{-m} + \cdots + b_{-1} p^{-1} + b_0 + b_1 p + b_2 p^2 + \cdots + b_n p^n + \cdots$$

$$= \sum_{n \geq -m} b_n p^n$$

with $0 \leq b_n \leq p - 1$ and $-m = v_p(x)$. This representation is unique.

PROOF: All that remains to be checked is the statement about $v_p(x)$, which is clear. □

This lands us right back in Chapter 1, and shows that one can think of an element of $\mathbb{Q}_p$, i.e., a p-adic number, as a power series in p. Or, if we prefer, we can think of them as "infinitely[4] long numbers in base p," writing $\ldots b_n \ldots b_2 b_1 b_0.b_{-1} \ldots b_{-m}$ for the series above.

Problem 114 Let $x \in \mathbb{Z}_p$. What condition on its p-adic expansion will guarantee that x is a p-adic unit?

Problem 115 One of the consequences of the fact that $\mathbb{Z}_p$ is compact is the fact that every infinite sequence of elements of $\mathbb{Z}_p$ has a convergent subsequence. Use the p-adic expansion to show this directly.

---

[3] Should they be called pigits? Or pits?
[4] Allowing for infinite strings of zeros, of course.

As we noted in passing, the coefficients b_n must be taken in a set of representatives of the classes modulo p. The numbers between 0 and $p-1$ are the most obvious choice for these representatives, but there are situations where other choices are expedient. For example, we will soon prove that $\mathbb{Z}_p$ contains the $(p-1)$-st roots of unity and that they are all different mod p. That means those roots of unity, together with 0, form a set of representatives of the congruence classes mod p, and so might be used as digits. For future use, then, let's record the fact that any set of representatives can serve as the p-adic digits.

Corollary 4.3.5 *Choose $A \subset \mathbb{Z}_p$ to be a set of representatives of $\mathbb{Z}/p\mathbb{Z}$. Every $x \in \mathbb{Q}_p$ can be written in the form*

$$x = b_{-m}p^{-m} + \cdots + b_{-1}p^{-1} + b_0 + b_1 p + b_2 p^2 + \cdots + b_n p^n + \cdots$$

$$= \sum_{n \geq -m} b_n p^n$$

with $b_n \in A$ for each n and $-n_0 = v_p(x)$. This representation is unique.

PROOF: Suppose first that $x \in \mathbb{Z}_p$ and look at its image in $\mathbb{Z}_p/p\mathbb{Z}_p = \mathbb{Z}/p\mathbb{Z}$. By our choice of A there is a unique element $b_0 \in A$ such that $x - b_0 \in p\mathbb{Z}_p$. Then $x - b_0 = px_1$ for some $x_1 \in \mathbb{Z}_p$. As before, there exists a unique $b_1 \in A$ such that $x_1 - b_1 \in p\mathbb{Z}_p$, so that $x = b_0 + b_1 p + p^2 x_2$ for some $x_2 \in \mathbb{Z}_p$. Continuing in this way, we obtain for each n:

$$x = b_0 + b_1 p + b_2 p^2 + \cdots + b_n p^n + p^{n+1}x_{n+1}$$

with $x \in \mathbb{Z}_p$, and so

$$\left| x - \left(b_0 + b_1 p + b_2 p^2 + \cdots + b_n p^n \right) \right| \leq p^{-(n+1)}.$$

This shows that the series

$$b_0 + b_1 p + b_2 p^2 + \cdots + b_n p^n + \cdots$$

converges to x.

If $x \notin \mathbb{Z}_p$, write $x = p^{-m}x_0$ with $x_0 \in \mathbb{Z}_p$, expand x_0 as before and multiply by p^{-m} to get our series. $\square$

Notice that the elements of A don't even need to be in $\mathbb{Z}$. Of course, the resulting expansion depends very strongly on the choice of the set A of digits. The next problem dramatizes this.

Problem 116 Let $p = 3$ and let $A = \{5, 6, 7\}$ be our set of digits. Find the p-adic expansion of $x = 0$ with these digits.

Returning to the "standard" digits, we now have a fairly close analogy between the real numbers $\mathbb{R}$ and the various $\mathbb{Q}_p$. Each is obtained from $\mathbb{Q}$ by completing with respect to an absolute value, and together they cover all possible (nontrivial) absolute values. Any real number x can be expressed (alas, not uniquely) as a decimal expansion

$$x = b_{-n}10^n + b_{-n-1}10^{n-1} + \cdots + b_{-1}10^1 + b_0 + b_1 10^{-1} + b_2 10^{-2} + \cdots$$

with $0 \leq b_i \leq 9$. Any $x \in \mathbb{Q}_p$ can be expressed (uniquely) as a p-adic expansion

$$x = b_{-n}p^{-n} + b_{-n-1}p^{-(n-1)} + \cdots + b_{-1}p^{-1} + b_0 + b_1 p^1 + b_2 p^2 + \cdots$$

with $0 \leq b_i \leq p - 1$. When we work with real numbers, we do so only up to a certain degree of precision; we need to do the same for p-adic numbers. In both cases, expansions that are eventually periodic correspond to rational numbers. Both $\mathbb{R}$ and the $\mathbb{Q}_p$ are locally compact topological fields, but the topology is very different, because the $\mathbb{Q}_p$ are totally disconnected.

4.4 What Does $\mathbb{Q}_p$ Look Like?

It's natural to wonder whether we can visualize $\mathbb{Q}_p$ (or $\mathbb{Z}_p$, should that be easier). After all, we have a very nice visual representation for the real numbers as the points on a line. For a compact subset of $\mathbb{R}$, say the unit ball $[-1, 1]$, the image is even clearer: a line segment. Our pictures of $\mathbb{R}$ are actually finite unions of blown-up versions of that interval, which we extrapolate in our minds to the complete line.

Things cannot be that easy for the p-adics, especially due to the fact that nontrivial ultrametric spaces are totally disconnected but not discrete. It's easy to draw a (compact) connected set: an interval or a blob. It's also easy to draw a (finite) discrete set: a bunch of points. How do we draw a compact set whose points are not discretely spaced out but also not connected to each other?

Rather than giving a complete answer, this section gives a few hints of what can be done. In particular, all the "proofs" in this section are either sketches or references. For the details, see [39, §2.2, 2.3], [53, I.2], and especially [22]. See [37] for a different approach.

Before we start, we should consider what we mean by a visual representation. The real line reflects not only the topology of $\mathbb{R}$, but also the order and the metric: bigger numbers are to the right of smaller ones; the distance between two points in the picture is proportional to the distance in $\mathbb{R}$. One can even see how adding works along the line by thinking of appending one segment to the end of another. Achieving all that for $\mathbb{Z}_p$ is unlikely. Instead, we will try to find ways to represent $\mathbb{Z}_p$ that preserve the *topology* but not necessarily the metric. That means, in particular, that the distance between two points in our images will not be proportional to the p-adic distance. On the other hand, convergent sequences in $\mathbb{Z}_p$ will look like convergent sequences in our pictures.

Remember that two metric spaces X and Y are *homeomorphic* if there exists a continuous invertible function $f : X \longrightarrow Y$ whose inverse is also continuous. Homeomorphic sets are "topologically the same." So what we are looking for are homeomorphic images of $\mathbb{Z}_p$; to be visualizable they should be contained in the line, in $\mathbb{R}^2$, or in $\mathbb{R}^3$.

There is one totally disconnected compact set that shows up in many Real Analysis courses: the Cantor "middle thirds" set. It is created by an iterative process. Start with the closed interval $C_0 = [0, 1]$, and delete its middle third to get $C_1 = [1, \frac{1}{3}] \cup [\frac{2}{3}, 1]$. Repeat this procedure with both of the closed intervals to get

$$C_2 = \left[0, \frac{1}{9}\right] \cup \left[\frac{2}{9}, \frac{1}{3}\right] \cup \left[\frac{2}{3}, \frac{7}{9}\right] \cup \left[\frac{8}{9}, 1\right].$$

Keep doing this forever, i.e., let

$$C = C_0 \cap C_1 \cap C_2 \cap \ldots$$

The resulting set C is compact and totally disconnected. See, for example, [55, §2.44].

The Cantor set has an interesting feature in common with $\mathbb{Z}_2$: there is a hierarchical structure in which the main interval is broken into two subsets, then each of those into two subsets, and so on. $\mathbb{Z}_2$ is like that: every element is congruent to either 0 or 1 (mod 2), and if you know what it is (mod 2) there are two choices for what it can be (mod 4), and so on. So maybe you will not be surprised by our next theorem.

Theorem 4.4.1 $\mathbb{Z}_2$ *with the 2-adic norm is homeomorphic to the middle thirds Cantor set C with the norm it inherits from $\mathbb{R}$.*

PROOF: Just as we can write real numbers as decimal expansions, any real number $y \in [0, 1]$ can be represented as

$$y = \frac{a_1}{3} + \frac{a_2}{3^2} + \cdots + \frac{a_n}{3^n} + \cdots$$

with $a_i \in \{0, 1, 2\}$. (The representation is not unique, as in the case of decimals, but this can be controlled.) It's easy to see that y is in the Cantor set if and only if we can represent y without using the digit 1, i.e., we have $a_i \in \{0, 2\}$ for all i.

On the other hand, any element $z \in \mathbb{Z}_2$ has a 2-adic expansion

$$y = b_0 + b_1 2 + b_2 2^2 + \cdots + b_n 2^n + \cdots$$

with $b_i \in \{0, 1\}$. There is a bijection $\{0, 1\} \longrightarrow \{0, 2\}$: just multiply by 2. So we can make a function $f : \mathbb{Z}_2 \longrightarrow C$ like this

$$f(b_0 + b_1 2 + \cdots + b_n 2^n + \cdots) = \frac{2b_0}{3} + \frac{2b_1}{3^2} + \cdots + \frac{2b_n}{3^{n+1}} + \cdots$$

This is clearly a bijection, and it turns out to be easy to prove that it is continuous with continuous inverse. See the references for the details. $\qquad\square$

It's easy to generalize this observation. It turns out that there are many "Cantor sets" obtained by tweaking the construction above. They are all compact, totally disconnected, and "perfect," which means[5] that any point in the set is the limit of some nonstationary sequence contained in the set. For each p we can imitate the function above to create a Cantor set that is homeomorphic to $\mathbb{Z}_p$. So we can visualize $\mathbb{Z}_p$ as a Cantor set contained in $\mathbb{R}$ if we want to. But there is a surprise.

Theorem 4.4.2 *For any p, $\mathbb{Z}_2$ is homeomorphic to $\mathbb{Z}_p$.*

This follows at once from a theorem about subsets of $\mathbb{R}$: any compact, perfect, and totally disconnected subset of $\mathbb{R}$ is homeomorphic to the Cantor set C. See [39, Theorem 2.29].

Of course, Cantor sets are hard to draw too! One gets slightly better results by going to $\mathbb{R}^2$ or $\mathbb{R}^3$. The idea is to use "vector digits": choose a point $\mathbf{x}_i$ in $\mathbb{R}^n$ for each digit $b_i \in \{0, 1, \ldots, p-1\}$. Choose a base b (so $b \in \mathbb{Z}$, $b \geq 2$). Then we can create a function

$$\sum_{i=0}^{\infty} a_i p^i \mapsto \sum_{i=0}^{\infty} \frac{1}{b^{i+1}} \mathbf{x}_i.$$

It can be shown that if the $\mathbf{x}_i$ are well chosen and b is large enough this is a homeomorphism. The image of $\mathbb{Z}_p$ will be a disjoint union of self-similar sets, i.e., a fractal. For example, one can use this idea to show that $\mathbb{Z}_3$ is homeomorphic to the Sierpinski gasket.

Sage contains an implementation of this. It uses the vertices of a regular p-gon for the vector digits, so the results are only interesting when $p > 2$. The command `Zp(5).plot()` produces something like Figure 4.1. A better choice might be to use 0 and the $(p-1)$-th roots of unity in $\mathbb{C}$ as the vector digits.

The self-similar structure makes it possible to "see" all of $\mathbb{Q}_p$ as well. Suppose the image of $\mathbb{Z}_p$ is a fractal where the full image contains p isomorphic but smaller subsets, each of which contain p isomorphic but smaller subsets, etc. We can arrange this so that there will be a scaling factor $\theta < 1$ that relates each bigger set to its smaller copies. Then to see $\frac{1}{p}\mathbb{Z}_p$ we scale the entire thing by θ^{-1}. If we want $\frac{1}{p^2}\mathbb{Z}_p$ we do that again, and so on. The union of all of these scaled copies is an image of $\mathbb{Q}_p$.

---

[5]See, for example, [55, Def. 2.18(h)]. I don't know what's so perfect about such sets.

Figure 4.1: Sage's depiction of $\mathbb{Z}_5$

Notice, however, that these are *topologically* equivalent to $\mathbb{Z}_p$ and $\mathbb{Q}_p$, but not metrically equivalent. And these images also do not encode the algebraic structure in any good way. So while they are fun to create and think about, they do not really help us much when we work in the p-adic world.

4.5   Hensel's Lemma

The theorem[6] known as "Hensel's Lemma" is probably the most important algebraic property of the p-adic numbers (and of other fields like $\mathbb{Q}_p$, which are complete with respect to a non-archimedean valuation). It says that in many circumstances one can decide quite easily whether a polynomial has roots in $\mathbb{Z}_p$.

When we work with polynomials over $\mathbb{R}$, it is often possible to decide on the existence of roots by looking at signs. For example, $x^2 + 1$ cannot have

---

[6]To be honest one should say "theorems": as we will see, there are several different ones that go by the same generic name.

any real roots because $x^2 + 1 > 0$ for every x. When we look for roots in $\mathbb{Z}_p$, what replaces sign considerations is reduction mod p.

In order to state our theorem we need to work with the formal derivative of a polynomial $F(X)$. This makes sense for polynomials with coefficients in any ring.

Definition 4.5.1 *Let* $F(X) = a_0 + a_1 X + a_2 X^2 + \cdots + a_n X^n$ *be a polynomial with coefficients* $a_i \in R$, *where* R *is a ring. The formal derivative of* $F(X)$ *is*

$$F'(X) = a_1 + 2a_2 X + \cdots + na_n X^{n-1}.$$

Notice that this definition does not involve any limit processes—or at least not yet. (We're saving that for the next chapter.)

Theorem 4.5.2 (Hensel's Lemma) *Let* $F(X) = a_0 + a_1 X + a_2 X^2 + \cdots + a_n X^n$ *be a polynomial whose coefficients are in* $\mathbb{Z}_p$. *Suppose that there exists a* p-*adic integer* $\alpha_1 \in \mathbb{Z}_p$ *such that*

$$F(\alpha_1) \equiv 0 \pmod{p\mathbb{Z}_p}$$

and

$$F'(\alpha_1) \not\equiv 0 \pmod{p\mathbb{Z}_p},$$

where $F'(X)$ *is the formal derivative of* $F(X)$. *Then there exists a unique* p-*adic integer* $\alpha \in \mathbb{Z}_p$ *such that* $\alpha \equiv \alpha_1 \pmod{p\mathbb{Z}_p}$ *and* $F(\alpha) = 0$.

PROOF: We will show that the root α exists by constructing a Cauchy sequence of integers converging to it. The idea is essentially what is known as "Newton's method" in real analysis. The attentive reader will recognize an idea that we have been using repeatedly since the first chapter.[7]

What we will construct is a sequence of integers $\alpha_1, \alpha_2, \ldots, \alpha_n, \ldots$ such that, for all $n \geq 1$, we have

i) $F(\alpha_n) \equiv 0 \pmod{p^n}$,

ii) $\alpha_{n+1} \equiv \alpha_n \pmod{p^n}$.

It is easy to see that such a sequence will be Cauchy (in fact, it is a "coherent sequence" in our terms above), and that its limit α will satisfy $F(\alpha) = 0$ (by continuity) and $\alpha \equiv \alpha_1 \pmod{p}$ (by construction). Conversely, a root α will determine such a sequence α_n. Thus, once we have the α_n the theorem will be proved.

The main assumption in the theorem is that α_1 exists. To find α_2, we note that condition (ii) requires that

$$\alpha_2 = \alpha_1 + b_1 p$$

---

[7]I hope. If not, please reread Section 3 of that chapter.

for some $b_1 \in \mathbb{Z}_p$. Plugging this expression into the polynomial $F(X)$ and expanding, we get

$$F(\alpha_2) = F(\alpha_1 + b_1 p)$$

$$= F(\alpha_1) + F'(\alpha_1)b_1 p + \text{terms in } p^n, \ n \geq 2$$

$$\equiv F(\alpha_1) + F'(\alpha_1)b_1 p \pmod{p^2}.$$

(This is easy to check directly, but it is probably best to think of it as a kind of formal Taylor expansion—see Problem 117.) To show that one can find α_2, we have to show that one can find b_1 so that

$$F(\alpha_1) + F'(\alpha_1)b_1 p \equiv 0 \pmod{p^2}.$$

Now, we know that $F(\alpha_1) \equiv 0 \pmod{p}$, so that $F(\alpha_1) = px$ for some x. The equation then becomes

$$px + F'(\alpha_1)b_1 p \equiv 0 \pmod{p^2},$$

which gives (after we divide by p)

$$x + F'(\alpha_1)b_1 \equiv 0 \pmod{p}.$$

To solve this, notice that $F'(\alpha_1)$ is not divisible by p, and hence is *invertible* in $\mathbb{Z}_p$, so that we can (and must) take

$$b_1 \equiv -x(F'(\alpha_1))^{-1} \pmod{p}.$$

In fact, we can choose such a b_1 in $\mathbb{Z}$, with $0 \leq b_1 \leq p - 1$, and then b_1 is uniquely determined. For this choice of b_1, we set $\alpha_2 = \alpha_1 + b_1 p$, which will have the stated properties.

This shows that one can take the first step: given α_1, find α_2. But a careful inspection shows that exactly the same calculation works to get α_{n+1} from α_n. Hence, we can construct the whole sequence, and it is uniquely determined at each step. This proves the theorem. $\square$

Problem 117 Let $F(X)$ be a polynomial with coefficients in a field $\Bbbk$ of characteristic zero. Show that the Taylor formula is true for $F(X)$, i.e., that

$$F(x + h) = F(x) + F'(x)h + \frac{1}{2!}F''(x)h^2 + \frac{1}{3!}F'''(x)h^3 + \dots$$

for any $x, h \in \Bbbk$.

Problem 118 Check that the calculation given in the proof does indeed work to get α_{n+1} from α_n and that α is indeed unique.

We said above that the calculation in Hensel's Lemma is analogous to Newton's Method for numerically approximating roots. Let's work that out in detail. In the classical Newton's method, we start with an initial guess x_0 and then compute what we hope are better and better approximations using the formula

$$x_{n+1} = x_n - \frac{f(x_n)}{f'(x_n)}.$$

In our setup, we found α_{n+1} by setting it equal to $\alpha_n + pb_n$, and computed b_n by setting $F(\alpha_n) = px$ and $b_n = -x(F'(\alpha_n))^{-1}$. Plugging everything in gives

$$\alpha_{n+1} = \alpha_n - p\frac{F(\alpha_n)}{p}(F'(\alpha_n))^{-1} = \alpha_n - \frac{F(\alpha_n)}{F'(\alpha_n)}.$$

In other words, the sequence produced in Hensel's Lemma is given by exactly the same formula.

There are differences between the classical and the p-adic case, of course. Mostly, $\mathbb{Z}_p$ is nicer. First of all, we checked that this procedure never leaves $\mathbb{Z}_p$ (in other words, the division in the formula can always be performed in $\mathbb{Z}_p$). Next, we checked that in the p-adic case the method *always* works, provided only that $F'(\alpha_1) \not\equiv 0 \pmod{p}$. This is far from true in the classical case. Finally, we get an extra bit of information, that the limit α is the unique root such that $\alpha \equiv \alpha_1 \pmod{p}$, which can be read as saying that the root we get is not too far from the initial estimate (this too is not true in the classical case).

It's worth giving the theorem again in this language:

Theorem 4.5.3 (Hensel's Lemma) *Let $F(X) = a_0 + a_1 X + a_2 X^2 + \cdots + a_n X^n$ be a polynomial whose coefficients are in $\mathbb{Z}_p$. Suppose that there exists a p-adic integer $\alpha_1 \in \mathbb{Z}_p$ such that $|F(\alpha_1)| < 1$ and $|F'(\alpha_1)| = 1$. Setting, for each $n \geq 1$,*

$$\alpha_{n+1} = \alpha_n - \frac{F(\alpha_n)}{F'(\alpha_n)}$$

defines a convergent sequence whose limit $\alpha \in \mathbb{Z}_p$ is the unique p-adic integer such that $|\alpha - \alpha_1| < 1$ and $F(\alpha) = 0$.

The theory of *discrete dynamical systems* studies what happens when we take a function $G(x)$ and produce a sequence by computing $x_1 = G(x_0)$, $x_2 = G(x_1)$, etc. That is what we are doing here, so Hensel's Lemma can be understood using that theory, which gives another way to prove the sequence converges and even allows us to investigate how fast the convergence happens. A (fairly advanced) textbook on that topic is [9].

Problem 119 What happens to the calculation if we do *not* assume that we have $F'(\alpha_1) \not\equiv 0 \pmod{p}$? Can you give an example where the theorem fails because this condition does not hold? (Hint: look back at our games with the polynomial $X^2 - m$ in Chapter 1.)

It's worth emphasizing that Hensel's Lemma is both an existence and a uniqueness result: there is a root, and it is the only root satisfying the congruence condition. Both facts are used often.

There are many different versions of this theorem, all of which tend to be referred to as "Hensel's Lemma." The next problem, for example, gives a version that can be used when the hypothesis on $F'(\alpha_1)$ does not hold. It turns out to be very useful as well.

Problem 120 Show that in Hensel's Lemma we can weaken the conditions $F(\alpha) \equiv 0$, $F'(\alpha_1) \not\equiv 0$ by replacing them with the condition $|F(\alpha_1)| < |F'(\alpha_1)|^2$. What should replace the conclusion that $\alpha \equiv \alpha_1 \pmod{p}$? Why is this version of Hensel's Lemma more general than the first? Can you give an example where this version can be used but the original version cannot?

(Hint: Instead of using congruences, start from the formula in Theorem 4.5.3 and prove that the sequence converges.)

Because our proof is completely explicit, it is not hard to implement it on a computer. In GP, there is a function `padicappr(f(x),a)` that finds a root of the polynomial $f(x)$ close to the p-adic number a. So, for example, we can find a 2-adic root of $x^2 + x + 6$ close to $x = 0$ like this:

```
gp > padicappr(x^2+x+6,0+O(2^20))
%1 = [2 + 2^2 + 2^4 + 2^9 + 2^10 + 2^12 + 2^13 + 2^15
        + 2^18 + 2^19 + O(2^20)]~
```

The result is a column vector (the tilde means "transpose", so it's printed as a row) with the p-adic roots of the polynomial that are congruent to a mod p. In this case there is only one root. The algorithm can handle cases where we need Problem 120:

```
padicappr(x^2-17,1+O(2^20))
%1 = [1 + 2^3 + 2^5 + 2^6 + 2^7 + 2^9 + 2^10 + 2^13 +
        2^16 + 2^17 + O(2^20), 1 + 2 + 2^2 + 2^4 + 2^8 +
        2^11 + 2^12 + 2^14 + 2^15 + 2^18 + 2^19 + O(2^20)]~
```

In this case two roots are found, one congruent to 1 (mod 8) and the other congruent to 7 (mod 8).

In Sage, we need to first create $\mathbb{Q}_2$ and the polynomial ring $\mathbb{Q}_2[x]$ in which $g = x^2 + x + 6$ is going to live

```
K=Qp(2)
R=PolynomialRing(K,'x')
x=R.gen(0)
g=x^2+x+6
print g
```

The output is

```
(1 + O(2^20))*x^2 + (1 + O(2^20))*x + 2 + 2^2 + O(2^21)
```

So we have the right polynomial over $\mathbb{Q}_2$. We know 2 is an approximate root. To apply Hensel's Lemma we proceed with

```
a=g.hensel_lift(2)
print a
print a^2+a+6
```

That gives (line breaks added for clarity as usual)

```
2 + 2^2 + 2^4 + 2^9 + 2^10 + 2^12 + 2^13
  + 2^15 + 2^18 + 2^19 + O(2^21)
O(2^21)
```

For the square root of 17, we see slightly different behavior in Sage.

```
K=Qp(2)
R=PolynomialRing(K,'x')
x=R.gen(0)
g=x^2-17
a=g.hensel_lift(1)
print a
```

gives

```
1 + 2^3 + 2^5 + 2^6 + 2^7 + 2^9 + 2^10 + 2^13 + 2^16 + O(2^17)
```

In other words, Sage gives us only one root. Asking for `a=g.hensel_lift(3)` gets us the other one.

4.6   Using Hensel's Lemma

This section considers two nice applications of Hensel's Lemma. The first is to determine which roots of unity can be found in $\mathbb{Q}_p$. The second finds all the squares in $\mathbb{Q}_p$.

Recall that an element ζ of a field is called an *m-th root of unity* if $\zeta^m = 1$; it is called a *primitive m-th root of unity* if in addition $\zeta^n \neq 1$ for $0 < n < m$. In $\mathbb{R}$, there are only two roots of unity, 1 and -1. On the other hand, we have already checked (in a problem long ago...) that the equation $X^2 + 1 = 0$ has a root in $\mathbb{Q}_5$, and it is easy to see that its root will be a fourth root of unity. So it is interesting to try to determine which roots of unity exist.

Hensel's Lemma finds roots that are in $\mathbb{Z}_p$, so let's notice that any root of unity must be a p-adic integer. That's easy: since $|\zeta|^m = |\zeta^m| = 1$ and absolute values are positive real numbers, any root of unity must have absolute value 1.

To use Hensel's Lemma, we need a polynomial. Since we are looking for roots of unity, we will use $F(X) = X^m - 1$. Notice that $F'(X) = mX^{m-1}$. The value $F'(\lambda) = m\lambda^{m-1}$ will be congruent to zero modulo p if either p

divides λ (in which case λ will not satisfy $\lambda^m - 1 \equiv 0$ anyway, so this isn't a problem) or p divides m. Thus, the second condition in the theorem will hold provided m is not divisible by p. For the first condition, we need to find an approximate root, and it is actually quite easy to decide when that can be done:

Problem 121 Fix a prime p and a number k not divisible by p. Show that there exists an integer α_1 such that $\alpha_1^k \equiv 1 \pmod{p}$ but $\alpha_1 \not\equiv 1 \pmod{p}$ if and only if $\gcd(k, p-1) > 1$. Show that for any such α_1 the least positive integer m such that $\alpha_1^m \equiv 1$ must be a divisor of $p-1$. (Hint: $\mathbb{Z}/p\mathbb{Z}$ is a field, and the set of its invertible elements is a *cyclic* group.)

Then Hensel's Lemma yields:

Proposition 4.6.1 *For any prime p and any positive integer m not divisible by p, there exists a primitive m-th root of unity in $\mathbb{Q}_p$ if and only if m divides $p-1$.*

Problem 122 Prove the proposition. (You will need both parts of Hensel's Lemma: the part that finds a root and the part that says that the root satisfying a certain condition is unique.)

If m divides $p-1$, then any m-th root of unity is also a $(p-1)$-st root of unity, so that the upshot is that the roots of unity in $\mathbb{Q}_p$ of order prime to p are exactly the $(p-1)$-st roots. This determines all the roots of unity in $\mathbb{Q}_p$, except for the possibility of there existing p^n-th roots of unity in $\mathbb{Q}_p$. These are harder[8] to analyze by this method. It turns out, as we will show later, that they are *not* in $\mathbb{Q}_p$ (except when $p = 2$, in which case ± 1—but no fourth roots of 1—do belong to $\mathbb{Q}_2$). Hence, we have determined *all* the roots of unity belonging to $\mathbb{Q}_p$, though we will only be able to *prove* that this is the case later on. (If you can't wait, look at page 144.)

Problem 123 Show that the set of roots of unity in $\mathbb{Q}_p$ is a subgroup of the group $\mathbb{Z}_p^\times$ of p-adic units. Show that the set of $(p-1)$-st roots of unity in $\mathbb{Q}_p$ is a cyclic group of order $(p-1)$. (The main content of the last statement is that there are $(p-1)$ p-adic roots of the polynomial $X^{p-1} - 1$. Use Hensel's Lemma.)

Another interesting application is to determine the squares in $\mathbb{Q}_p$. This is something we essentially did in Chapter 1. First we do the p-adic units:

Proposition 4.6.2 *Let $p \neq 2$ be a prime, and let $b \in \mathbb{Z}_p^\times$ be a p-adic unit. If there exists an $\alpha_1 \in \mathbb{Z}_p$ such that $\alpha_1^2 \equiv b \pmod{p\mathbb{Z}_p}$, then b is the square of an element of $\mathbb{Z}_p^\times$.*

---

[8]But not impossible—see [18, Theorem 3.1].

PROOF: Apply Hensel's Lemma to $X^2 - b$, and notice that $p \neq 2$ and $b \in \mathbb{Z}_p^\times$ are enough to make sure that $2\alpha_1 \not\equiv 0 \pmod{p}$. $\qquad\square$

Then we extend to all of $\mathbb{Q}_p$, by noticing that any $x \in \mathbb{Q}_p$ can be written as $x = p^{v_p(x)}x'$ with $x' \in \mathbb{Z}_p^\times$ (in fact, that is pretty much the definition of $v_p(x)$). What the next result says is that x will be a square if $v_p(x)$ is even and x' is a square.

Corollary 4.6.3 *Let $p \neq 2$ be a prime. An element $x \in \mathbb{Q}_p$ is a square if and only if it can be written as $x = p^{2n}y^2$ with $n \in \mathbb{Z}$ and $y \in \mathbb{Z}_p^\times$ a p-adic unit. The quotient group $\mathbb{Q}_p^\times/(\mathbb{Q}_p^\times)^2$ has order four. If $c \in \mathbb{Z}_p^\times$ is any element whose reduction modulo p is not a quadratic residue, then the set $\{1, p, c, cp\}$ is a complete set of coset representatives.*

PROOF: The first statement is essentially obvious (because powers of p and p-adic units "do not mix"). Applying the proposition and standard properties of quadratic residues and non-residues gives the rest. $\qquad\square$

It is interesting to compare this result to its analogue in $\mathbb{R}$, which says that a real number is a square if it is positive, and that the quotient $\mathbb{R}^\times/(\mathbb{R}^\times)^2$ is of order two, with coset representatives $\{1, -1\}$. From this point of view, the corollary can be thought of as a p-adic version of the "rule of signs" for multiplying real numbers.

We still need to consider $p = 2$. For that, we need to use the stronger form of Hensel's Lemma given in Problem 120, since $F'(\alpha_1) = 2\alpha_1$ will of course always be divisible by 2.

Problem 124 Show that if $b \in \mathbb{Z}_2$, and $b \equiv 1 \pmod{8\mathbb{Z}_2}$ (so that in particular b is a 2-adic unit), then b is a square in $\mathbb{Z}_2$. Conversely, show that any 2-adic unit which is a square is congruent to 1 modulo 8. Conclude that the group $\mathbb{Q}_2^\times/(\mathbb{Q}_2^\times)^2$ has order 8, and is generated by the classes of -1, 5, and 2, so that a complete set of coset representatives is $\{1, -1, 5, -5, 2, -2, 10, -10\}$.

4.7 Hensel's Lemma for Polynomials

In this section we prove another form of Hensel's lemma, more general than the first. The idea is to interpret the first form of Hensel's Lemma as saying that if a polynomial factors modulo p and one of the factors is of the form $(X - \alpha)$, so that

$$f(X) \equiv (X - \alpha)g(X) \pmod{p},$$

then (under some extra condition) there is a similar factorization in $\mathbb{Z}_p[X]$. The obvious generalization is to consider arbitrary factorizations. The condition $f'(\alpha) \not\equiv 0 \pmod{p}$ says that α is not a double root mod p, that is, that the second factor $g(X)$ is not divisible by $(X - \alpha)$. For general factorizations, then, the assumption should be that the factors are relatively prime (as polynomials) modulo p. Let's make this precise:

Definition 4.7.1 *Let $g(X)$ and $h(X)$ be polynomials in $\mathbb{Z}_p[X]$. Let $\bar{g}(X)$ and $\bar{h}(X) \in \mathbb{F}_p[X]$ be the polynomials obtained by reducing the coefficients modulo p. We say $g(X)$ and $h(X)$ are relatively prime modulo p if $\gcd(\bar{g}, \bar{h}) = 1$ in $\mathbb{F}_p[X]$, or, equivalently, if there exist polynomials $a(X), b(X) \in \mathbb{Z}_p[X]$ such that*

$$a(X)g(X) + b(X)h(X) \equiv 1 \pmod{p},$$

where we understand congruence coefficient-by-coefficient, i.e., we say two polynomials are congruent modulo p if each coefficient of one is congruent modulo p to the corresponding coefficient of the other.

Problem 125 Is being relatively prime modulo p weaker or stronger than being relatively prime in $\mathbb{Z}_p[X]$?

The next theorem says that this idea does work.

Theorem 4.7.2 (Hensel's Lemma for Polynomials) *Let $f(X) \in \mathbb{Z}_p[X]$ be a polynomial with coefficients in $\mathbb{Z}_p$, and assume that there exist polynomials $g_1(X)$ and $h_1(X)$ in $\mathbb{Z}_p[X]$ such that*

   i) $g_1(X)$ is monic,[9]

   ii) $g_1(X)$ and $h_1(X)$ are relatively prime modulo p, and

   iii) $f(X) \equiv g_1(X)h_1(X) \pmod{p}$ (understood coefficient-by-coefficient).

Then there exist polynomials $g(X), h(X) \in \mathbb{Z}_p[X]$ such that

   i) $g(X)$ is monic,

   ii) $g(X) \equiv g_1(X) \pmod{p}$ and $h(X) \equiv h_1(X) \pmod{p}$, and

   iii) $f(X) = g(X)h(X)$.

PROOF: This is just like the original version: we start from the "approximate" factorization, and improve the approximation more and more until, in the limit, we get a factorization over $\mathbb{Z}_p$. Notice that requiring $g(X)$ to be monic implies that $\deg g(X) = \deg g_1(X)$.

Let d be the degree of $f(X)$, and m be the degree of $g_1(X)$ (remember that g_1 is monic). Then we can assume that $\deg(h_1) \leq d - m$ (it could be less, because the top coefficient of f could be divisible by p). We want to construct two sequences of polynomials $g_n(X)$ and $h_n(X)$ such that

   i) each g_n is monic and of degree m,

   ii) $g_{n+1} \equiv g_n \pmod{p^n}$ and $h_{n+1} \equiv h_n \pmod{p^n}$, and

   iii) $f(X) \equiv g_n(X)h_n(X) \pmod{p^n}$.

---

[9]This means that the coefficient of the term of highest degree is one.

(As always, we take the congruences coefficient-by-coefficient.) If we can find such sequences, we are clearly done, since going to the limit gives the desired polynomials $g(X)$ and $h(X)$. In other words, the coefficients of, say, $g(X)$ will be the limits of the corresponding coefficients of the $g_n(X)$. (Can you see why it's important to know that the degrees of the g_n are not changing?)

We already have $g_1(X)$ and $h_1(X)$; let's describe how to get $g_2(X)$ and $h_2(X)$. Since the g's are to be congruent, we must have

$$g_2(X) = g_1(X) + p\, r_1(X)$$

for some polynomial $r_1(X) \in \mathbb{Z}_p[X]$; similarly, we must have

$$h_2(X) = h_1(X) + p\, s_1(X).$$

To show that g_2 and h_2 exist, we simply have to show that it is possible to find r_1 and s_1 such that the desired conditions are satisfied. For that, we need to solve the equation

$$f(X) \equiv g_2(X)h_2(X) \pmod{p^2},$$

which we expand to

$$f(X) \equiv (g_1(X) + p\, r_1(X))(h_1(X) + p\, s_1(X)) \pmod{p^2}.$$

Multiplying out, we get

$$f(X) \equiv g_1(X)h_1(X) + p\, r_1(X)h_1(X) + p\, s_1(X)g_1(X) + p^2\, r_1(X)s_1(X)$$

$$\equiv g_1(X)h_1(X) + p\, r_1(X)h_1(X) + p\, s_1(X)g_1(X) \pmod{p^2}.$$

Now remember that $f(X) \equiv g_1(X)h_1(X) \pmod{p}$, so that we have

$$f(X) - g_1(X)h_1(X) = p\, k_1(X)$$

for some $k_1(X) \in \mathbb{Z}_p[X]$. Rearranging, we get

$$p\, k_1(X) \equiv p\, r_1(X)h_1(X) + p\, s_1(X)g_1(X) \pmod{p^2}.$$

Dividing through by p, we get

$$k_1(X) \equiv r_1(X)h_1(X) + s_1(X)g_1(X) \pmod{p}.$$

This is the equation we need to solve to determine r_1 and s_1.

The first step towards doing so is to recall that we have assumed that g_1 and h_1 are relatively prime modulo p. This means that we know that there exist $a(X), b(X) \in \mathbb{Z}_p[X]$ such that $a(X)g_1(X) + b(X)h_1(X) \equiv 1 \pmod{p}$. Consider, then, the two polynomials

$$\tilde{r}_1(X) = b(X)k_1(X) \qquad \text{and} \qquad \tilde{s}_1(X) = a(X)k_1(X).$$

These will almost do the trick: they clearly will make all the congruence conditions true. The only problem is that we have no control over the degree of $\tilde{r}_1(X)$; if that degree is bigger than m, then $g_1(X) + p\tilde{r}_1(X)$ will not be monic of degree m.

To remedy that, only a slight change is needed. We already know that

$$\tilde{r}_1(X)h_1(X) + \tilde{s}_1(X)g_1(X) \equiv k_1(X) \pmod{p}.$$

Now divide $\tilde{r}_1(X)$ by $g_1(X)$, and let $r_1(X)$ be the remainder:

$$\tilde{r}_1(X) = g_1(X)q(X) + r_1(X).$$

And now we know that $\deg r_1(X) < \deg g_1(X)$. If we set

$$s_1(X) = \tilde{s}_1(X) + h_1(X)q(X),$$

it all works out:

$$
\begin{aligned}
r_1(X)&h_1(X) + s_1(X)g_1(X) \\
&\equiv (\tilde{r}_1(X) - g_1(X)q(X))h_1(X) + (\tilde{s}_1(X) + h_1(X)q(X))g_1(X) \\
&\equiv \tilde{r}_1(X)h_1(X) - g_1(X)h_1(X)q(X) + \tilde{s}_1(X)g_1(X) + g_1(X)h_1(X)q(X) \\
&\equiv \tilde{r}_1(X)h_1(X) + \tilde{s}_1(X)g_1(X) \\
&\equiv k_1(X) \pmod{p},
\end{aligned}
$$

so that our congruence conditions are satisfied, and the fact that the degree of $r_1(X)$ is smaller than the degree of $g_1(X)$ is enough to guarantee that $g_1(X) + pr_1(X)$ is monic, and we are done.

This shows that g_2 and h_2 exist. Since they are congruent to g_1 and h_1 modulo p, they are also relatively prime modulo p, so that there will be no difficulty in going on to the next step.

Now we repeat the argument changing the indices and exponents to find g_3 and h_3. It is an easy exercise to show that this can always be done, and that produces the sequence whose convergence proves the theorem. □

Problem 126 To make sure you understand that final twist in the proof, work out the details for the following example. Let $p = 2$, and consider the polynomial $f(X) = 2X^2 + X + 2$. Modulo 2, this is easy to factor: just take $g_1(X) = X$ and $h_1(X) = 1$. Follow the steps in the proof to find $g_2(X)$ and $h_2(X)$, and discuss what happens if we try to use $\tilde{r}_1$ and $\tilde{s}_1$ instead of r_1 and s_1.

Problem 127 Work out the construction of g_3 and h_3 in full detail, to convince yourself that you understand the process.

Problem 128 Fill in the last step of the proof by giving a full proof, by induction, that the g_n and h_n exist for every n.

The reader will have noticed that this argument is essentially identical to the one we gave for the first version of Hensel's Lemma. It might be interesting to check whether one can formulate a stronger version that is analogous to the one in Problem 120. That would likely involve the p-adic valuation of the resultant of the two approximate factors.

Both Sage and GP implement this. In Sage, just create a polynomial over $\mathbb{Q}_p$ as above and use `g.factor()` or `factor(g)`. In GP the command is `factorpadic(g,2,20)`, where the second argument tells GP that $p = 2$ and the third says to work with precision `+O(2^20)`.

4.8   Local and Global

One of the consequences of Hensel's Lemma is that, given a polynomial with integer coefficients, it is usually not too hard to decide whether it has roots in $\mathbb{Z}_p$, since it is enough to find roots modulo p. The "same" is true for $\mathbb{R}$, where we can usually decide whether there are roots by sign considerations (for example, if the polynomial has different signs at $x = a_1$ and $x = a_2$, there must be a root between these two numbers).

Suppose, however, that we want to look for roots in $\mathbb{Q}$. At least this much is easy to see: if there are roots in $\mathbb{Q}$, then there are also roots in $\mathbb{Q}_p$ for every $p \leq \infty$ (i.e., in all the $\mathbb{Q}_p$ and in $\mathbb{R}$). Hence we can certainly conclude that there are *no* rational roots if there is some $p \leq \infty$ for which there are no p-adic[10] roots. For example:

- $X^2 + 1 = 0$ has no roots in $\mathbb{R}$, hence has none in $\mathbb{Q}$.

- $X^2 - 2 = 0$ has no roots in $\mathbb{Q}_2$, hence has none in $\mathbb{Q}$.

- The only solution of $X^2 - 37Y^2 = 0$ in $\mathbb{Q}_5$ is $X = Y = 0$, so that is also the only solution in $\mathbb{Q}$.

These are easy examples, of course, but they do point the way.

The way to think about this situation is following Hensel's original analogy: the p-adic fields (including $\mathbb{R}$) are analogous to fields of Laurent expansions, and correspond to "local" information "near" the prime p. The fact that roots in $\mathbb{Q}$ automatically are roots in $\mathbb{Q}_p$ for every p means that a "global" root is also a "local" root at each p, i.e., "locally everywhere."

Much more interesting would be a converse: that "local" roots could be "patched together" to give a "global" root. This would be very useful, since deciding on the existence of local roots is very easy. Here is an (easy) example of such a converse.

---

[10] Here, of course, "∞-adic" means "real." In general, this section will constantly refer to all the absolute values taken together, and thus will constantly use the convention that the usual absolute value corresponds to the "prime" ∞, so that we will write $\mathbb{Q}_\infty = \mathbb{R}$ for the real numbers.

Proposition 4.8.1 *A number $x \in \mathbb{Q}$ is a square if and only if it is a square in every $\mathbb{Q}_p$, $p \leq \infty$.*

PROOF: This is really very easy: for any $x \in \mathbb{Q}$, we have

$$x = \pm \prod_{p < \infty} p^{v_p(x)}.$$

If x is a square at infinity, it is positive. If it is a square at a prime p, then $v_p(x)$ is even. It follows (just write out the prime factorization) that such an x is a square in $\mathbb{Q}$.   □

This very important idea goes back to Hensel, but it is known as the "Hasse principle" because it was Hensel's student Helmut Hasse who proved the first really important theorem along those lines. We could state the principle like this: *putting together local information at all $p \leq \infty$ should give global information*. Exactly in what sense this is true (if it is) depends on each specific problem, but there are many situations in which this principle plays a central role.

A very interesting example of this sort of method is the theory of diophantine equations, in which we are given an equation for which we want to find solutions in $\mathbb{Q}$, or at least to decide if any exist. This is in general an extremely difficult (and absolutely fascinating) subject, but in some cases the question can be decided by the local-global game. Consider, for example, the equation

$$X^2 + Y^2 + Z^2 = 0.$$

One sees at once that the only solution is the trivial one $X = Y = Z = 0$, because this is the only solution in $\mathbb{R}$ (and any other solution in $\mathbb{Q}$ would also be a solution in $\mathbb{R}$). Similarly, it doesn't take too much to see that the equation

$$X^2 + Y^2 - Z^2 = 0$$

does have a solution in $\mathbb{Q}$, and therefore in all of the $\mathbb{Q}_p$.

What one would hope for in this context is that one would have a perfect correspondence between "global" properties and "local" properties that hold "locally everywhere." In this example, it is clear that if a global solution (i.e., one in $\mathbb{Q}$) exists, then local solutions exist for all primes (of course, since the solution in $\mathbb{Q}$ belongs to all the $\mathbb{Q}_p$). One would also like the converse to be true, i.e., that the *lack* of a global solution could always be detected locally. To put it in other words, one would like it if the existence of a local solution for every p would guarantee the existence of a global solution. This is far from clear, however, because the local solutions in each $\mathbb{Q}_p$ live in different fields, and there seems to be no compelling reason why they should "glue together" somehow to provide a solution over $\mathbb{Q}$.

For equations like the ones above (of degree 2, homogeneous), a few experiments begin to convince us that the hope is indeed plausible, because for every equation that does not have solutions one can quickly find a prime so that the equation has no solutions in $\mathbb{Q}_p$:

i) $X^2 + Y^2 + Z^2 = 0$ has no nontrivial solutions in $\mathbb{R}$;

ii) $3X^2 + 2Y^2 - Z^2 = 0$ has no nontrivial solutions in $\mathbb{Q}_3$ (check!);

iii) $X^2 - 3Y^2 = 0$ has no nontrivial solutions in $\mathbb{Q}_7$ (check!).

This suggests the following bold statement:

Local-Global Principle: *The existence or non-existence of solutions in $\mathbb{Q}$ (global solutions) of a diophantine equation can be detected by studying, for each $p \leq \infty$, the solutions of the equation in $\mathbb{Q}_p$ (local solutions).*

Of course, as stated, this is too vague to be a "theorem," but the local-global principle has proved to be a valuable guide for the study of diophantine problems. What it has suggested is a "plan of attack" on any given equation (or type of equation): first think locally, then try to put together the local information to obtain global information.

The most naïve version of the principle would be the one we suggested above: the statement that an equation has solutions in $\mathbb{Q}$ if and only if it has solutions in all the $\mathbb{Q}_p$. This sounds wonderful, since it says that "solvable locally everywhere" is the same as "solvable globally." Unfortunately, it is false:

Problem 129 Show that the equation

$$(X^2 - 2)(X^2 - 17)(X^2 - 34) = 0$$

has a root in $\mathbb{Q}_p$ for all $p \leq \infty$, but has no roots in $\mathbb{Q}$.

Problem 130 (This is quite hard.) Show that $X^4 - 17 = 2Y^2$ is solvable locally everywhere, but is not solvable in $\mathbb{Q}$. (The existence of local solutions is easily checked; the non-existence of rational solutions is the hard part.)

One might try to salvage the principle in various ways, for example:

Problem 131 Decide whether it is true that a polynomial in one variable with coefficients in $\mathbb{Z}$ is irreducible in $\mathbb{Q}[X]$ if and only if it is irreducible in $\mathbb{Q}_p[X]$ for every $p \leq \infty$. (Recall that a polynomial is irreducible if it does not factor into a product of polynomials of lower degree.)

Finally, here is an example where the principle is gloriously successful:

Theorem 4.8.2 (Hasse–Minkowski) *Let*

$$F(X_1, X_2, \ldots X_n) = \sum_{i,j} c_{ij} X_i X_j \in \mathbb{Q}[X_1, X_2, \ldots X_n]$$

be a quadratic form (that is, a homogeneous polynomial of degree 2 in n variables). The equation

$$F(X_1, X_2, \ldots X_n) = 0$$

has non-trivial solutions in $\mathbb{Q}$ if and only if it has non-trivial solutions in $\mathbb{Q}_p$ for each $p \leq \infty$.

The proof is just a little out of our reach in this book, since it requires a more thorough study of quadratic forms and their properties than we are prepared to spend time on. A very good account of the proof can be found in [60], where it is the culmination of the first half of the book.

One should notice that this theorem completely solves the problem of deciding whether a quadratic form has non-trivial zeros, since the local question can be decided rather easily in each case. In fact, for each prime p, an appropriate version of Hensel's Lemma shows that there is a finite procedure for deciding whether the equation is solvable in $\mathbb{Q}_p$ (so that a computer could do it). It is a little worrying that there are infinitely many primes to consider, but it turns out that the whole problem can be sufficiently broken down so that one gets a finite procedure for checking for local solutions at *all* primes, so that (given the Hasse–Minkowski theorem) the whole problem gets reduced to a finite procedure.

In lieu of a proof of the Hasse–Minkowski theorem (or that one can check all primes with a finite amount of work), it might be fun to work out in detail an example of its application. So let a, b, and c be rational numbers, and consider the equation

$$aX^2 + bY^2 + cZ^2 = 0.$$

We want to use the Hasse–Minkowski theorem to settle completely when it is that such an equation has non-trivial rational solutions ("non-trivial" just means "other than $X = Y = Z = 0$"). We will do this by checking what conditions on a, b, and c are needed so that the equation has a nontrivial solution in $\mathbb{Q}_p$ for all p. It will take less than four pages, i.e., a finite amount of work.

To start off, if any of a, b, and c is equal to zero, there certainly is a nontrivial solution (with one variable non-zero, and the other two equal to zero). Next, we can clear denominators to make a, b, and c integers. We can also assume that they have no common factors (which we could cancel). Finally, we can assume that a, b, and c are square-free (i.e., they have no factors which are squares), by absorbing any square factor into one of the unknowns.

Problem 132 Suppose that $a = a'n^2$. Check that any rational solution (x, y, z) of $aX^2 + bY^2 + cZ^2 = 0$ corresponds to a rational solution (nx, y, z) of $a'X^2 + bY^2 + cZ^2 = 0$. Explain why this means that we can assume that a, b, and c are square-free.

Problem 133 We have already observed that we may assume that a, b, and c have no common factors. Show that in fact we can go farther, and assume that no two of these three numbers have any common factors. In other words, we may assume that the product abc is square-free.

Very well, we are set up now as follows: we have an equation

$$aX^2 + bY^2 + cZ^2 = 0$$

where a, b, and c are pairwise relatively prime integers with no square factors. What Hasse–Minkowski tells us is that we can decide whether this equation has non-trivial rational solutions by looking at each $\mathbb{Q}_p$ in turn. So let's:

1. Suppose $p = \infty$, so that $\mathbb{Q}_p = \mathbb{R}$. Then it's all about signs: as long as we can get something positive, we can take a square root to find a real solution. So the equation will have a non-trivial solution exactly when a, b, and c are not all positive or all negative. (If you have any doubts, work it out!)

2. Suppose p is an odd prime that does not divide any of the coefficients. We'll need to use Hensel's Lemma, so the first step towards a solution in $\mathbb{Q}_p$ is to study the solutions modulo p.

Proposition 4.8.3 *Let p be an odd prime, and let a, b, c be pairwise relatively prime integers not divisible by p. Then there exist integers x_0, y_0, and z_0, not all divisible by p, such that*

$$ax_0^2 + by_0^2 + cz_0^2 \equiv 0 \pmod{p}.$$

PROOF: This is a special case of a famous theorem due to Chevalley and Warning. It could be proved more directly, but we give a proof that works in the general case, which makes it somehow the "right" proof. It also involves a neat trick worth knowing.

As x, y, and z run over the integers between 0 and $p - 1$ (which, since we are working modulo p, is all we need to worry about), there are p^3 different triples (x, y, z). Let's try to count how many of these are solutions of

$$aX^2 + bY^2 + cZ^2 \equiv 0 \pmod{p}.$$

For that, we use a dastardly trick: notice that

$$(ax^2 + by^2 + cz^2)^{p-1} \equiv \begin{cases} 1 \pmod{p} & \text{if } (x, y, z) \text{ is not a solution} \\ 0 \pmod{p} & \text{if } (x, y, z) \text{ is a solution} \end{cases}$$

This is because, by Fermat's Little Theorem, we have $n^{p-1} \equiv 1 \pmod{p}$ whenever $n \not\equiv 0 \pmod{p}$. This means that if we let N be the total number of *non*-solutions, then

$$N \equiv \sum_{(x,y,z)} (ax^2 + by^2 + cz^2)^{p-1} \pmod{p},$$

where each of x, y, and z ranges through the numbers from 0 to $p-1$. Now, when we expand these powers, we are going to get an equation representing N as a sum of a bunch of sums, each of which is of the form

$$\sum_{(x,y,z)} \alpha x^{2i} y^{2j} z^{2k}$$

with $2i + 2j + 2k = 2(p-1)$ and $\alpha \in \mathbb{Z}$. We claim that each one of these sums is zero modulo p. To see this, note that we must certainly have that one of $2i$, $2j$, and $2k$ is less than $p-1$ (if they were all $\geq p-1$, then the sum would be $\geq 3(p-1)$, which it isn't). Say $2i < p-1$ (the argument is the same in the other cases). Then we can rewrite our sum as

$$\sum_{(y,z)} \left(\alpha y^{2j} z^{2k} \sum_{x} x^{2i} \right).$$

Now we invoke a little lemma:

Lemma 4.8.4 *Let n be an integer, $0 \leq n < p-1$. Then*

$$\sum_{x=0}^{p-1} x^n \equiv 0 \pmod{p}.$$

Assuming the lemma for now (the proof will come later), we see that the inner sum in the last formula is always congruent to zero modulo p. It follows that $N \equiv 0 \pmod{p}$. In other words, the number of triples that are *not* solutions is divisible by p. Since the total number of triples is p^3, we also get that the number of triples that are solutions is divisible by p.

But we already know one solution: $x = y = z = 0$! In other words, the number of triples which are solutions is at the same time divisible by p and at least 1. That means there must be more than one solution, which means there must be a solution (x, y, z) where not all three components are divisible by p, which is what we claimed.   □

To be completely happy, we just need to prove the lemma, which[11] we'll let the reader have some fun with.

---

[11] You saw this coming, no?

Problem 134 Prove the lemma. (Hint: remember that the integers modulo p form a field, with all sorts of nice properties. Note: if $n = 0$, the sum seems to refer to 0^0; read this as simply a synonym for 1.) What is the sum congruent to for other exponents n?

What we know, then, after the proposition, is that when $p \nmid 2abc$ there always are "good" solutions (i.e., solutions that are "non-trivial mod p") of the congruence

$$aX^2 + bY^2 + cZ^2 \equiv 0 \pmod{p}.$$

Once we know that, it's easy to settle the question in $\mathbb{Q}_p$ for those primes p: let (x_0, y_0, z_0) be a "solution mod p" as in the proposition; we know x_0, y_0, and z_0 are not all divisible by p; suppose $p \nmid x_0$ (otherwise, permute the names). Look at the equation

$$aX^2 + by_0^2 + cz_0^2 = 0$$

(in other words, replace the variable Y by the integer y_0, and similarly Z by z_0). This is now a polynomial in one variable, and we know that x_0 is a solution modulo p. Given our assumptions, Hensel's Lemma now tells us that there is an $x \in \mathbb{Z}_p$ which is a root of this equation. But then we've done it: (x, y_0, z_0) is a non-trivial solution in $\mathbb{Q}_p$ of the original equation. The upshot:

Corollary 4.8.5 *If p is an odd prime that does not divide abc, then the equation*

$$aX^2 + bY^2 + cZ^2 = 0$$

has a non-trivial solution in $\mathbb{Q}_p$.

Problem 135 Work out the details of the application of Hensel's lemma which we breezed by above.

Problem 136 At which points in the above argument did we use the assumption that $p \nmid abc$?

That handles almost all the primes, but we still have to look at what happens when $p = 2$ and when p divides one of the coefficients (we agreed above that we can assume that no one prime divides two of the coefficients).

3. Suppose $p = 2$, and a, b, and c are all odd. In this case, we will need some special condition to guarantee that there are solutions in $\mathbb{Q}_2$. Suppose a nontrivial solution (x, y, z) with x, y, $z \in \mathbb{Q}_2$ exists. We can clearly assume that $\max\{|x|_2, |y|_2, |z|_2\} = 1$, i.e., that x, y, and z are 2-adic integers which are not all in $2\mathbb{Z}_2$. (Given a solution, multiply by a positive or negative power of 2 to get this to hold.)

Reducing mod $2\mathbb{Z}_2$, and remembering that the coefficients are all odd, we see that exactly two of x, y, and z will be 2-adic units, and the other will be divisible by 2. Suppose that y and z are units. The square of a 2-adic unit always belongs to $1 + 4\mathbb{Z}_2$, while the square of an element in $2\mathbb{Z}_2$ will belong to $4\mathbb{Z}_2$. So, looking mod $4\mathbb{Z}_2$, we get that

$$b + c \equiv 0 \pmod 4.$$

If instead of x being the non-unit, either y or z is, then we get a similar condition involving two other coefficients of the equation.

In other words, if $p = 2$, $2 \nmid abc$, and there is a solution in $\mathbb{Q}_2$, then the sum of two of the coefficients of the equation must be divisible by 4.

It turns out that this is also sufficient:

Problem 137 Suppose a, b, and c are all odd, and the sum of two of them is divisible by 4. Show that the equation

$$aX^2 + bY^2 + cZ^2 = 0$$

has a non-trivial solution in $\mathbb{Q}_2$. (Hints: You need to use the result of Problem 120. Look for solutions modulo 8. Notice that $a + b \equiv 0 \pmod 4$ breaks into two cases when you work mod 8.)

4. Suppose $p = 2$, and one of the coefficients is even. We'll leave this and the next one to the reader:

Problem 138 Suppose $p = 2$, and one of a, b, and c is even. Show that if there exists a non-trivial solution of $aX^2 + bY^2 + cZ^2 = 0$ in $\mathbb{Q}_2$, then either the sum of two coefficients or the sum of all three coefficients will be divisible by 8. Show that this condition is also sufficient to guarantee that a non-trivial solution exists.

5. Suppose $p \neq 2$ and a is divisible by p.

Problem 139 Suppose $p \neq 2$ and a is divisible by p. Show that if there exists a non-trivial solution of $aX^2 + bY^2 + cZ^2 = 0$ in $\mathbb{Q}_p$, then there must exist an integer $r \in \mathbb{Z}$ such that

$$b + r^2 c \equiv 0 \pmod p.$$

(Another way of putting this is: $-b/c$ is a quadratic residue modulo p.) Show that this condition is also sufficient.

Putting all of this information together, we now have conditions that guarantee, for each p, that there are nontrivial solutions in $\mathbb{Q}_p$. Using the Hasse–Minkowski Theorem, we get:

Proposition 4.8.6 *Let a, b, and c be pairwise relatively prime square-free integers. The equation*

$$aX^2 + bY^2 + cZ^2 = 0$$

has non-trivial solutions in $\mathbb{Q}$ if and only if the following conditions are satisfied:

i) *a, b, and c are not all positive or all negative.*

ii) *For each odd prime dividing a, there exists an integer $r \in \mathbb{Z}$ such that $b + r^2 c \equiv 0 \pmod{p}$, and similarly for the odd primes dividing b and c.*

iii) *If a, b, and c are all odd, then there are two of them whose sum is divisible by 4.*

iv) *If a is even, then either $b + c$ or $a + b + c$ is divisible by 8 (and similarly if b or c is even).*

A direct proof (without using Hasse–Minkowski) of this special case of the theorem can be found in chapters 3–5 of [15]. The strategy of the proof is to use conditions (*ii*), (*iii*), and (*iv*) and Minkowski's "geometry of numbers," to show that one can find a solution (x, y, z) that satisfies the inequality

$$|a|x^2 + |b|y^2 + |c|z^2 < 4|abc|.$$

(Here $| \; | = | \; |_\infty$ is the "usual" absolute value.) This equation defines an ellipsoid in $\mathbb{R}^3$, and the number of triples (x, y, z) of integers satisfying this condition is finite, so that we can easily run through all of them (on a computer, probably) and find a solution. In other words, Cassels' argument in [15] goes further than merely giving an existence result: it actually gives us the means to find the solution.

Problem 140 The reader who was very attentive to the wording of that last paragraph may have noticed one other feature of Cassels' proof that is worth remarking on: condition (*i*) is never used in the proof. This is rather surprising. For example, it means that if we know that the equation has a solution in $\mathbb{Q}_p$ for every prime $p < \infty$, then it has a solution in $\mathbb{R}$. Or, in more elementary and more dramatic terms, it says that three integers a, b, and c satisfying conditions (*ii*), (*iii*), and (*iv*) cannot all have the same sign. Would you have guessed that something like that was true?

Can you speculate about what might be going on here? (Comment: these are deep waters, but it's always worth the effort to think a little about things like this.)

For equations of degree higher than two, it is unlikely that anything as strong as the Hasse–Minkowski Theorem can be true. In fact, in many cases one has counterexamples (see Problem 130 for one) that show that one may have local solutions everywhere and still have no global solutions. Still, even in situations where this strong form of the local-global principle is false, the basic idea that getting local information everywhere should give global information often remains useful. In the case of cubic equations, for instance, it is *not* true that the existence of local solutions everywhere guarantees the existence of global solutions; nevertheless, there are still strong connections (or at least one suspects so). For example, there is a conjecture, due to Birch and Swinnerton-Dyer, that says, when looked at from this angle, that the quantity of global solutions can be determined in terms of local information.

The Birch and Swinnerton-Dyer conjecture is widely believed to be true, and offers one example of how the local-global principle remains one of the fundamental ideas of modern number theory.

5   Elementary Analysis in $\mathbb{Q}_p$

The field of p-adic numbers is in many ways analogous to the field of real numbers: it is a field with an absolute value, and it is complete with respect to the metric given by that absolute value. In fact, the similarities go deeper: $\mathbb{R}$ and the various $\mathbb{Q}_p$ are completions of $\mathbb{Q}$, hence contain $\mathbb{Q}$ as a dense subset; they are all locally compact; none of them are algebraically closed.

These similarities all suggest that much of what is usually done in $\mathbb{R}$ can be extended to $\mathbb{Q}_p$. In particular, the basic structures of the calculus should all extend. The goal of this chapter is to examine what form these basic ideas take in the p-adic context. The central theme will be the theory of infinite series, which we will use to construct a number of different functions on $\mathbb{Q}_p$ which imitate the classical transcendental functions.

The reader will probably remark on the fact that our "elementary analysis" focuses on power series, touching only lightly on the derivatives and integrals that played such a large role in everyone's calculus classes. As far as derivatives are concerned, the main reason for this is simply that derivatives are much less interesting in a p-adic context than they are in real analysis. In particular, the fact that the mean value theorem does not hold means that simply working with differentiable functions will usually not be good enough. Functions defined by power series are nicer.

Integration is a different story entirely. It is certainly possible to construct a p-adic theory of integration (indeed, more than one such theory). But things get complicated. We give a brief sketch of the situation in Section 5.3.

Before we go on, we should also note that while there are many similarities between $\mathbb{R}$ and the $\mathbb{Q}_p$, there are also rather large differences; noticing them at this point will prepare us for the changes to come later. To begin with, $\mathbb{R}$ is an *ordered* field: there is a well-defined notion of "bigger than" that is nicely compatible with the operations. This is certainly not true for the $\mathbb{Q}_p$. Secondly, $\mathbb{R}$ is archimedean (more precisely, the absolute value on $\mathbb{R}$ is), while the $\mathbb{Q}_p$ are all non-archimedean. This means, in particular, that $\mathbb{R}$ is *connected* as a metric space, while $\mathbb{Q}_p$, as we saw above, is *totally disconnected*. It follows, for example, that there is nothing in $\mathbb{Q}_p$ analogous to an interval in $\mathbb{R}$. Nor is there any analogue of the notion of a curve, because any continuous function $[0, 1] \longrightarrow \mathbb{Q}_p$ will be constant. It is these contrasts that will cause most of the differences between real and p-adic analysis.

© Springer Nature Switzerland AG 2020
F. Q. Gouvêa, *p-adic Numbers*, Universitext, https://doi.org/10.1007/978-3-030-47295-5_5

5.1   Sequences and Series

We begin by studying the basic convergence properties of sequences and series. The most important fact has already been noted: $\mathbb{Q}_p$ is a complete field, so that every Cauchy sequence converges. Furthermore, notice that all of the axioms that hold for the absolute value in $\mathbb{R}$ still hold in $\mathbb{Q}_p$ (being non-archimedean is an *extra* property). Hence, most of the basic theorems still hold in the p-adic context, *with the same proofs!* We will leave it to the reader to look over the basic theory in her real analysis text,[1] and emphasize rather the points where the non-archimedean property introduces serious differences from the real case. Perhaps the most important such difference is the fact, also noted above, that in a non-archimedean context it is easier to test for the Cauchy property.

Lemma 5.1.1 *A sequence* (a_n) *in* $\mathbb{Q}_p$ *is a Cauchy sequence, and therefore convergent, if and only if it satisfies*

$$\lim_{n \to \infty} |a_{n+1} - a_n| = 0.$$

PROOF: This is the case $\Bbbk = \mathbb{Q}_p$ of Lemma 3.2.2.   □

Except for this important difference, the theory of sequences and their convergence properties is pretty much identical to the theory over $\mathbb{R}$. The basic definition of convergence is the same. Of course, some sequences will converge in $\mathbb{Q}_p$ that don't converge in $\mathbb{R}$, and vice versa.

Let's look at an example. Let $a_1 = 1+p$ and define a sequence recursively by $a_n = (a_{n-1})^p$. Notice that

$$(1+p)^p = 1 + p^2 + \binom{p}{2}p^2 + \binom{p}{3}p^3 + \cdots + p^p.$$

Since $\binom{p}{k}$ is divisible by p when $0 < k < p$, we see that $(1+p)^p - 1$ is divisible by p^2, i.e., $a_2 \equiv 1 \pmod{p^2}$. Repeating the argument, we conclude that for every n we have $a_n \equiv 1 \pmod{p^n}$, i.e.,

$$|a_n - 1| \le p^{-n},$$

and we see that a_n converges to 1 as $n \to \infty$.

Problem 141 Decide if the following sequences converge in $\mathbb{Q}_p$, and find the limit of those that do:

- $a_n = n!$ (As n grows, $n!$ gets more and more divisible by p, so the limit should be zero. Check that it is.)

- $a_n = n$ (This one should not converge to anything; can you see why?)

---

[1] For example, [55, Ch. 3] contains all the basics. See also [39], where the real versus p-adic comparison is emphasized.

- $a_n = 1/n$
- $a_n = p^n$
- $a_n = (1 + px)^{p^n}$, where $x \in \mathbb{Z}_p$.

Problem 142 Let (a_n) be a convergent sequence in $\mathbb{Q}_p$. Show that either $\lim |a_n| = 0$ or there exists an integer M such that $|a_n| = |a_M|$ for every $n \geq M$. In words: the sequence of absolute values of a convergent sequence either tends to zero or becomes constant for large enough n.

As for sequences, so for series: the classical theory still holds. For example, the following is still true:

Problem 143 Let $a_n \in \mathbb{Q}_p$. Show that absolute convergence implies convergence, i.e., that if the series of absolute values $\sum |a_n|$ converges (in $\mathbb{R}$), then the series $\sum a_n$ converges in $\mathbb{Q}_p$.

This is an important and useful result in real analysis. In the p-adic context, however, Lemma 5.1.1 gives us something much better:

Corollary 5.1.2 *An infinite series* $\displaystyle\sum_{n=0}^{\infty} a_n$ *with* $a_n \in \mathbb{Q}_p$ *is convergent if and only if*

$$\lim_{n \to \infty} a_n = 0,$$

in which case we also have

$$\left| \sum_{n=0}^{\infty} a_n \right| \leq \max_n |a_n|.$$

PROOF: A series converges when the sequence of partial sums converges. Now, the n-th term a_n is exactly the difference between the n-th and the $(n-1)$-st partial sums; if it tends to zero, it follows from Lemma 5.1.1 that the sequence of partial sums is a Cauchy sequence, hence is convergent.

Finally, the estimate for the sum is a straight extension of the non-archimedean inequality, and we leave its verification to the reader. Note that since $|a_n| \to 0$ the maximum will be attained, i.e., we do not have to say "sup" instead. □

Problem 144 Check the inequality for the absolute value of the sum of a convergent series.

Problem 145 The corollary flies in the face of many admonitions in calculus class: in $\mathbb{R}$, the fact that the general term tends to zero is *not* a sufficient condition for convergence. In other words, the corollary is *false* in $\mathbb{R}$. Give an example of a series in $\mathbb{R}$ whose general term tends to zero, but which does not converge. Give another.

Problem 146 The inequality

$$\left| \sum_{n=0}^{\infty} a_n \right| \leq \max_n |a_n|$$

is also false for convergent series in $\mathbb{R}$. Give a counterexample.

The upshot is that it is much easier to decide on the convergence of an infinite series in the p-adic context than over $\mathbb{R}$. This has the effect of making the theory of series in $\mathbb{Q}_p$ generally a lot simpler than the classical theory. For example, when we work in $\mathbb{R}$ the sum of a conditionally convergent series depends on the order of its terms. In $\mathbb{Q}_p$, that never happens. Proving this is a good warm-up for the proofs to follow.

Problem 147 Show that the sum of a convergent series in $\mathbb{Q}_p$ does not change when we reorder the terms. (Hint: if the terms tend to zero, for any $\varepsilon > 0$ we know that all but finitely many have absolute value $< \varepsilon$. Compare two partial sums that are long enough to include all such terms.)

Another example of a situation where things are easier in the non-archimedean case is a theorem[2] about double series and reversing the order of summation. We want to consider a "double sequence" b_{ij} of p-adic numbers and ask about the two series we get by summing either first in i, then in j, or the other way around. In other words, we want to consider

$$\sum_{i=0}^{\infty} \left(\sum_{j=0}^{\infty} b_{ij} \right)$$

and

$$\sum_{j=0}^{\infty} \left(\sum_{i=0}^{\infty} b_{ij} \right)$$

and decide whether the two are equal.

For this to make sense, we need that the b_{ij} tend to zero when we fix one index and let the other go to infinity (otherwise the inner series won't converge). We'll say that

$$\lim_{i \to \infty} b_{ij} = 0 \qquad \text{uniformly in } j$$

if given any positive number ε we can find an integer N *which does not depend on j* such that

$$i \geq N \implies |b_{ij}| < \varepsilon.$$

In other words, for each j the sequence b_{ij} tends to zero when $i \to \infty$, and the convergence is "at the same rate" for all j. The first thing we need is a lemma:

---

[2]This is a variant of a theorem given in [14]; I learned it from Keith Conrad. See also [39, Theorem 3.8].

Lemma 5.1.3 *Let $b_{ij} \in \mathbb{Q}_p$. The following are equivalent:*

i) *For every i, $\lim_{j \to \infty} b_{ij} = 0$, and $\lim_{i \to \infty} b_{ij} = 0$ uniformly in j.*

ii) *Given any $\varepsilon > 0$ there exists an integer N depending only on ε such that*
$$\max\{i, j\} \geq N \implies |b_{ij}| < \varepsilon.$$

iii) $\lim_{i \to \infty} b_{ij} = 0$ *uniformly in j and* $\lim_{j \to \infty} b_{ij} = 0$ *uniformly in i.*

PROOF: It's clear that (*ii*) implies (*iii*) and even clearer that (*iii*) implies (*i*). What we need to prove, then, is that the apparently weaker (*i*) implies (*ii*).

Assume (*i*) and let $\varepsilon > 0$ be given. The uniformity in j says that we can choose N_0, depending on ε but not on j, such that $|b_{ij}| < \varepsilon$ if $i \geq N_0$. The convergence in i is weaker: it says that for each i we can find $N_1(i)$ (the notation emphasizes that it *does* depend on i) such that if $j \geq N_1(i)$ we have $|b_{ij}| < \varepsilon$. Now take

$$N = N(\varepsilon) = \max\{N_0, N_1(0), N_1(1), \ldots, N_1(N_0 - 1)\}.$$

This N does the trick: if $\max(i, j) \geq N$, then either $i \geq N_0$, and we know $|b_{ij}| < \varepsilon$ regardless of what j is, or $i < N_0$ and $j \geq N$, in which case i must be equal to one of $0, 1, \ldots, N_0 - 1$ and j will be bigger than the appropriate N_1, giving $|b_{ij}| < \varepsilon$ again. □

The crucial point is that the fact that $b_{ij} \to 0$ uniformly in j allows us to restrict to only a finite number of cases in which we have to use the other condition. Now we can go on to prove our theorem on double series.

Proposition 5.1.4 *Let $b_{ij} \in \mathbb{Q}_p$, and suppose that*

i) *for every i, $\lim_{j \to \infty} b_{ij} = 0$, and*

ii) $\lim_{i \to \infty} b_{ij} = 0$ *uniformly in j.*

Then both series

$$\sum_{i=0}^{\infty} \left(\sum_{j=0}^{\infty} b_{ij} \right) \qquad and \qquad \sum_{j=0}^{\infty} \left(\sum_{i=0}^{\infty} b_{ij} \right)$$

converge, and their sums are equal.

PROOF: From the lemma, we know that given ε we can choose N such that if $\max\{i, j\} \geq N$ then $|b_{ij}| < \varepsilon$. In particular, b_{ij} tends to zero for every i when $j \to \infty$ and vice versa, which means that the internal sums

$$\sum_{j=0}^{\infty} b_{ij} \qquad and \qquad \sum_{i=0}^{\infty} b_{ij}$$

converge (the first for all i, and the second for all j). In addition, for $i \geq N$ we have, by Corollary 5.1.2,

$$\left| \sum_{j=0}^{\infty} b_{ij} \right| \leq \max_{j} \{ |b_{ij}| \} < \varepsilon;$$

similarly, for $j \geq N$ we have

$$\left| \sum_{i=0}^{\infty} b_{ij} \right| < \varepsilon.$$

In particular, we see that

$$\lim_{i \to \infty} \sum_{j=0}^{\infty} b_{ij} = 0 \qquad \text{and} \qquad \lim_{j \to \infty} \sum_{i=0}^{\infty} b_{ij} = 0,$$

so that both double series converge.

It remains to check that the sums are equal. For that, we continue to use N and ε chosen as above, so that $|b_{ij}| < \varepsilon$ when either i or j is $\geq N$, and we use over and over the fact that in a non-archimedean field a bound on each term in a sum gives a bound on the sum itself; this is just the ultrametric inequality $|x + y| \leq \max\{|x|, |y|\}$, as generalized to series in Corollary 5.1.2.

Begin by noticing that

$$\left| \sum_{i=0}^{\infty} \left(\sum_{j=0}^{\infty} b_{ij} \right) - \sum_{i=0}^{N} \left(\sum_{j=0}^{N} b_{ij} \right) \right| = \left| \sum_{i=0}^{N} \left(\sum_{j=N+1}^{\infty} b_{ij} \right) + \sum_{i=N+1}^{\infty} \left(\sum_{j=0}^{\infty} b_{ij} \right) \right|.$$

Now, if $j \geq N+1$, we have $|b_{ij}| < \varepsilon$ for every i; by the ultrametric inequality, it follows that $\left| \sum_{j=N+1}^{\infty} b_{ij} \right| < \varepsilon$ for every i, and then (use the ultrametric inequality again!)

$$\left| \sum_{i=0}^{N} \left(\sum_{j=N+1}^{\infty} b_{ij} \right) \right| < \varepsilon.$$

Similarly, we get an estimate for the other summand:

$$\left| \sum_{i=N+1}^{\infty} \left(\sum_{j=0}^{\infty} b_{ij} \right) \right| < \varepsilon,$$

and one more application of the ultrametric inequality allows us to conclude that

$$\left| \sum_{i=0}^{\infty} \left(\sum_{j=0}^{\infty} b_{ij} \right) - \sum_{i=0}^{N} \left(\sum_{j=0}^{N} b_{ij} \right) \right| < \varepsilon.$$

Of course, reversing i and j we get a similar inequality for the other double sum. Finally, since clearly one can reverse the order of summation in the finite sum, we can use the ultrametric inequality once again to conclude that

$$\left| \sum_{i=0}^{\infty} \left(\sum_{j=0}^{\infty} b_{ij} \right) - \sum_{j=0}^{\infty} \left(\sum_{i=0}^{\infty} b_{ij} \right) \right| < \varepsilon.$$

Since this is true for any $\varepsilon > 0$ it follows that the two sums must be equal.□

What this result says is that if the b_{ij} tend to zero in a sufficiently uniform way, then their sum can be taken in any order. This result will prove quite useful in our later applications.

Problem 148 How important is the ultrametric inequality in the proof? What is the best result along these lines in real analysis?

The next two problems show that series can be added and multiplied in the "natural" way.

Problem 149 Show that if $a = \sum a_n$ and $b = \sum b_n$ are convergent, and we set

$$c_n = a_n + b_n,$$

then the series $\sum c_n$ is convergent and has sum $a + b$. (This is easy.)

Problem 150 Show that if $a = \sum a_n$ and $b = \sum b_n$ are convergent, and we set

$$c_n = \sum_{i=0}^{n} a_i b_{n-i},$$

then the series $\sum c_n$ is convergent and has sum ab. (This is not easy. In the archimedean case, it is only true under the assumption that one of the series converges absolutely; otherwise, $\sum c_n$ may not converge at all.)

5.2   Functions, Continuity, Derivatives

The basic ideas about functions and continuity remain unchanged when we go to the p-adics, since after all they depend only on the metric structure. There are no intervals to work with (in fact, no non-trivial connected sets at all), so usually our functions will be defined in (open or closed) balls. Recall that we write $B(a, r)$ for the open ball with center a and radius r and $\overline{B}(a, r)$ for the closed ball with center a and radius r. We defined continuity and uniform continuity on page 41, but it doesn't hurt to recall the definition:

Definition 5.2.1 *Let $U \subset \mathbb{Q}_p$. A function $f : U \to \mathbb{Q}_p$ is said to be continuous at $a \in U$ if for every $\varepsilon > 0$ there exists a $\delta > 0$ (possibly depending on a) such that, for every $x \in U$,*

$$|x - a| < \delta \Longrightarrow |f(x) - f(a)| < \varepsilon.$$

Let $U \subset \mathbb{Q}_p$. A function $f : U \to \mathbb{Q}_p$ is said to be uniformly continuous on U if for every $\varepsilon > 0$ there exists a $\delta > 0$ such that, for all $x, y \in U$,

$$|x - y| < \delta \Longrightarrow |f(x) - f(y)| < \varepsilon.$$

The basic results about continuity are true in all metric spaces, and hence are true here too. For example, if U is compact (and remember, in $\mathbb{Q}_p$ it's perfectly possible for a set to be both open *and* compact) and f is continuous at every point of U, then f is uniformly continuous on U.

Problem 151 Is there a p-adic analogue of the intermediate value theorem?

Derivatives are a bit more interesting, if only because it'll turn out that they don't work as well as in the classical case. It certainly makes perfect sense to define derivatives of functions $f : \mathbb{Q}_p \longrightarrow \mathbb{Q}_p$ in the usual way:

Definition 5.2.2 *Let $U \subset \mathbb{Q}_p$ be an open set, and let $f : U \longrightarrow \mathbb{Q}_p$ be a function. We say f is differentiable at $x \in U$ if the limit*

$$f'(x) = \lim_{h \to 0} \frac{f(x + h) - f(x)}{h}$$

exists. If $f'(x)$ exists for every x in U we say f is differentiable in U, and we write $f' : U \longrightarrow \mathbb{Q}_p$ for the function $x \mapsto f'(x)$.

To some extent, the derivative works as expected. For example, we can show that differentiable functions are continuous, in exactly the same way as we do it over $\mathbb{R}$ or $\mathbb{C}$. Along the same vein:

Problem 152 Let $n \in \mathbb{Z}$. What is the derivative of the function $\mathbb{Q}_p \longrightarrow \mathbb{Q}_p$ given by $x \mapsto x^n$?

Inevitably, there are some surprising things. Recall, for example, that the closed unit ball in $\mathbb{Q}_p$ is the union of disjoint open unit balls:

$$\overline{B}(0, 1) = B(0, 1) \cup B(1, 1) \cup B(2, 1) \cup \cdots \cup B(p - 1, 1).$$

If we define a function to be constant in each of the open balls, say, $f(x) = i$ when $x \in B(i, 1)$, we get a function which is *locally constant*, i.e., each $x \in \overline{B}(0, 1)$ has an open neighborhood on which f is constant. Clearly this function must have derivative zero. Since $\mathbb{Q}_p$ is totally disconnected, we are often going to run into such functions. In fact, one can show that any bounded continuous function is the uniform limit of locally constant functions; see [56, 26.2] or [53, IV.3.3].

But things are actually much worse.

Problem 153 Consider the function (stolen from [56]) $f : \mathbb{Z}_p \longrightarrow \mathbb{Z}_p$ which maps

$$x = a_0 + a_1 p + a_2 p^2 + a_3 p^3 + \cdots + a_n p^n + \cdots$$

to

$$f(x) = a_0 + a_1 p^2 + a_2 p^4 + a_3 p^6 + \cdots + a_n p^{2n} + \cdots$$

Show that f is injective and that $f'(x) = 0$ for all x.

Since the function in the problem is injective, it cannot be locally constant in any neighborhood of any $x \in \mathbb{Z}_p$. But its derivative is still zero!

This is one of the reasons why the derivative seems to play such a minor role in p-adic analysis: having zero derivative does not imply that a function is locally constant.

What breaks in the proof we saw in calculus? The usual proof relies on the mean value theorem, which is the linchpin of the elementary theory of differentiable functions. It says that given a and b in an interval where $f(x)$ is differentiable, there exists a number ξ between a and b such that

$$f(b) - f(a) = f'(\xi)(b - a).$$

If we know $f'(\xi) = 0$ for all ξ, we conclude at once that $f(a) = f(b)$ and that f is constant on that interval.

In thinking about what a p-adic version of that theorem might say, we might initially think that the problem is that "between": since $\mathbb{Q}_p$ is not ordered it makes no sense in our case. That, however, turns out not to be the core problem. After all, if we had an inequality we could still reach our conclusion, since if an absolute value is ≤ 0 then it is 0. So we could hope for a theorem that said that if the derivative of $f(x)$ is bounded by a number M, then $|f(b) - f(a)| \leq M|b - a|$. (This version of the mean value theorem holds in the multivariable case over $\mathbb{R}$, for example.) So here is an attempt at a minimal p-adic version of the mean value theorem:

What a p-adic mean value theorem might say: *If a function $f(X)$ is differentiable with (continuous?) derivative on $U \subset \mathbb{Q}_p$ and if $|f'(x)| \leq M$ for all $x \in U$, then, for any two numbers a and b in U we have*

$$|f(b) - f(a)| \leq M|b - a|.$$

Unfortunately, things aren't that simple. We know this must be false, but we might guess that it only fails for complicated functions. So let's give a counterexample where $f(x)$ is a very nice function.

Proposition 5.2.3 *The "p-adic mean value theorem" we just stated is false.*

PROOF: Take $U = \mathbb{Z}_p$, $f(x) = x^p$, $a = 1$, $b = 0$. Then $f'(x) = p x^{p-1}$, so we have $|f'(x)| \leq p^{-1}$ for all $x \in U$. But $f(1) - f(0) = 1$, so

$$|f(1) - f(0)| = 1 > \frac{1}{p} = \frac{1}{p}|1 - 0|. \qquad \square$$

So the lack of an MVT is what allows functions to have zero derivative without being locally constant. Some authors call such functions "pseudo-constant" or "almost constant." And there are many of them!

Problem 154 Show that the chain rule is still true in $\mathbb{Q}_p$. Use this fact to show that if f has zero derivative everywhere and g is any continuously differentiable function, then both $f \circ g$ and $g \circ f$ have zero derivative everywhere. Explain why this means that there are a great many "pseudo-constant" functions.

In particular, it follows that two functions which have the same derivative do *not* need to differ by a constant. So the theory of anti-derivatives becomes much more messy than it was over $\mathbb{R}$.

The effect of all this is that knowing that a function is differentiable isn't as useful in the p-adic context as it is classically. In fact, since the zero function is continuous, even having a continuous derivative doesn't help.

There is still another property of derivatives that fails in the p-adic setting. Over $\mathbb{R}$, if a function is continuously differentiable and has nonzero derivative at a point then it will be injective on an interval containing that point. Alas, there exists a function $f : \mathbb{Z}_p \longrightarrow \mathbb{Q}_p$ such that $f'(x) = 1$ for all x but $f(p^n) = f(p^n - p^{2n})$ for every $n \geq 0$, so that f is not injective in *any* neighborhood of zero. See [56, 26.6] for the construction.

The solution, as usual, is to come up with a stricter notion of differentiability. We won't go into the details; the basic idea is to consider difference quotients

$$\frac{f(u) - f(v)}{u - v}$$

as functions of *two* variables and take the limit as $(u, v) \to (a, a)$. Over $\mathbb{R}$, the existence of this limit boils down to f being continually differentiable. Over $\mathbb{Q}_p$, the limit can fail to exist even if f' is continuous, and so this becomes a stronger condition. Readers interested in going deeper into this might look at [45], [56, §26–29], and especially [53, Ch. V].

In this book we will concentrate on functions defined by power series instead. Of course, the function we used in the example above is a polynomial, so nothing will rescue the "p-adic mean value theorem" we tried to formulate. Nevertheless, it is possible to prove an analogue of the mean value theorem for functions defined by power series, provided one restricts to the case when $|b - a|$ is small enough. See Alain Robert's account of this in [53, V.3].

5.3   Integrals

This short section is an introduction to the problems surrounding p-adic integration. Rather than giving definitions and theorems, the goal is to explain how and why things get complicated, and then provide pointers to the literature. We mostly follow [56, §30].

When we set out to generalize integration to a p-adic setting, the first thing to decide is what exactly we want to generalize. Over $\mathbb{R}$, integrals and anti-derivatives are closely related: if $F'(x) = f(x)$ and $f(x)$ is continuous, then

$$\int_a^b f(x)\,dx = F(b) - F(a).$$

For this formula to make sense, it is crucial that $F(b) - F(a)$ doesn't depend on the choice of anti-derivative, which follows from the fact that a function with derivative identically zero is constant. Since that fact is false in p-adic calculus, the "fundamental theorem of calculus" cannot be true. So in our world integrals and anti-derivatives are unrelated.

There are a few more things that can be said about anti-derivatives in the p-adic setting, but the upshot is that there is no good general theory (see [56, 30.2 and 30.3], for example). Things are better for functions defined by power series, so we will return to this issue in the next section.

How about a theory of integration? That is usually done for bounded functions on a compact set, but let's consider a concrete example to see how it might go. Suppose we have a continuous (and therefore bounded) function $f : \mathbb{Z}_p \longrightarrow \mathbb{Q}_p$. To compute an integral, we would need to write $\mathbb{Z}_p$ as a union of smaller subsets U_i and then consider limits of sums like

$$\sum_i f(x_i)\,\text{size}(U_i).$$

Since f takes values in $\mathbb{Q}_p$, we would need the size of U_i to be a p-adic number of some kind. A function that assigns to each (nice) subset a certain number is called a *measure*.

What properties should a measure have? Here's a possible set of requirements. First, for each compact subset $U \subset \mathbb{Z}_p$ we would like to have a number $m(U) \in \mathbb{Q}_p$. Ideally, this should have the following properties:

  i) (additivity) if $U \cap V = \varnothing$, $m(U \cup V) = m(U) + m(V)$;

  ii) (translation invariance) if $a \in \mathbb{Z}_p$, $m(a + U) = m(U)$.

Suppose such a function exists and let $m(\mathbb{Z}_p) = a$. Consider the closed balls $U_i = i + p\mathbb{Z}_p$, $i = 0, 1, \ldots, p - 1$. These are compact and disjoint, their union is $\mathbb{Z}_p$. By translation invariance, they must all have the same measure: $m(U_i) = m(p\mathbb{Z}_p)$. By additivity, the sum of the $m(U_i)$ must be a. So we get $p\,m(p\mathbb{Z}_p) = a$, so $m(p\mathbb{Z}_p) = \frac{a}{p}$. So the measure of the subset $p\mathbb{Z}_p \subset \mathbb{Z}_p$ has *bigger* p-adic absolute value than the measure of $\mathbb{Z}_p$.

The same argument shows that $m(p^n\mathbb{Z}_p) = \frac{a}{p^n}$. So as n grows we get smaller neighborhoods of zero $p^n\mathbb{Z}_p$ whose measures get p-adically *bigger*! Indeed

$$\lim_{n \to \infty} |m(p^n\mathbb{Z}_p)| = +\infty.$$

The technical slogan for what we have just shown is *there is no translation-invariant bounded $\mathbb{Q}_p$-valued measure on $\mathbb{Z}_p$*. What that means is that there is no perfect theory of integration either.

There are two options from here, leading to two different theories of integration. One can accept unbounded measures, or one can drop translation invariance. Each leads to a theory of integration that is of some value. The first option leads to the Volkenborn integral, which is closely related to Mahler's theory of continuous p-adic functions; see [56, §50–57] and [53, V.5]. The second is related to p-adic interpolation problems; see [42, Chapter II].

There is still another possible variant. We could try to integrate *real-valued* functions on $\mathbb{Q}_p$. For that, we would need a measure as above, but with $m(U) \in \mathbb{R}$. That turns out to yield a much nicer theory: $\mathbb{Q}_p$ is a locally compact abelian (additive) group, so it carries a Haar measure, which is exactly a translation-invariant real-valued measure. That makes it easy to do integration for functions $f : \mathbb{Q}_p \longrightarrow \mathbb{R}$. Something similar works for $f : \mathbb{Q}_p \longrightarrow \mathbb{Q}_q$ for $q \neq p$. See [46, Ch. VI] for a discussion of such integrals.

5.4   Power Series

In real analysis, power series

$$\sum_{n=0}^{\infty} a_n (X - \alpha)^n$$

offer a convenient way of representing functions, and in particular can be used to *define* several important functions, such as the exponential and trigonometric functions. As one might expect, the p-adic theory turns out to be quite similar to the classical version, except that some of the tricky points become a lot simpler to handle. On the other hand, the non-archimedean property does introduce a few surprises. The biggest of these surprises is the fact that the relation between the formal composition of power series and the composition of the functions they define becomes *more* complicated in the p-adic context than it is in the classical situation. Because this is such an unexpected development, we spend quite a bit of time on it.

The next few sections explore the main ideas about power series and functions defined by power series, focusing, in the end, on the p-adic versions of the logarithm and the exponential. The main influences on our treatment are [14], [34], and [19]. We have stated most of our results for power series in X, but of course they remain true for power series in $(X - \alpha)$ if we replace all conditions $|x| < k$ by $|x - \alpha| < k$.

Consider, then, a power series

$$f(X) = \sum_{n=0}^{\infty} a_n X^n.$$

Given $x \in \mathbb{Q}_p$, we want to consider $f(x)$, which is[3] the series $\sum a_n x^n$; we already know that this converges if and only if $|a_n x^n| \to 0$. As in the classical case, the set of all such x (which we call the *region of convergence*) is a disk whose radius can be computed directly.

Proposition 5.4.1 *Let* $f(X) = \sum_{n=0}^{\infty} a_n X^n$, *and define*

$$\rho = \frac{1}{\limsup_{n \to \infty} \sqrt[n]{|a_n|}},$$

where we use the usual conventions when the limit is zero or infinity, so that $0 \le \rho \le \infty$.

  i) If $\rho = 0$, *then* $f(x)$ *converges only when* $x = 0$.

  ii) If $\rho = \infty$, *then* $f(x)$ *converges for every* $x \in \mathbb{Q}_p$.

  iii) If $0 < \rho < \infty$ *and* $\lim_{n \to \infty} |a_n| \rho^n = 0$, *then* $f(x)$ *converges if and only if* $|x| \le \rho$.

  iv) If $0 < \rho < \infty$ *and* $|a_n| \rho^n$ *does not tend to zero as* n *goes to infinity, then* $f(x)$ *converges if and only if* $|x| < \rho$.

PROOF: (This requires you to understand what lim sup means; see [55, pp. 55–57] for a short explanation. If $\limsup b_n = B$ then for any $\varepsilon > 0$ two things hold: first, $b_n < B + \varepsilon$ for all but finitely many n; second, $b_n > B - \varepsilon$ for infinitely many n.)

We already know that the region of convergence is

$$\left\{ x \in \mathbb{Q}_p : \lim_{n \to \infty} |a_n x^n| = 0 \right\},$$

so that the point of the theorem is to translate this into more precise information. First of all, it is clear that $f(0)$ converges.

Next, if $|x| > \rho$, it is easy to see that $|a_n||x|^n$ cannot tend to zero when n tends to infinity: the definition of ρ implies that for infinitely many values of n, $|a_n|$ is close to $1/\rho^n$, and, since $|x| > \rho$, $(|x|/\rho)^n$ gets arbitrarily large as n grows.

Similarly, if $|x| < \rho$, choose a ρ_1 such that $|x| < \rho_1 < \rho$. Then $|x|/\rho_1 < 1$ and for all but finitely many n we have $|a_n| < 1/\rho_1^n$, so $|a_n x^n| \le |x|^n/\rho_1^n$ and

---

[3]In this section and the following ones, we adopt the convention that X represents an indeterminate, while x usually represents a p-adic number. Hence, $f(X)$ in this statement is to be thought of as the formal power series itself, while $f(x)$ is the numerical series we obtain by substituting x for X. It makes no sense to discuss the convergence of $f(X)$: the series is just *there*. It does make sense to discuss the convergence of $f(x)$; whether it converges or not will depend on x. When it does converge, we will write $f(x)$ for *both* the numerical series and its value; this should normally not cause any confusion.

so $|a_n x^n| \to 0$. Finally, the statements about what happens when $|x| = \rho$ are immediate from Corollary 5.1.2.   □

As in the archimedean case, the number ρ is called the *radius of convergence* of the series.

It's worth noting that our theorem is simpler than the classical version. Over $\mathbb{R}$, the region of convergence may include none, both, or only one of the endpoints of the interval $-\rho < x < \rho$. Over $\mathbb{C}$, it is worse: $|x| < \rho$ is a disk, and the set of points on the boundary for which the series converges can be pretty complicated. By contrast, in the p-adic case what happens at the points on the "boundary" of the region of convergence (i.e., the points with $|x| = \rho$) is rather simple: either the series is convergent at *all* such points or at none of them. (On the other hand, recall that the points such that $|x| = \rho$ are not really the boundary of the open disk!)

Problem 155 Find the region of convergence of the following p-adic power series:

   i) $\sum p^n X^n$

   ii) $\sum p^{-n} X^n$

   iii) $\sum n! X^n$

One of the nice things about starting from formal power series is the fact that several of the operations we want to do with power series make sense at the formal level. Let's look at these formal operations, and then ask the important question: how do the formal properties translate to properties of the functions defined by the power series?

We start with the easiest operations: given two formal power series $f(X)$ and $g(X)$, we can consider their sum and their product. If

$$f(X) = \sum_{n=0}^{\infty} a_n X^n \qquad \text{and} \qquad g(X) = \sum_{n=0}^{\infty} b_n X^n,$$

then we define

$$(f + g)(X) = \sum_{n=0}^{\infty} (a_n + b_n) X^n$$

and

$$(fg)(X) = \sum_{n=0}^{\infty} \left(\sum_{k=0}^{n} a_k b_{n-k} \right) X^n.$$

Notice that if we want to think of these definitions as the result of actually adding or multiplying the series, they imply a lot of reordering and recombining of terms!

As we have defined it, this is a formal operation only. Of course, we'd like to know that it actually works when we plug in numbers for X. It's not too hard to see everything works as expected:

Proposition 5.4.2 *Let $f(X)$ and $g(X)$ be formal power series, and suppose $x \in \mathbb{Q}_p$. If $f(x)$ and $g(x)$ both converge, then:*

i) $(f+g)(x)$ converges and is equal to $f(x) + g(x)$, and

ii) $(fg)(x)$ converges and is equal to $f(x)g(x)$.

It follows that the radii of convergence of $f + g$ and of fg are each greater than or equal to the smaller of the radii of convergence of f and g.

PROOF: Basically, all that's needed is an appeal to the results you proved for numerical series in Problems 149 and 150.   □

Problem 156 Fill in the details of the proof.

Having had such success with adding and multiplying series, we can be more ambitious, and consider the composition of formal series. Suppose we have two formal series

$$f(X) = \sum_{n=0}^{\infty} a_n X^n \qquad \text{and} \qquad g(X) = \sum_{n=0}^{\infty} b_n X^n,$$

and that $b_0 = 0$ (another way of saying that would be to say $g(0) = 0$). We want to check that it makes sense to define the composition $h(X) = f(g(X))$ of the two series. This should be

$$h(X) = a_0 + a_1 g(X) + a_2 g(X)^2 + \cdots + a_n g(X)^n + \ldots$$

That looks like an awful mess, but in fact we can (working formally, of course) reorganize it into a well-behaved power series. The idea is this: since $g(X)$ has no independent term, $g(X)^2$ starts with a term of degree 2, $g(X)^3$ starts with a term of degree 3, and so on. So, when we try to work out what the coefficients of $h(X) = f(g(X)) = \sum c_n X^n$ should be, each coefficient only requires a finite amount of work.

- The zeroth coefficient is just $c_0 = a_0$.

- The first coefficient only requires that we look at the first two terms $a_0 + a_1 g(X)$, and therefore $c_1 = a_1 b_1$.

- The second coefficient requires that we look at the first three terms

$$a_0 + a_1 g(X) + a_2 g(X)^2 = a_0 + a_1(b_1 X + b_2 X^2 + \ldots) \\ + a_2(b_1^2 X^2 + \ldots),$$

and so $c_2 = a_1 b_2 + a_2 b_1^2$.

- For the third coefficient, look at

$$a_0 + a_1 g(X) + a_2 g(X)^2 + a_3 g(X)^3 = a_0 + a_1(b_1 X + b_2 X^2 + b_3 X^3 \dots)$$
$$+ a_2(b_1^2 X^2 + 2b_1 b_2 X^3 + \dots)$$
$$+ a_3(b_1^3 X^3 + \dots)$$

so that $c_3 = a_1 b_3 + 2a_2 b_1 b_2 + a_3 b_1^3$.

- And so on! One can clearly find c_n for every n.

Problem 157 Can you find a general formula for the c_n?

So we know that given two formal power series $f(X)$ and $g(X)$ with $g(0) = 0$, we have a formal power series $h(X) = f(g(X))$ which is their formal composition. Now we need to ask questions about convergence, and those are not as easy to answer. The point is that plugging a number x into the power series $h(X)$ might give a different answer from what one gets by first plugging x into $g(X)$ and then plugging the result into $f(X)$. One might suspect that there are problems simply by contemplating the amount of rearranging that's going on in our definition of the composite series $h(X)$. In fact, it turns out we need to be very careful, as the following computation shows.

Problem 158 Consider the following example in $\mathbb{Q}_2$. Let

$$f(X) = 1 + X + \frac{X^2}{2!} + \dots + \frac{X^n}{n!} + \dots$$

be the usual formal series for the exponential, let

$$g(X) = 2X^2 - 2X,$$

and let

$$h(X) = f(g(X)).$$

In GP, the function defined by $f(X)$ is called exp. This GP code creates both the function $\exp(2x^2 - 2x)$ and a truncation of the formal power series $h(X)$.

```
funct(x)=exp(2*x^2-2*x)
hseries=truncate(exp(2*x^2-2*x))
```

Check that `funct(1+O(2^20))` and `subst(hseries,x,1+O(2^20))` give completely different answers. In other words, plugging 1 into the series $h(X)$ does not give the same answer as computing $f(g(1))$!

Clearly we need an extra condition to guarantee that the formal composed series converges to the composite function. Here's the theorem:

Theorem 5.4.3 Let $f(X) = \sum a_n X^n$ and $g(X) = \sum b_n X^n$ be formal power series with $g(0) = 0$, and let $h(X) = f(g(X))$ be their formal composition. Suppose that *i)* *g(x) converges,*

ii) *f(g(x)) converges (this means: plugging the number to which g(x) converges into f(X) gives a convergent series),*

iii) *for every n, we have $|b_n x^n| \leq |g(x)|$ (in other words, no term of the series converging to g(x) is bigger than the sum).*

Then h(x) also converges, and f(g(x)) = h(x).

PROOF: (Following [34, Ch. 17].) We have

$$f(X) = \sum_{n=0}^{\infty} a_n X^n \qquad \text{and} \qquad g(X) = \sum_{n=1}^{\infty} b_n X^n.$$

Let

$$g(X)^m = \sum_{n=m}^{\infty} d_{m,n} X^n.$$

It's not hard to work out the $d_{m,n}$ by using the formula for the product of formal power series: $d_{m,n} = 0$ if $n < m$, and, for any $n \geq m$,

$$d_{m,n} = \sum_{i_1 + i_2 + \cdots + i_m = n} b_{i_1} b_{i_2} \ldots b_{i_m}$$

(that formula looks uglier than it really is: basically, take all the products of m-tuples of b_i's whose indices add up to n). This (and solving Problem 157) allows us to write $h(X) = f(g(X))$ explicitly:

$$h(X) = a_0 + \sum_{n=1}^{\infty} \left(\sum_{m=1}^{n} a_m d_{m,n} \right) X^n.$$

Now let's start thinking about convergence. First of all, since $g(x)$ converges, we can use Proposition 5.4.2 to conclude that the formal series $g(X)^m$ converges when we plug in $X = x$, and in fact converges to $g(x)^m$; in other words,

$$g(x)^m = \sum_{n=m}^{\infty} d_{m,n} x^n.$$

More interesting is the fact that the special assumption we made about $g(X)$ is still true for the series $g(X)^m$: for every n, we have

$$|d_{m,n} x^n| \leq |g(x)^m|.$$

To see this, note first that $|g(x)^m| = |g(x)|^m$. Now look at the general term $|d_{m,n} x^n|$. If $n < m$, then $|d_{m,n} x^n| = 0$ and there is nothing to prove. On the other hand, if $n \geq m$, then the ultrametric inequality gives

$$|d_{m,n}x^n| = \left| \sum_{i_1+i_2+\cdots+i_m=n} b_{i_1}x^{i_1}b_{i_2}x^{i_2}\ldots b_{i_m}x^{i_m} \right|$$
$$\leq \max\left\{ |b_{i_1}x^{i_1}| \cdot |b_{i_2}x^{i_2}| \cdots |b_{i_m}x^{i_m}| \right\},$$

where the maximum is once again taken over all m-tuples $(i_1, i_2, \ldots, i_m)$ such that $i_1 + i_2 + \cdots + i_m = n$. But we know, from the hypothesis on $g(X)$, that $|b_{i_j}x^{i_j}| \leq |g(x)|$ for every i_j; multiplying all these inequalities gives $|d_{m,n}x^n| \leq |g(x)|^m$, which is the inequality we want.

So now we know that $g(x)$ converges, that powers of $g(x)$ converge, and that both the series for $g(x)$ and for $g(x)^m$ satisfy the extra condition that no term is larger than the final sum. We also know, from our assumptions, that $f(g(x))$ converges, that is, that $a_m(g(x))^m$ tends to zero as m grows. We have

$$f(g(x)) = a_0 + \sum_{m=1}^{\infty} a_m g(x)^m = a_0 + \sum_{m=1}^{\infty} a_m \left(\sum_{n=m}^{\infty} d_{m,n}x^n \right)$$
$$= a_0 + \sum_{m=1}^{\infty}\sum_{n=m}^{\infty} a_m d_{m,n}x^n$$

(where the order of the summations is crucial, of course), and, on the other hand,

$$h(x) = a_0 + \sum_{n=1}^{\infty} \left(\sum_{m=1}^{n} a_m d_{m,n} \right) x^n = a_0 + \sum_{n=1}^{\infty}\sum_{m=1}^{n} a_m d_{m,n}x^n.$$

These series are obtained from each other by reversing the order of the summation, so what we need to do is check that this is legal and that both series will have the same sum. That's what Proposition 5.1.4 is for!

To apply the proposition, we need to show that the general term $a_m d_{m,n}x^n$ tends to zero sufficiently uniformly. So let's study that general term. The crucial thing is to notice that we can use the fact that $g(x)^m$ is larger than any term of the series to get a uniform bound:

$$|a_m d_{m,n}x^n| \leq |a_m g(x)^m|,$$

where the important thing is that the right-hand side is independent of n. Given ε, we can, since $a_m g(x)^m \to 0$, choose N such that $|a_m g(x)^m| < \varepsilon$ if $m \geq N$. This shows part of what we want:

$$\lim_{m\to\infty} a_m d_{m,n}x^n = 0, \qquad \text{uniformly in } n.$$

On the other hand, for each m we know that the series

$$g(x)^m = \sum_{n=0}^{\infty} d_{m,n}x^n$$

converges, and it follows (after multiplying by a_m) that, for every m,

$$\lim_{n \to \infty} a_m d_{m,n} x^n = 0.$$

That's what we need to be able to apply Proposition 5.1.4 and conclude both that the series for $h(x)$ converges and that its sum is equal to $f(g(x))$, which is what we needed to prove. □

Notice that we know in general that

$$|g(x)| \leq \max_{n \geq 1}\{|b_n x^n|\}$$

by the ultrametric inequality. The extra condition in the theorem implies that in fact this is an equality. It's worth noticing a case when the extra condition is guaranteed to fail: if $g(x) = 0$ then we would want $\max\{|b_n x^n|\} = 0$, which for $x \neq 0$ is true only if all the b_n are zero.

In the literature one finds other conditions that are sufficient to guarantee that the formal composition of power series agrees with the composition of the functions they define. In [56, 41.2], for example, Schikhof gives the following. With notations as in the theorem, we can set $r = \max\{|b_n x^n|\}$ and replace condition iii with the assumption that $\lim |a_n| r^n = 0$, i.e., that $f(x)$ converges when $|x| = r$. Then $h(x) = f(g(x))$. One advantage of this version is that when $g(x) = 0$ we have no chance to satisfy the conditions in Theorem 5.4.3, but we might be able to satisfy Schikhof's condition.

Problem 159 Let $f(X) = \sum a_n X^n$ and $g(X) = \sum b_n X^n$ be formal power series with $g(0) = 0$, and let $h(X) = f(g(X))$ be their formal composition. Suppose that

i) $g(x)$ converges,

ii) if $r = \max_{n \geq 1}\{|b_n x^n|\}$, we have $\lim |a_n| r^n = 0$.

Then $f(g(x))$ and $h(x)$ converge and $f(g(x)) = h(x)$.

Problem 160 Suppose $f(X)$ and $g(X)$ satisfy the assumptions of Theorem 5.4.3. Suppose $g(x)$ converges and let $r = \max\{|b_n x^n|\}$. Show that $\lim |a_n| r^n = 0$.

In [53, VI.1.5] we find the same condition expressed in terms of the radii of convergence of $f(X)$ and $g(X)$. Let ρ_f and ρ_g be the radii of convergence. Then Robert shows that if $|x| < \rho_g$ and $\max\{|b_n x^n|\} < \rho_f$, then we can conclude that plugging x into the composed series $f(g(X))$ gives $f(g(x))$.

As the computation we did above shows, the extra assumption on $g(X)$ is essential! In other words, an equality of formal power series that involves composition does not need to imply equality of the functions unless we can check this extra condition.

Problem 161 Let's return to our example in $\mathbb{Q}_2$. Let

$$f(X) = 1 + X + \frac{X^2}{2!} + \cdots + \frac{X^n}{n!} + \cdots$$

be the usual formal series for the exponential, let

$$g(X) = 2X^2 - 2X,$$

and let

$$h(X) = f(g(X)).$$

We will show later that $f(x)$ converges for every $x \in 4\mathbb{Z}_2$ and diverges otherwise. Since $g(X)$ is a polynomial, $g(x)$ converges for every x. In particular, suppose we take $x = 1$. Then $g(1) = 0$ and so $f(g(1)) = 1$.

It's harder to figure out when $h(X)$ converges. Let $h(X) = \sum a_n X^n$. It's possible (but not easy) to prove that for all $n \geq 2$ we have $v_2(a_n) \geq 1 + n/4$. Therefore, $h(x)$ converges for all $x \in \mathbb{Z}_2$. In particular, it converges when $x = 1$.

   i) Check that the first two conditions in the Theorem 5.4.3 are satisfied, but the third is not. Check that the condition in Problem 159 is also not satisfied.

   ii) By computing out the first few terms of $h(X)$ and using the estimate for the valuation of the a_n, show that $h(1) \equiv 3 \pmod 4$.

   iii) Conclude that $h(1) \neq f(g(1))$.

The counterexample is a composition $f(g(X))$ where g is a polynomial. If, on the other hand, f is a polynomial, there is never a problem. This is because when f is a polynomial all that we need to do to compute $f(g(x))$ is to multiply $g(x)$ by itself, scale those products, and add. Hence Proposition 5.4.2 suffices to show that $h(X) = f(g(X))$ converges whenever $g(X)$ does and that $h(x) = f(g(x))$ for all x in the region of convergence of $g(X)$.

The behavior of a composition of formal power series is one of the most striking differences between p-adic and classical analysis. It is interesting to remark that in this case classical analysis is actually easier: if the radius of convergence of $f(X)$ is ρ and $|g(x)| < \rho$, then $h(x)$ converges and we have $f(g(x)) = h(x)$. See, for example, Proposition 5.1 in Section 2 of Chapter 1 of [13]. (For the ambitious reader: look through the proof in the classical case to see if you can spot where it fails for non-archimedean absolute values.)

Problem 162 One other operation with power series which we didn't mention is differentiation. Given a power series $f(X) = \displaystyle\sum_{n=0}^{\infty} a_n X^n$, we define its formal derivative to be $f'(X) = \displaystyle\sum_{n=1}^{\infty} n a_n X^{n-1}$. Show that this has the usual properties of a derivative:

   i) $(f + g)'(X) = f'(X) + g'(X)$.

   ii) $(fg)'(X) = f'(X)g(X) + f(X)g'(X)$.

   iii) If $h(X) = f(g(X))$, then $h'(X) = f'(g(X))g'(X)$.

Notice that these are equalities of formal series! We'll check that formal differentiation does give the derivative in the next section.

5.5   Functions Defined by Power Series

We will use power series to define functions. In other words, given a power series $f(X)$ we will think of it as defining a function whose domain is the set of x for which $f(x)$ converges. Just as in the classical case, the functions that are defined by power series have nice properties. The simplest one is continuity.

Lemma 5.5.1 *Let $f(X) = \sum a_n X^n$ be a power series with coefficients in $\mathbb{Q}_p$. If $f(x)$ converges when $|x| \leq r$, then the function $f : \overline{B}(0, r) \longrightarrow \mathbb{Q}_p$ defined by $x \mapsto f(x)$ is bounded and uniformly continuous.*

PROOF: Notice that we are not assuming that the closed ball is the region of convergence, just that the series converges on the closed ball. The proof we give follows [19].

Since f converges on $\overline{B}(0, r)$, we know that $|a_n| r^n$ tends to 0 as $n \to \infty$. It follows that $M_r = \max\limits_{n \geq 0} |a_n| r^n$ is finite.

Let's first show $f(x)$ is bounded on $\overline{B}(0, r)$. That's very easy: if $|x| \leq r$ we have

$$|f(x)| = \left| \sum_{n=0}^{\infty} a_n x^n \right| \leq \max\{|a_n x^n|\} \leq \max\{|a_n| r^n\} = M_r,$$

which shows that $f(x)$ is bounded by M_r when $x \in \overline{B}(0, r)$.

For uniform continuity, suppose $x, y \in \overline{B}(0, r)$. If we subtract $f(y)$ from $f(x)$, the constant terms cancel and we can factor out $(x - y)$ from the remaining sum:

$$f(x) - f(y) = \sum_{n=1}^{\infty} a_n (x^n - y^n)$$

$$= \sum_{n=1}^{\infty} a_n (x - y)(x^{n-1} + x^{n-2}y + \cdots + y^{n-1})$$

$$= (x - y) \sum_{n=1}^{\infty} a_n (x^{n-1} + x^{n-2}y + \cdots + y^{n-1}).$$

But

$$\left| x^{n-1} + x^{n-2}y + \cdots + y^{n-1} \right| \leq \max_{0 \leq i \leq n-1} \{|x|^{n-1-i}|y|^i\} \leq r^{n-1},$$

and this allows us to estimate the sum with the ultrametric inequality:

$$\left| \sum_{n=1}^{\infty} a_n (x^{n-1} + x^{n-2}y + \cdots + y^{n-1}) \right| \leq \max\{|a_n| r^{n-1}\} = \frac{1}{r} M_r.$$

So we get

$$|f(x) - f(y)| \leq \frac{1}{r} M_r |x - y|,$$

which shows that f is uniformly continuous on $\overline{B}(0, r)$.   □

When thinking about this lemma, one should keep in mind that while we are working over $\mathbb{Q}_p$ we are trying to give arguments that apply as generally as possible. Closed balls in $\mathbb{Q}_p$ are compact, and recall that continuous functions on a compact set are always bounded and uniformly continuous. So as long as we stay in $\mathbb{Q}_p$ the lemma is not really telling us more than that the function is continuous. The thing to notice, however, is that all we used in the proof was the non-archimedean property, and so the result will be true in a complete non-archimedean field even if we are in a context where closed balls are *not* compact (and such calamities do happen).

Another thing worth noting is that in $\mathbb{Q}_p$ every open ball is also a closed ball, since $B(0, p^k) = \overline{B}(0, p^{k-1})$. This is because the p-adic valuation is discrete. But if we pass to an extension of $\mathbb{Q}_p$ whose valuation is not discrete, there will be power series whose region of convergence is an open ball and that are not bounded or uniformly continuous on their region of convergence.

Corollary 5.5.2 *Let* $f(X) = \sum a_n X^n$ *be a power series with coefficients in* $\mathbb{Q}_p$, *and let* $\mathcal{D} \subset \mathbb{Q}_p$ *be its region of convergence, i.e., the set of* $x \in \mathbb{Q}_p$ *for which* $f(x)$ *converges. The function*

$$f : \mathcal{D} \to \mathbb{Q}_p$$

defined by $x \mapsto f(x)$ *is continuous on* $\mathcal{D}$.

Problem 163 Prove the corollary. In $\mathbb{R}$, continuity at the endpoints of the interval of convergence is a problem. Make sure that your p-adic proof handles those points as well.

As in the classical case, we can change the center of the series expansion, i.e., re-write our function as a power series in $(X - \alpha)$ for any α in the region of convergence. In the classical case, the resulting series can (and usually does) have a different region of convergence than the original series, and this fact is one of the ways to obtain "analytic continuations." Surprisingly, in the p-adic case changing the center never helps with analytic continuation.

Proposition 5.5.3 *Let* $f(X) = \sum a_n X^n$ *be a power series with coefficients in* $\mathbb{Q}_p$, *and let* $\alpha \in \mathbb{Q}_p$, $\alpha \neq 0$, *be a point for which* $f(\alpha)$ *converges. For each* $m \geq 0$, *define*

$$b_m = \sum_{n \geq m} \binom{n}{m} a_n \alpha^{n-m},$$

and consider the power series

$$g(X) = \sum_{m=0}^{\infty} b_m (X - \alpha)^m.$$

i) *The series defining b_m converges for every m, so that the b_m are well-defined.*

ii) *The power series $f(X)$ and $g(X)$ have the same region of convergence, that is, $f(\lambda)$ converges if and only if $g(\lambda)$ converges.*

iii) *For any λ in the region of convergence, we have $g(\lambda) = f(\lambda)$.*

PROOF: Claim (i) is easy to see: since binomial coefficients are integers and α belongs to the region of convergence for $f(X)$, we get (for fixed m)

$$\left| \binom{n}{m} a_n \alpha^{n-m} \right| \le |a_n \alpha^{n-m}| = |\alpha|^{-m} \cdot |a_n \alpha^n| \to 0,$$

which gives the desired convergence by Lemma 5.1.2.

To show (ii) and (iii), take any λ in the region of convergence of $f(X)$, and compute

$$f(\lambda) = \sum_n a_n (\lambda - \alpha + \alpha)^n = \sum_n \sum_{m \le n} \binom{n}{m} a_n \alpha^{n-m} (\lambda - \alpha)^m.$$

The last sum looks a lot like a partial sum for $g(\lambda)$, except that it needs to be re-ordered. For that, we use Proposition 5.1.4. To check that the condition is satisfied, set

$$\beta_{nm} = \begin{cases} \binom{n}{m} a_n \alpha^{n-m} (\lambda - \alpha)^m & \text{if } m \le n \\[2ex] 0 & \text{if } m > n \end{cases}$$

We need to check that the sequence β_{nm} satisfies the conditions in Proposition 5.1.4. Note first that

$$|\beta_{nm}| = \left| \binom{n}{m} a_n \alpha^{n-m} (\lambda - \alpha)^m \right| \le |a_n \alpha^{n-m} (\lambda - \alpha)^m|,$$

so that the problem is bounding this last expression. To do that, recall that the region of convergence is a (closed or open) disk of some radius ρ; since both λ and α are in the region of convergence, there exists a radius ρ_1 such that

• the closed disk of radius ρ_1 is contained in the region of convergence, and

- we have both $|\lambda| \leq \rho_1$ and $|\alpha| \leq \rho_1$.

(This dodge takes care of the question of whether the region of convergence is the open or the closed disk: if it is the closed disk, we can take $\rho_1 = \rho$; if the open, take ρ_1 to be the larger of the absolute values of α and λ, so that $\rho_1 < \rho$. The second choice actually works in both cases.) Then we have

- $|\alpha|^{n-m} \leq \rho_1^{n-m}$ by construction, and

- $|\lambda - \alpha|^m \leq \max\{|\lambda|, |\alpha|\}^m \leq \rho_1^m$ by the non-archimedean property.

Going back to the terms we want to estimate, we get

$$|\beta_{nm}| \leq |a_n \alpha^{n-m} (\lambda - \alpha)^m| \leq |a_n| \rho_1^n,$$

which is independent of m and tends to zero as $n \to \infty$. This means that given any $\varepsilon > 0$ there exists an N for which $|\beta_{nm}| < \varepsilon$ if $n \geq N$ and for any m. This shows that β_{nm} tends to zero uniformly in m. The other condition is easy: if $m > n$, we have $\beta_{nm} = 0$, hence it's certainly true that for every n we have $\beta_{nm} \to 0$ when $m \to \infty$. Thus, the conditions in Proposition 5.1.4 are satisfied, and we can reverse the order of summation.

Changing the order of summation in the expression for $f(\lambda)$ gives the expression for $g(\lambda)$, so that applying Proposition 5.1.4 allows us to conclude that $g(\lambda)$ converges and is equal to $f(\lambda)$. This shows that g converges whenever f does, and in that case their values are equal. To conclude, notice that we can switch the roles of g and f in the argument, which shows that in fact the regions of convergence are identical.   □

Problem 164 Before you relax from that long proof, are you sure that the smoke-and-mirrors phrase "switch the roles of g and f" is really justified? Does anything further need to be checked?

Problem 165 Prove the following relative of the proposition: Let $f(X)$ be a power series such that $f(x)$ converges for $|x| < \rho$, and suppose $|a| = 1$ and $|b| < \rho$. Then the function $g(x) = f(ax + b)$ is given by a power series $g(X)$ which converges for $|x| < \rho$.

As in the classical theory, functions which can be expressed as power series in a disk $B(a, r)$ or $\overline{B}(a, r)$ around each a in their domain are called *analytic*. Functions of this kind in general have very nice properties, and this is also true in $\mathbb{Q}_p$. Unfortunately, the theory is not so nice as, for example, the theory over the complex numbers. One of the crucial reasons is the proposition we have just proved: we cannot get an "analytic continuation" for a function by choosing another center and expanding in a power series. Doing so in $\mathbb{Q}_p$ produces a power series with exactly the same region of convergence, which therefore does not allow us to "continue" the function to a larger domain.

Also unpleasant is the fact that many functions are "locally analytic" for trivial reasons having to do with the fact that $\mathbb{Q}_p$, since it is non-archimedean, is totally disconnected. In fact, consider the function given by

$$f(x) = \begin{cases} 1 & \text{if } x \in \mathbb{Z}_p \\ 0 & \text{if } x \notin \mathbb{Z}_p \end{cases}$$

Since both $\mathbb{Z}_p$ and its complement are open sets, around any point one can find a ball in which $f(x)$ is constant, and hence can be written as a (constant) power series! One would not want to think of such a function as "analytic." Hence, while the set of analytic functions on a closed ball behaves well, it isn't clear how to move from that "local" theory to a "global" notion of analytic function.

It turns out that one can get around such difficulties, and come up with a good concept of "analytic" functions and of "analytic continuation." Unfortunately, this requires quite a sophisticated approach. The resulting theory is developed in what is called *Rigid Analytic Geometry*; its foundations are due to John Tate, and it has become a very important branch of modern number theory. For an introduction to this rather difficult subject, the reader might look at [7] or [11].

We will stick to simpler things, mostly by focusing on how a function defined by a power series behaves in a closed ball contained in the region of convergence. First of all, if a function is given by a power series it completely determines that power series. As in the classical case, this can be shown by using derivatives (see below), but we prove something stronger. Let's say a sequence (x_m) converging to a limit L is *stationary* if there exists an n such that $x_m = L$ for all $m \geq n$.

Proposition 5.5.4 *Let $f(X)$ and $g(X)$ be formal power series, and suppose there is a non-stationary sequence $x_m \in \mathbb{Q}_p$ converging to zero in $\mathbb{Q}_p$ and such that $f(x_m) = g(x_m)$ for every m. Then $f(X) = g(X)$, i.e., $f(X)$ and $g(X)$ have the same coefficients.*

PROOF: (This is identical to the classical proof.) Replacing the sequence (x_m) by a subsequence if necessary, we can assume $x_m \neq 0$ for all m. If we consider the difference $h(X) = f(X) - g(X) = \sum a_n X^n$, then we have $h(x_m) = 0$ for every m, and we want to show that $a_n = 0$ for every n. Suppose not; then let r be the least index for which $a_r \neq 0$, so that

$$h(X) = a_r X^r + a_{r+1} X^{r+1} + a_{r+2} X^{r+2} \cdots$$
$$= X^r (a_r + a_{r+1} X + a_{r+2} X^2 + \cdots)$$
$$= X^r h_1(X),$$

where $h_1(0) = a_r \neq 0$. Since h_1 is a function defined by a power series, it is continuous, so $h_1(x_m) \to a_r$ as $m \to \infty$ (remember that our assumption is that $x_m \to 0$); in particular, $h_1(x_m)$ is non-zero for large enough m. Since we know $x_m \neq 0$, it follows that $h(x_m) = x_m^r h_1(x_m)$ is non-zero for large enough m, which is a contradiction. ☐

Problem 166 Suppose $f(X)$ and $g(X)$ are formal power series, and suppose that x_m is a non-stationary sequence in $\mathbb{Q}_p$ converging to a point x such that both $f(x)$ and $g(x)$ converge. Show that if $f(x_m) = g(x_m)$ for every m, then $f(X) = g(X)$.

In Problem 162, we considered a "formal derivative" operation on formal power series. If a function is defined by a power series, we would like its derivative to correspond to the formal derivative of the power series, and it does. Before we give a proof, note that there is a classical theorem about when a series of functions can be differentiated term-by-term; see [55, Theorem 7.17], for example. Unfortunately, the proof uses the mean value theorem, which we already know is false in the p-adic setting. (Is the theorem even true in the p-adic setting?) So we need a different approach.

Proposition 5.5.5 *Let $f(X) = \sum a_n X^n$ be a power series with non-zero radius of convergence and let $f'(X)$ be its formal derivative. Let $x \in \mathbb{Q}_p$. If $f(x)$ converges, then so does $f'(x)$, and we have*

$$f'(x) = \lim_{h \to 0} \frac{f(x+h) - f(x)}{h}.$$

PROOF: (Following [34]. See [19] for a proof based on Theorem 5.1.4.)

Note first that there are indeed elements $h \to 0$ for which $f(x+h)$ converges, since the region of convergence is a (closed or open) ball centered at the origin. In fact, let ρ be the radius of convergence. If $x = 0$, any h with $|h| < \rho$ works; if $x \neq 0$, then any h with $|h| < |x|$ works. (Remember that if $|h| < |x|$, then $|x + h| = |x|$.) In particular, the limit that appears in the proposition does make sense.

Suppose, then, that $f(x)$ converges, which is equivalent to saying that $a_n x^n \to 0$. If $x = 0$ then it is clear that $f'(x)$ converges. If $x \neq 0$, notice that since the absolute value of an integer is at most 1, as $n \to \infty$ we have

$$|na_n x^{n-1}| \leq |a_n x^{n-1}| = \frac{1}{|x|}|a_n x^n| \to 0,$$

and again we see[4] that $f'(x)$ converges.

Recall that either $f(x)$ converges in the closed ball $\overline{B}(0, \rho)$ or in the open ball $B(0, \rho)$. In the first case, set $\rho_1 = \rho$. In the second case, choose ρ_1 such that $|x| \leq \rho_1 < \rho$. The point is that in either case the series will converge when $|x| \leq \rho_1$.

---

[4]This is one of those places where Corollary 5.1.2 really simplifies our life!

Since we only care about h close to zero, we may assume, if $x \neq 0$, that $|h| < |x| \leq \rho_1$. Otherwise, $x = 0$ and we can simply assume $|h| \leq \rho_1$. Now,

$$f(x + h) = \sum_{n=0}^{\infty} a_n(x + h)^n = \sum_{n=0}^{\infty} a_n \sum_{m=0}^{n} \binom{n}{m} x^{n-m} h^m.$$

Subtracting $f(x)$ and dividing by h, we get

$$\frac{f(x + h) - f(x)}{h} = \sum_{n=1}^{\infty} \sum_{m=1}^{n} a_n \binom{n}{m} x^{n-m} h^{m-1}.$$

It remains to take the limit as $h \to 0$ on both sides. On the left we will get $f'(x)$. On the right, the question is whether we can take the limit of a sum by computing the sum of the limits.

Since we have $|x| \leq \rho_1$ and $|h| \leq \rho_1$, we have

$$\left| a_n \binom{n}{m} x^{n-m} h^{m-1} \right| \leq |a_n| \rho_1^{n-1},$$

and the series converges when $|x| = \rho_1$ we have $|a_n| \rho_1^n \to 0$. Given $\varepsilon > 0$, we can find an M so that $m \geq M$ implies $|a_n| \rho_1^{n-1} < \varepsilon$, and this implies that for our fixed $|x| \leq \rho_1$ and all $|h| \leq \rho_1$ we have

$$\left| a_n \binom{n}{m} x^{n-m} h^{m-1} \right| \leq |a_n| \rho_1^{n-1} < \varepsilon$$

for *all* n. In other words, the inner terms tend to zero uniformly in h.

This implies[5] that we can take the limit term-by-term. In the case of the inner terms, that amounts to setting $h = 0$ in the polynomial

$$\sum_{m=1}^{n} a_n \binom{n}{m} x^{n-m} h^{m-1} = n a_n x^{n-1} + \binom{n}{2} a_n x^{n-2} h + \cdots,$$

which gives $n a_n x^{n-1}$. So we get

$$f'(x) = \sum_{n=1}^{\infty} n a_n x^{n-1},$$

which is what we want.   □

Problem 167 Prove the result we needed for that crucial last step. Suppose that for all $|h| \leq r$ we have

$$f(h) = \sum_{n=0}^{\infty} f_n(h)$$

---

[5]Can you see a problem coming?

and that $\lim_{n\to\infty} f_n(h) = 0$ uniformly in h. Show that if all the limits exist we have

$$\lim_{h\to 0} f(h) = \sum_{n=0}^{\infty} \left(\lim_{h\to 0} f_n(h) \right).$$

The theorem says that whenever $f(x)$ converges so does $f'(x)$. In particular, the radius of convergence of $f'(X)$ is at least as big as the radius of convergence of $f(X)$. In fact they are the same.

Problem 168 Suppose that the formal power series $f(X)$ has radius of convergence ρ. Show that the radius of convergence of $f'(X)$ is also ρ.

Problem 169 Let $f(X) = \sum a_n X^n$ be a formal power series with radius of convergence $\rho > 0$, and suppose $f(x)$ converges. Show that for every k the k-th derivative $f^{(k)}(x)$ exists, and is given by

$$f^{(k)}(x) = \sum_{n\geq k} \frac{n!}{(n-k)!} a_n(x-a)^{n-k} = k! \sum_{n\geq k} \binom{n}{k} a_n(x-a)^{n-k}.$$

This series has radius of convergence ρ, and we have

$$a_k = \frac{f^{(k)}(0)}{k!}.$$

Problem 170 Can you think of any reason why one would want to write the derivative as above, with $k!$ factored out?

To demonstrate that we've actually proved quite a bit, here's an easy consequence of our results. As we pointed out above, it is possible for two p-adic functions to have the same derivative without it being the case that their difference is constant. That doesn't happen for functions defined by power series.

Corollary 5.5.6 *Suppose $f(X)$ and $g(X)$ are power series, and suppose that both series converge for $|x| < \rho$. If $f'(x) = g'(x)$ for all $|x| < \rho$, then there exists a constant $c \in \mathbb{Q}_p$ such that $f(X) = g(X) + c$ as power series. In particular, $f(X)$ and $g(X)$ have the same disk of convergence, and we have $f(x) = g(x) + c$ for all x in the disk of convergence.*

PROOF: Let $f(X) = \sum a_n X^n$, $g(X) = \sum b_n X^n$, and let $f'(X)$ and $g'(X)$ be the formal derivatives. By Proposition 5.5.5 we know that whenever $|x| < \rho$ we have

$$\sum_{n=1}^{\infty} na_n x^{n-1} = \sum_{n=1}^{\infty} nb_n x^{n-1}.$$

By Proposition 5.5.4, we can conclude that $a_n = b_n$ for all $n \geq 1$, and the conclusion follows.   □

The upshot is that functions defined by power series behave well. In particular, pseudo-constant functions cannot be given by power series on any ball where they are not actually constant.

The fact that functions given by power series cannot be pseudo-constant means that we can hope for a fairly good theory of anti-derivatives for such functions. Of course, the formal theory is easy: if

$$f(X) = a_0 + a_1 X + a_2 X^2 + \cdots + a_n X^n + \cdots,$$

then any anti-derivative of $f(X)$ must look like

$$F(X) = C + a_0 X + \frac{a_1}{2} X^2 + \frac{a_2}{3} X^3 + \cdots + \frac{a_n}{n+1} X^{n+1} + \cdots$$

Clearly the function defined by $F(X)$ will have derivative given by $f(X)$ whenever it converges, but the fact that $F(X)$ has denominators means that it might not converge for the same range of x. Problem 168 says that the radius of convergence of the two series is the same, but the behavior at points where $|x| = \rho$ can be completely different. Since the sphere given by $|x| = \rho$ is an open set, there are many such points!

For example, consider

$$f(x) = \sum_{k=1}^{\infty} p^k X^{p^k - 1}.$$

This clearly converges for all $x \in \mathbb{Z}_p$. But the formal anti-derivative is

$$F(x) = \sum_{k=1}^{\infty} X^{p^k},$$

which diverges at any x with $|x| = 1$. So while $f(x)$ is an analytic function on $\mathbb{Z}_p$, its anti-derivative is not; we only get a good analytic anti-derivative on $p\mathbb{Z}_p$.

One can show, however, that the anti-derivative of a function defined by a power series converging in $\mathbb{Z}_p$ is always *locally* analytic, i.e., one can cover $\mathbb{Z}_p$ with smaller balls and find a power series for the anti-derivative on each of those balls. Unfortunately, this means we can't quite rescue the idea that the "integral" of $f(x)$ should be $F(b) - F(a)$, because b and a might not be in the same smaller ball. Fixing this problem leads to the Coleman integral.

5.6 Strassman's Theorem

The next theorem we want to look at is a fundamental result about the zeros of functions defined by power series.

Theorem 5.6.1 (Strassman) *Let*

$$f(X) = \sum_{n=0}^{\infty} a_n X^n = a_0 + a_1 X + a_2 X^2 + \cdots$$

be a non-zero power series with coefficients in $\mathbb{Q}_p$, and suppose that we have $\lim_{n\to\infty} a_n = 0$, so that $f(x)$ converges for all $x \in \mathbb{Z}_p$. Let N be the integer defined by the two conditions

$$|a_N| = \max_n |a_n| \qquad and \qquad |a_n| < |a_N| \quad for\ n > N.$$

Then the function $f : \mathbb{Z}_p \longrightarrow \mathbb{Q}_p$ defined by $x \mapsto f(x)$ has at most N zeros.

(The existence of N follows from the fact that the series is nonzero but the coefficients a_n tend to zero: there is a largest absolute value, and N is the index of the last coefficient for which the maximum is attained.)

Strassman's theorem is usually proved using a high-powered result known as the p-adic Weierstrass preparation theorem, which shows that any function on $\mathbb{Z}_p$ defined by a power series as above is equal to the product of a polynomial of degree N and a power series with no roots in $\mathbb{Z}_p$. We will eventually prove that (Theorem 7.2.6), but for now we forgo that approach and give the elementary proof found in [14, Ch. 4, Thm. 4.1].

PROOF: We use induction on N.

a) If $N = 0$, we must have $|a_0| > |a_n|$ for all $n \geq 1$, and what we want to prove is that in that case there are no zeros: $f(x) \neq 0$ for all $x \in \mathbb{Z}_p$. Indeed, if we had $f(x) = 0$, then

$$0 = f(x) = a_0 + a_1 x + a_2 x^2 + \cdots,$$

from which it would follow that

$$|a_0| = |a_1 x + a_2 x^2 + \cdots|$$

$$\leq \max_{n \geq 1} |a_n x^n|$$

$$\leq \max_{n \geq 1} |a_n|.$$

But this contradicts the assumption that $|a_0| > |a_n|$ for all $n \geq 1$.

b) To handle the induction step, we use an idea from the algebra of polynomials: a zero implies a factorization. Suppose that

$$|a_N| = \max_n |a_n| \qquad and \qquad |a_n| < |a_N| \quad for\ n > N,$$

and suppose that $f(\alpha) = 0$ for some $\alpha \in \mathbb{Z}_p$. Choose any $x \in \mathbb{Z}_p$. Then we have

$$f(x) = f(x) - f(\alpha) = \sum_{n \geq 1} a_n (x^n - \alpha^n)$$

$$= (x - \alpha) \sum_{n \geq 1} \sum_{j=0}^{n-1} a_n x^j \alpha^{n-1-j}.$$

By Proposition 5.1.4, we can re-order the series as a power series in x, which gives

$$f(x) = (x - \alpha) \sum_{j=0}^{\infty} b_j x^j = (x - \alpha) g(x),$$

where the coefficients b_j are given by

$$b_j = \sum_{k=0}^{\infty} a_{j+1+k} \alpha^k.$$

It is easy to see that $b_j \to 0$ as $j \to \infty$. It's also clear that if they were all zero then $f(X)$ would be the zero power series, contradicting one of our assumptions. So $g(X)$ satisfies the assumptions of the theorem.

To use the induction hypothesis we need to find the last $|b_j|$ with maximum absolute value. First, note that

$$|b_j| \leq \max_{k \geq 0} |a_{j+1+k}| \leq |a_N|$$

for every j, so all the $|b_j|$ are bounded by $|a_N|$. On the other hand, since $|\alpha| \leq 1$, for any $i \geq 1$ we have $|a_{N+i} \alpha^i| \leq |a_{N+i}| < |a_N|$, so the ultrametric inequality gives

$$|b_{N-1}| = |a_N + a_{N+1}\alpha + a_{N+2}\alpha^2 + \cdots| = |a_N|.$$

Finally, if $j \geq N$,

$$|b_j| \leq \max_{k \geq 0} |a_{j+k+1}| \leq \max_{j \geq N+1} |a_j| < |a_N|.$$

This shows that the magic number in Strassman's theorem when applied to $g(X)$ is $N - 1$. By induction, we can assume that $g(X)$ has at most $N - 1$ zeros in $\mathbb{Z}_p$, which implies that $f(X)$ has at most N zeros (those of $g(X)$, plus α). This proves the theorem. $\square$

Problem 171 Check that the application of Proposition 5.1.4 in the proof of Strassman's theorem is valid.

Strassman's theorem is only the first of several important theorems[6] about zeros of functions on $\mathbb{Q}_p$ defined by power series. Even so, it is a very powerful theorem. Here are some consequences.

Corollary 5.6.2 *Let* $f(X) = \sum a_n X^n$ *be a non-zero power series which converges on* $\mathbb{Z}_p$, *and let* $\alpha_1, \ldots, \alpha_m$ *be the roots of* $f(X)$ *in* $\mathbb{Z}_p$. *Then we can find a power series* $g(X)$ *which converges on* $\mathbb{Z}_p$ *but has no zeros in* $\mathbb{Z}_p$, *for which*

$$f(X) = (X - \alpha_1) \cdots (X - \alpha_m) g(X).$$

PROOF: Clear from the proof of the theorem and Proposition 5.5.4.   □

Since $\mathbb{Z}_p$ is just the closed unit ball in $\mathbb{Q}_p$, we can extend the result to other disks by simple scaling.

Corollary 5.6.3 *Let* $f(X) = \sum a_n X^n$ *be a non-zero power series which converges on* $p^m \mathbb{Z}_p$, *for some* $m \in \mathbb{Z}$. *Then* $f(X)$ *has a finite number of zeros in* $p^m \mathbb{Z}_p$.

PROOF: Define $g(X) = f(p^m X) = \sum a_n p^{mn} X^n$. Since $f(X)$ converges in $p^m \mathbb{Z}_p$, $g(x)$ converges for $x \in \mathbb{Z}_p$, and applying the theorem to $g(X)$ gives the finiteness.   □

Problem 172 Strassman's Theorem actually gives a bound for the number of roots in $\mathbb{Z}_p$. What is a bound for the number of roots in $p^m \mathbb{Z}_p$?

Problem 173 Say all you can about the zeros of the functions defined by the power series in Problem 155.

This result allows us to prove a variant of Proposition 5.5.4:

Corollary 5.6.4 *Let* $f(X) = \sum a_n X^n$ *and* $g(X) = \sum b_n X^n$ *be two p-adic power series which converge in a disk* $p^m \mathbb{Z}_p$. *If there exist infinitely many numbers* $\alpha \in p^m \mathbb{Z}_p$ *such that* $f(\alpha) = g(\alpha)$, *then* $a_n = b_n$ *for all* $n \geq 0$.

PROOF: Apply the previous corollary to $f(X) - g(X)$.   □

Notice that since $p^m \mathbb{Z}_p$ is compact, the existence of infinitely many α as above implies the existence of a convergent sequence of such α, so that (in $\mathbb{Q}_p$) this result could also be proved directly from Proposition 5.5.4. But if we move to a context where closed balls are not compact, this proof still works.

One consequence of this is something of a surprise:

Corollary 5.6.5 *Let* $f(X) = \sum a_n X^n$ *be a p-adic power series which converges in some disk* $p^m \mathbb{Z}_p$. *If the function* $p^m \mathbb{Z}_p \longrightarrow \mathbb{Q}_p$ *defined by* $x \mapsto f(x)$ *is periodic, that is, if there exists* $\pi \in p^m \mathbb{Z}_p$ *such that* $f(x + \pi) = f(x)$ *for all* $x \in p^m \mathbb{Z}_p$, *then* $f(X)$ *is constant.*

---

[6]For example, the p-adic Weierstrass Preparation Theorem and the theory of Newton polygons, which allow very detailed control of the zeros. See Chapter 7 for more details.

PROOF: The series $f(X) - f(0)$ has zeros at $n\pi$ for all $n \in \mathbb{Z}$. Since $\pi \in p^m\mathbb{Z}_p$ implies $n\pi \in p^m\mathbb{Z}_p$, this gives infinitely many zeros, and hence the series $f(X) - f(0)$ must be identically zero, i.e., $f(X)$ must be constant.   □

This offers an intriguing contrast to the classical case, where the sine and cosine functions are both periodic and "entire," i.e., they can each be expressed as a power series that converges everywhere. The crucial difference is that in the classical case it never happens that all the multiples of the period are in the same bounded interval, while in our case the non-archimedean property guarantees just that.

While periodicity is very different in the classical and the p-adic situations, the zeros of an entire function are distributed similarly in both cases:

Corollary 5.6.6 *Let $f(X) = \sum a_n X^n$ be a p-adic power series, and suppose that $f(X)$ is entire, i.e., that $f(x)$ converges for every $x \in \mathbb{Q}_p$. Then $f(X)$ has at most countably many zeros. Furthermore, if the set of zeros is not finite then the zeros form a sequence α_n with $|\alpha_n| \to \infty$.*

PROOF: This is clear, because the number of zeros in each bounded disk $p^m\mathbb{Z}_p$ is finite.   □

It is natural (and tempting) to conjecture from these results that there should be a representation of any entire function as an infinite product over the zeros; something like

$$f(X) = h(X) \prod (1 - \alpha^{-1}X),$$

where α ranges over the zeros of $f(X)$ and $h(X)$ is an entire function with no zeros. (Why it's best to write the expansion in terms of the inverses of the roots may be a little mysterious now; we will go back to this in Chapter 7.) It is easy to see that such a representation does exist, but it will not be very interesting unless we are ready to go to the algebraic closure of $\mathbb{Q}_p$, since even polynomials may fail to have roots in $\mathbb{Q}_p$. When we have the necessary machinery set up for working over the algebraic closure, we will be able to obtain a very precise description of entire functions in this spirit.

This brings out a rather embarrassing point: in the case of $\mathbb{R}$, the algebraic closure is an old friend, the field of complex numbers. By contrast, at this point we really know very little about the algebraic closure of $\mathbb{Q}_p$. In fact, we do not even know whether the p-adic absolute value on $\mathbb{Q}_p$ can be extended to its algebraic closure. It turns out that this extension is indeed possible (we will discuss this a little later), but that the algebraic closure *is not complete* with respect to this absolute value. (This is very different from the classical case, where $\mathbb{C}$ is just as complete as $\mathbb{R}$.) The obvious thing to do is to go through the completion process again. The resulting field, usually called $\mathbb{C}_p$, is both complete and algebraically closed, and is the p-adic analog of the complex numbers. Arguably the field $\mathbb{C}_p$ is the "correct" context in which to do p-adic analysis, and we will go through the process of constructing it and studying the results in Chapters 6 and 7 of this book. For now, we want to stay at a more intuitive level, and hence will continue working in $\mathbb{Q}_p$. What we will do, however, is to be careful to construct our arguments in such a way that they will be easy to generalize to other fields. This will save us a lot of work later on.

5.7   Logarithm and Exponential Functions

In this section, our goal is to use power series to define p-adic functions which are analogous to the classical exponential and logarithm functions. In contrast to the archimedean case, it is the logarithm that has the better convergence properties.

We begin with the usual power series for the logarithm:

$$\mathbf{f}(X) = \mathbf{log}(1 + X) = \sum_{n=1}^{\infty} (-1)^{n+1} \frac{X^n}{n} = X - \frac{X^2}{2} + \frac{X^3}{3} + \cdots$$

(We use **log**—rather than log—to emphasize that we are considering the formal power series, and not the function, which after all we have not yet defined in the p-adic context.) Since the coefficients of this power series are rational numbers, it makes sense to think of the series as a power series in $\mathbb{Q}_p$ (for any prime p). The first step towards understanding it is, of course, to compute its radius of convergence. Before we jump into the limit calculation, however, we should note another classical vs. p-adic contrast. In the classical case, all the integers in the denominators help the convergence, because they tend to make the terms of the series smaller. In the p-adic case, this is exactly reversed: integers in the denominator either do not change the absolute value (when they are not divisible by p) or make it *bigger* (when they are). What saves convergence in the case of this series is that "in general" n is not too divisible by p.

To compute the radius of convergence ρ, let $\mathbf{f}(X) = \sum_{n=1}^{\infty} (-1)^{n+1} \frac{X^n}{n}$, so that $a_n = \dfrac{(-1)^n}{n}$. Then

$$|a_n| = \left| \frac{1}{n} \right| = p^{v_p(n)}.$$

From this, we get

$$\sqrt[n]{|a_n|} = p^{v_p(n)/n} \to 1$$

as $n \to \infty$. (Check!) Hence, $\rho = 1$. This doesn't decide for us whether the convergence happens on the open or closed ball of radius 1. To decide, we need to look at what happens when $|x| = 1$. But it is clear that in that case the absolute value $|a_n x^n| = |a_n| = |1/n|$ does not tend to zero (it's equal to 1 whenever p doesn't divide n). So we get

Lemma 5.7.1 *The series*

$$\mathbf{f}(X) = \sum_{n=1}^{\infty}(-1)^{n+1}\frac{X^n}{n} = X - \frac{X^2}{2} + \frac{X^3}{3} - \frac{X^4}{4} + \cdots$$

converges for $|x| < 1$ (and diverges otherwise).

Problem 174 Check that

$$\lim_{n\to\infty} p^{v_p(n)/n} = 1.$$

(The main idea is to estimate $v_p(n)$ as a function of n.)

The conclusion is that $\mathbf{f}(X)$ defines a function on the open ball $B(0, 1)$ of radius 1 and center 0. This suggests that we should define the logarithm in the obvious way, so that $\mathbf{f}(x) = \log(1 + x)$.

Definition 5.7.2 *Let $U_1 = B(1, 1) = \{x \in \mathbb{Z}_p : |x - 1| < 1\} = 1 + p\mathbb{Z}_p$. We define the p-adic logarithm of $x \in U_1$ as*

$$\log_p(x) = \mathbf{log}(1 + (x - 1)) = \sum_{n=1}^{\infty}(-1)^{n+1}\frac{(x-1)^n}{n}.$$

Of course, if we want this function to deserve to be called a logarithm, we had better check that it satisfies the functional equation that characterizes logarithms.

Proposition 5.7.3 *Suppose $a, b \in 1 + p\mathbb{Z}_p$. Then*

$$\log_p(ab) = \log_p(a) + \log_p(b).$$

PROOF: In the literature, this is often proved by noting that there is an underlying identity of power series. The problem with this is that verifying condition (iii) of Theorem 5.4.3 is somewhat problematic. So instead we give a direct proof that mimics the classical proof. To simplify the notation, let $a = 1 + x$ and $b = 1 + y$.

For any $x \in p\mathbb{Z}_p$, let

$$f(x) = \log_p(1 + x) = \sum_{n=1}^{\infty}(-1)^{n+1}\frac{x^n}{n}.$$

Then, by our results on derivatives of functions defined by power series, we have

$$f'(x) = \sum_{n=0}^{\infty}(-1)^n x^n = \frac{1}{1+x}.$$

Now fix $y \in p\mathbb{Z}_p$ and define

$$g(x) = \log_p((1 + x)(1 + y)) = f(y + (1 + y)x).$$

By the result in Problem 165, this is a power series that converges for $|x| < 1$. Now use the chain rule to compute the derivative of g:

$$g'(x) = (1+y)f'(y + (1+y)x) = \frac{(1+y)}{1+y+(1+y)x} = \frac{1}{1+x} = f'(x).$$

Since both $f(x)$ and $g(x)$ are defined by power series that converge for $|x| < 1$, it follows by Corollary 5.5.6 that $g(x) = f(x) + c$ whenever $|x| < 1$. Plugging in $x = 0$ shows that $c = g(0) = f(y)$. Hence we've shown that $g(x) = f(x) + f(y)$; translating back to logarithms, this says

$$\log_p((1+x)(1+y)) = \log_p(1+x) + \log_p(1+y),$$

and we are done.   □

Problem 175 Show that if $p = 2$ then $-1 \in B$, so that it makes sense to compute $\log_p(-1)$. Show that $\log_p(-1) = 0$. Compare with the example in Chapter 1. Can you estimate the highest power of 2 that divides the n-th partial sum?

In the previous chapter, we used Hensel's Lemma to determine for which m there exist m-th roots of unity in $\mathbb{Q}_p$. Our method restricted us to the case where $p \nmid m$, so we left open the possibility of the existence of p^n-th roots of unity in $\mathbb{Q}_p$. It turns out, as we said then, that these do not exist, except for the trivial case when $p = 2$ and $n = 1$. The next three problems use the p-adic logarithm to prove this claim. The idea is that if x is a root of unity and $x \in 1 + p\mathbb{Z}_p$, then we must have $\log_p(x) = 0$, so that studying the zeros of the logarithm will give us a handle on the roots of unity.

Problem 176 Use Strassman's Theorem to show that for $p \neq 2$ we have $\log_p(x) = 0$ if and only if $x = 1$. If $p = 2$, show that $\log_p(x) = 0$ if and only if $x = \pm 1$. (Hint: one can't use Strassman's Theorem directly, because the series does not converge in $\mathbb{Z}_p$, but rather in $p\mathbb{Z}_p$. But that is easily handled with a change of variables.)

Problem 177 Let $p \neq 2$. Show that if $x \in \mathbb{Q}_p$ and $x^p = 1$, then $x = 1$. Conclude that there are no p-th roots of unity (and hence no p^n-th roots of unity) in $\mathbb{Q}_p$.

Problem 178 Let $p = 2$. Show that if $x \in \mathbb{Q}_2$ and $x^4 = 1$, then $x = \pm 1$. Conclude that there are no fourth roots of unity in $\mathbb{Q}_2$. (There are, of course, the square roots of unity ± 1.)

Since knowing the roots of unity in $\mathbb{Q}_p$ turns out to be very useful, we summarize all that we know about them:

- for $p = 2$, the only roots of unity in $\mathbb{Q}_p$ are ± 1

- for $p \neq 2$, $\mathbb{Q}_p$ contains all of the $(p-1)$-st roots of unity, and no others.

(Recall that the existence of the $(p-1)$-st roots of unity was proved as an application of Hensel's Lemma in the last chapter.)

Having obtained a logarithm, exponentials cannot be far behind. In the classical case, the series

$$\exp(X) = \sum_{n=0}^{\infty} \frac{X^n}{n!} = 1 + X + \frac{X^2}{2} + \frac{X^3}{6} + \cdots$$

converges for all $x \in \mathbb{R}$, because the coefficients $1/n!$ tend very quickly to zero with respect to the real absolute value. In the p-adic context, of course, this changes drastically, because $n!$ tends to zero, so that $1/n!$ becomes arbitrarily large as n grows. This means that we cannot expect to have a large radius of convergence. To determine what that radius will be, we have to work out exactly how fast the coefficients $1/n!$ grow, i.e., we have to work out how divisible $n!$ is by p.

Lemma 5.7.4 *Let p be a prime. Then*

$$v_p(n!) = \sum_{i=1}^{\infty} \left\lfloor \frac{n}{p^i} \right\rfloor < \frac{n}{p-1},$$

where $\lfloor \cdot \rfloor$ is the greatest integer (or "floor") function. In particular

$$|n!|_p > p^{-n/(p-1)}.$$

PROOF: The formula

$$v_p(n!) = \sum_{i=1}^{\infty} \left\lfloor \frac{n}{p^i} \right\rfloor$$

is well known and easy to prove. We leave it as the next problem. The inequality then follows, because $\lfloor x \rfloor \le x$, so that

$$v_p(n!) = \sum_{i=1}^{\infty} \left\lfloor \frac{n}{p^i} \right\rfloor < \sum_{i=1}^{\infty} \frac{n}{p^i} = \frac{n}{p-1}$$

by the usual formula for geometric series. □

Problem 179 Prove that

$$v_p(n!) = \sum_{i=1}^{\infty} \left\lfloor \frac{n}{p^i} \right\rfloor.$$

Here is another version of the same formula, which is sometimes useful:

Problem 180 Let n be a positive integer, and let $n = a_0 + a_1 p + a_2 p^2 + \cdots + a_k p^k$ be its expansion in base p. Let $s = a_0 + a_1 + \cdots + a_k$ be the sum of the digits in the expansion. Show that

$$v_p(n!) = \frac{n-s}{p-1}.$$

(Hint: work out the difference between n/p^i and its integral part in terms of the base p expansion.)

Now we use these estimates to work out the radius of convergence of the exponential.

Lemma 5.7.5 *Let*

$$\mathbf{g}(X) = \sum_{n=0}^{\infty} \frac{X^n}{n!} = 1 + X + \frac{X^2}{2!} + \frac{X^3}{3!} + \cdots$$

Then $\mathbf{g}(x)$ *converges if and only if* $|x| < p^{-1/(p-1)}$.

PROOF: Since

$$|a_n| = |1/n!| = p^{v_p(n!)} < p^{n/(p-1)}$$

by our first estimate, we get

$$\rho \geq p^{-1/(p-1)}.$$

Thus, the series certainly converges for $|x| < p^{-1/(p-1)}$.

On the other hand, let $|x| = p^{-1/(p-1)}$ and let $n = p^m$ be a power of p. In this case, we have

$$v_p(n!) = v_p(p^m!) = 1 + p + \cdots + p^{m-1} = \frac{p^m - 1}{p - 1}$$

(notice that this is a special case of the result in Problem 180). Then, since $v_p(x) = 1/(p-1)$,

$$v_p\left(\frac{x^n}{n!}\right) = v_p\left(\frac{x^{p^m}}{p^m!}\right) = \frac{p^m}{p-1} - \frac{p^m - 1}{p - 1} = \frac{1}{p - 1}.$$

This does not depend on m, hence $x^n/n!$ cannot tend to zero, and the series doesn't converge. Since we know that the region of convergence is a disk, this proves the lemma. □

REMARK: There is something a little strange about the inequality in the lemma. If $p \neq 2$ and $x \in \mathbb{Z}_p$, then the absolute value of x can either be equal to 1 (which is bigger than $p^{-1/(p-1)}$) or less than or equal to $1/p = p^{-1}$ (which is smaller): there are no values "in the middle." Thus, if $p \neq 2$,

$$|x| < p^{-1/(p-1)} \iff |x| \leq p^{-1} \iff x \in p\mathbb{Z}_p \iff |x| < 1,$$

so that the disk in the lemma is just the open disk of radius one!

This seems to suggest that all our care in working out the precise radius of convergence is wasted. That is not really the case. The point is that our estimates only depend on the absolute value of x, and so they will work in any field *containing* $\mathbb{Q}_p$ (with an absolute value extending the one on $\mathbb{Q}_p$). In such fields there may indeed be elements with

$$p^{-1/(p-1)} \leq |x| < 1.$$

This will be particularly important in our next two chapters when we will indeed be working in larger fields.

Meanwhile, the reader should keep an eye out for arguments that depend on knowing that $|x| < 1$ implies $x \in p\mathbb{Z}_p$; these are the arguments that do not generalize.

In any case, as long as we stay in $\mathbb{Q}_p$, things are pretty simple. If $p \neq 2$, $\mathbf{g}(x) = \exp(x)$ converges for $x \in p\mathbb{Z}_p$. If $p = 2$, $-1/(2-1) = -1$, so the lemma tells us that $\mathbf{g}(x) = \exp(x)$ converges when $|x| < 1/2$, which happens when $x \in 4\mathbb{Z}_2$.

Now we can define the p-adic exponential function using the formal series $\exp(X)$.

Definition 5.7.6 Let $D = B(0, p^{-1/(p-1)}) = \{x \in \mathbb{Z}_p : |x| < p^{-1/(p-1)}\}$. *The p-adic exponential is the function* $\exp_p : D \longrightarrow \mathbb{Q}_p$ *defined by*

$$\exp_p(x) = \sum_{n=0}^{\infty} \frac{x^n}{n!}.$$

Notice that $\exp_p(1)$ is not defined, so there is no natural p-adic analogue of e in $\mathbb{Q}_p$. Within its domain, however, the p-adic exponential does satisfy most of the formal properties of the classical exponential. Let's begin with the most famous one:

Proposition 5.7.7 *If* $x, y \in D$ *we have* $x + y \in D$ *and*

$$\exp_p(x + y) = \exp_p(x)\exp_p(y).$$

PROOF: This is essentially a formal manipulation of power series:

$$\exp_p(x+y) = \sum_{n=0}^{\infty} \frac{(x+y)^n}{n!} = \sum_{n=0}^{\infty} \frac{1}{n!} \sum_{k=0}^{n} \binom{n}{k} x^{n-k} y^k$$

$$= \sum_{n=0}^{\infty} \sum_{k=0}^{n} \frac{1}{n!} \frac{n!}{(n-k)!k!} x^{n-k} y^k$$

$$= \sum_{n=0}^{\infty} \sum_{k=0}^{n} \frac{x^{n-k}}{(n-k)!} \frac{y^k}{k!}$$

$$= \left(\sum_{m=0}^{\infty} \frac{x^m}{m!}\right)\left(\sum_{k=0}^{\infty} \frac{y^k}{k!}\right)$$

$$= \exp_p(x)\exp_p(y),$$

as claimed. $\qquad\square$

Problem 181 Are there convergence issues to check in the proof?

This shows that, apart from the smallish radius of convergence, we have obtained something that is a lot like the classical exponential.

There is of course one more formal property we would like to be true also in the p-adic context: the fact that the logarithm and the exponential are inverses, i.e., the relation

$$\exp(\log(1 + X)) = 1 + X$$

and its inverse. This is a formal equality of power series, so that we only need to check that the conditions in Theorem 5.4.3 hold.

Proposition 5.7.8 *Let* $x \in \mathbb{Z}_p$, $|x| < p^{-1/(p-1)}$. *Then we have*

$$|\exp_p(x) - 1| < 1$$

so that $\exp_p(x)$ *is in the domain of* $\log_p$, *and*

$$\log_p(\exp_p(x)) = x.$$

Conversely, if $|x| < p^{-1/(p-1)}$ *we have*

$$|\log_p(1 + x)| < p^{-1/(p-1)}$$

so that $\log_p(1 + x)$ *is in the domain of* $\exp_p$, *and*

$$\exp_p(\log_p(1 + x)) = 1 + x.$$

PROOF: We need to check the estimates to know that all the series converge, and we also need to check condition (iii) from Theorem 5.4.3. Note first that both identities are clearly true when $x = 0$, so that we can assume $x \neq 0$.

To compute $\log_p(\exp_p(x))$, we are actually plugging $\exp_p(x) - 1$ into the series $\log(1 + X)$, so that is the quantity we need to estimate. We start from

$$\left| \frac{x^n}{n!} \right| = |x|^n \cdot p^{v_p(n!)} < |x|^n p^{n/(p-1)},$$

which we get from Lemma 5.7.4. Since $|x| < p^{-1/(p-1)}$, this is less than 1, and it follows that

$$\left| \exp_p(x) - 1 \right| = \left| \sum_{n=1}^{\infty} \frac{x^n}{n!} \right| < 1,$$

as claimed.

But in fact we can do better by using the result in Problem 180. Suppose $n \geq 2$; to make the computation easier, let's use the additive valuation v_p instead of absolute values. Since $v_p(x) > 1/(p-1)$, we get

$$v_p\left(\frac{x^{n-1}}{n!}\right) = (n-1)v_p(x) - v_p(n!) > \frac{n-1}{p-1} - \frac{n-s}{p-1} = \frac{s-1}{p-1} \geq 0,$$

where, as in Problem 180, s is the sum of the digits in the expansion of n in base p (so that $s \geq 1$). It follows that

$$\left|\frac{x^{n-1}}{n!}\right| < 1,$$

and so

$$\left|\frac{x^n}{n!}\right| < |x|.$$

So we have shown that in the series $\exp_p(x)$ all terms with $n \geq 2$ are smaller than the leading term x. This implies that $|\exp_p(x) - 1| = |x|$ and also that $|\exp_p(x)| > |x^n/n!|$ for all $n \geq 2$, so that condition (iii) in Theorem 5.4.3 is satisfied. Applying Theorem 5.4.3, we can conclude from the formal equality of power series that if $|x| < p^{-1/(p-1)}$ we have

$$\log_p(\exp_p(x)) = x.$$

Now let's consider the composition in the opposite order. This time we're plugging $\log_p(1 + x)$ into $\exp(X)$, so we need to estimate the valuation of $\log_p(1+x)$. We can actually take advantage of the estimates we already have by noticing that $|n!| \leq |n|$, so that

$$\left|\frac{x^n}{n}\right| \leq \left|\frac{x^n}{n!}\right|.$$

So if $|x| < p^{-1/(p-1)}$, the estimates above say

$$\left|\frac{x^n}{n}\right| \leq \left|\frac{x^n}{n!}\right| < |x|.$$

As above, we get

$$|\log_p(1 + x)| = |x| < p^{-1/(p-1)},$$

which shows both that $\log_p(1 + x)$ is in the domain of the exponential and that condition (iii) in Theorem 5.4.3 is satisfied. Hence the formal equality of power series implies what we want:

$$\exp_p(\log_p(1 + x)) = 1 + x.$$

This finishes the proof. □

Problem 182 When we work hard at a proof it is worthwhile to see if along the way we haven't proved something that might be interesting in itself. Notice that we proved that if $|x| < p^{-1/(p-1)}$ then $|\exp_p(x) - 1| = |x|$. Since $\exp_p(0) = 1$, this actually says that

$$|\exp_p(x) - \exp_p(0)| = |x - 0|.$$

Generalize it to show that if $|x| < p^{-1/(p-1)}$ and $|y| < p^{-1/(p-1)}$, then

$$|\exp_p(x) - \exp_p(y)| = |x - y|.$$

In other words, $\exp_p$ is an isometry.

The hypotheses of the theorem are indeed necessary, for two reasons. The first, and less crucial one, is that if $|x| < 1$ but $|x| \geq p^{-1/(p-1)}$, it can very well be that $\log_p(1 + x)$ does not belong to the domain of the exponential. But much more serious is the fact that it can happen that we have $|x| < 1$, $|x| \geq p^{-1/(p-1)}$, and also

$$|\log_p(1 + x)| < p^{-1/(p-1)},$$

so that all the series involved converge, but

$$\exp_p(\log_p(1 + x)) \neq 1 + x.$$

This is due to the fact that the extra condition in Theorem 5.4.3 really does matter. To see this concretely, consider what happens[7] when we take $p = 2$ and $x = -2$: in that case, $1 + x = 1 - 2 = -1$, so that

$$\log_p(1 + x) = \log_p(-1) = 0.$$

Then, when we plug into the series for the exponential, we get

$$\exp_p(\log_p(-1)) = \exp(0) = 1 \neq -1.$$

In other words, the p-adic exponential and logarithm are inverses only within the restricted domains specified in the proposition.

Problem 183 Why doesn't Theorem 5.4.3 apply in this situation?

Notice that the theorem also tells us something about the images of $\exp_p$ and $\log_p$. Recall that we defined $D = \{x \in \mathbb{Z}_p : |x| < p^{-1/(p-1)}\}$. Using this notation, the theorem tells us that $\exp_p$ is an isometric isomorphism from D to $1 + D$ and that $\log_p$ is its inverse.

Problem 184 Use power series to define p-adic analogues of the sine and cosine functions, and determine their regions of convergence. Show that if $p \equiv 1 \pmod{4}$ then there exists $i \in \mathbb{Q}_p$ such that $i^2 = -1$, and the classical relation

$$\exp_p(ix) = \cos_p(x) + i\sin_p(x)$$

holds for any x in the common region of convergence. The classical trigonometric functions are periodic; are the p-adic versions periodic?

---

[7] This is the example we referred to when we discussed Theorem 5.4.3.

5.8   The Structure of $\mathbb{Z}_p^\times$

As an application of the p-adic logarithm and exponential, we can study the group of p-adic units $\mathbb{Z}_p^\times$ a little more carefully. We've already shown, using Hensel's Lemma, that $\mathbb{Z}_p^\times$ contains the $(p-1)$-st roots of unity. We also showed, using Strassman's Theorem, that when $p \neq 2$ there are no other roots of unity. Now we want to know what "the rest of $\mathbb{Z}_p^\times$" looks like. The idea is to look carefully at the domains and images of the logarithm and exponential functions.

As we noted above, when we are working in $\mathbb{Z}_p$ the region D on which the exponential converges is the same as $p\mathbb{Z}_p$ unless $p = 2$. To simplify the notation, let's introduce a parameter q as follows:

- if p is an odd prime, then $q = p$;

- if $p = 2$, then $q = 4$.

The point is that then the p-adic exponential $\exp_p(x)$ will be defined for $x \in q\mathbb{Z}_p$, and $\log_p(x)$ will be defined for $x \in 1 + p\mathbb{Z}_p$. Notice also that $\mathbb{Q}_p$ contains the $(p-1)$-st roots of unity when p is odd, and contains the square roots of unity when $p = 2$. If we use Euler's φ function, defined by $\varphi(n) = $ the number of integers between 1 and n which are relatively prime to n, then the number of roots of unity in $\mathbb{Q}_p$ is always $\varphi(q)$, since $\varphi(p) = p - 1$ for any prime, and $\varphi(4) = 2$.

Problem 185 Another way to define the φ function is to say that $\varphi(n)$ is the number of elements in $(\mathbb{Z}/n\mathbb{Z})^\times$, i.e., the number of invertible elements in the ring $\mathbb{Z}/n\mathbb{Z}$. Check that the two definitions are equivalent.

Let's define two subsets of $\mathbb{Z}_p^\times$:

$$U_1 = \{x \in \mathbb{Z}_p^\times : |x - 1| < 1\} = 1 + p\mathbb{Z}_p$$

$$U_p = \{x \in \mathbb{Z}_p^\times : |x - 1| < p^{-1/(p-1)}\} = 1 + q\mathbb{Z}_p.$$

Notice that

- $U_p \subset U_1 \subset \mathbb{Z}_p^\times$,

- $U_p = U_1$ except if $p = 2$,

- if $p = 2$, then $U_1 = \mathbb{Z}_p^\times$, and

- U_1 and U_p are *subgroups* of $\mathbb{Z}_p^\times$.

The elements of U_1 are often called the "1-units" (which comes from "the units which are congruent to 1 $(\bmod\ p\mathbb{Z}_p)$").

Problem 186 Check that U_1 and U_p are indeed subgroups of $\mathbb{Z}_p^\times$.

We can now determine the structure of $\mathbb{Z}_p^\times$ quite precisely:

Proposition 5.8.1 *Let* U_1 *and* U_p *be as above, and let* $\mathbb{Z}_p^+$ *denote the additive group of* $\mathbb{Z}_p$. *Let*

$$W = \{x \in \mathbb{Z}_p : |x| < p^{-1/(p-1)}\} = q\mathbb{Z}_p,$$

considered as a subgroup of $\mathbb{Z}_p^+$.

i) *The p-adic logarithm* $\log_p$ *defines a homomorphism of groups*

$$\log_p : U_1 \longrightarrow \mathbb{Z}_p^+,$$

   whose image is contained in the valuation ideal $\wp = p\mathbb{Z}_p$.

ii) *The p-adic logarithm* $\log_p$ *defines an isometric isomorphism of groups*

$$\log_p : U_p \xrightarrow{\ \sim\ } W,$$

   with inverse $\exp_p$. *In particular,* $U_p \cong W \cong \mathbb{Z}_p^+$ *is torsion-free.*

PROOF: Most of this is a straight translation of the discussion above into the language of groups. Proposition 5.7.3 says that $\log_p$ is a homomorphism, and Proposition 5.7.7 does the same for $\exp_p$. Proposition 5.7.8 says that the function in (ii) is an isomorphism.

When $p \neq 2$, $U_1 = U_p$ and $W = p\mathbb{Z}_p$, so part (i) is equivalent to part (ii). For $p = 2$, however, $U_1 \neq U_p$, so we know that the image of U_p is W, but we have not yet determined the image of U_1. As the next problem shows, it turns out that $\log_2(U_1) = W = 4\mathbb{Z}_2$ as well. In particular, the image is contained in $2\mathbb{Z}_2$, as (i) claims.

Recall that a group is torsion-free if there exist no elements $x \neq 1$ such that $x^m = 1$ for some m, i.e., no roots of unity. In fact, (ii) gives another proof that there are no roots of unity in U_p, since the *additive* group $\mathbb{Z}_p^+$ is torsion-free.   □

Problem 187 Let $p = 2$ and let $x \in 1 + 2\mathbb{Z}_2$. Show that $\log_2(x) \in 4\mathbb{Z}_2$. (You can do a brute-force estimate, but there's a trick that will avoid all that work.)

Corollary 5.8.2 *For any prime p, we have an isomorphism* $\mathbb{Z}_p^\times \cong V \times U_p$, *where* $U_p \cong \mathbb{Z}_p^+$ *is a torsion-free pro-p-group and V is the torsion part of* $\mathbb{Z}_p^\times$. *Furthermore:*

i) *V is the set of roots of unity in* $\mathbb{Q}_p$, *which is a subgroup of* $\mathbb{Z}_p^\times$, *and*

ii) $V \cong (\mathbb{Z}/q\mathbb{Z})^\times$, *so that V is a cyclic group of order* $\varphi(q)$.

PROOF: It is easy to see that there is an exact sequence

$$1 \longrightarrow U_p \longrightarrow \mathbb{Z}_p^\times \xrightarrow{\pi} (\mathbb{Z}/q\mathbb{Z})^\times \longrightarrow 0.$$

(Remember that this means that the kernel of each homomorphism in the sequence is equal to the image of the previous one, so that this is basically just the *definition* of U_p.) In fancy language, what we want to prove is that the exact sequence "splits," but we will just give a direct proof.

We already know (from a combination of Hensel's Lemma and Strassman's Theorem) that $\mathbb{Z}_p^\times$ contains a group V of roots of unity. It is a cyclic group of order $p-1$ when p is odd, and of order 2 when $p = 2$, so in any case it has $\varphi(q)$ elements, i.e., just as many elements as $(\mathbb{Z}/q\mathbb{Z})^\times$ does. We know that any two of these elements are distinct modulo q. (If in doubt, review the discussion starting on page 93. Or prove it again by solving the next problem.) Suppose ζ_1 and ζ_2 are roots of unity in $\mathbb{Z}_p^\times$ that have the same image under the map π. That implies $\zeta_1 \zeta_2^{-1} \in U_p$, but there are no roots of unity in U_p, so that $\zeta_1 \zeta_2^{-1} = 1$, i.e., $\zeta_1 = \zeta_2$. In other words, π induces an isomorphism between V and $(\mathbb{Z}/q\mathbb{Z})^\times$, and the other assertions in the theorem follow easily.   □

Problem 188 Prove that two different roots of unity (of order prime to p) cannot be congruent modulo $q\mathbb{Z}_p$. (This means: their difference cannot belong to $q\mathbb{Z}_p$. There are a whole lot of ways to do this.)

Problem 189 Fill in whatever is missing in the proof of the Corollary.

One thing that follows from this result is that π gives an isomorphism between V and $(\mathbb{Z}/q\mathbb{Z})^\times$. If p is odd, the inverse of this isomorphism gives an inclusion

$$\omega : \mathbb{F}_p^\times \cong V \hookrightarrow \mathbb{Z}_p^\times,$$

where $\mathbb{F}_p$ is the field with p elements. We can extend ω to $\mathbb{F}_p$ by setting $\omega(0) = 0$. The function ω is called *the Teichmüller character*, and it appears quite frequently in many different guises. The word "character" indicates that ω is multiplicative, i.e., $\omega(ab) = \omega(a)\omega(b)$ for all $a, b \in \mathbb{F}_p$. When $p = 2$ we can just define $\omega(0) = 0$ and $\omega(1) = 1$ to get the same effect.

If we compose ω with the "reduction modulo p" map from $\mathbb{Z}$ to $\mathbb{F}_p$,

$$\mathbb{Z} \xrightarrow{\ (\mathrm{mod}\ p)\ } \mathbb{F}_p \xrightarrow{\ \omega\ } \mathbb{Z}_p,$$

we get a Dirichlet character[8] with values in $\mathbb{Z}_p$, which is also usually called the Teichmüller character and denoted by ω. With this version of ω we see that $\omega(n) \equiv n \pmod{p}$, so that ω gives a new way of choosing coset representatives for the elements of $\mathbb{Z}/p\mathbb{Z}$.

---

[8]Basically, a multiplicative function on $\mathbb{Z}$.

To complete the confusion, when $p \neq 2$ one often also uses ω to denote the projection from $\mathbb{Z}_p^\times$ onto its direct factor V, so that every $x \in \mathbb{Z}_p^\times$ is written uniquely as

$$x = \omega(x) \cdot x_1$$

with $x_1 \in 1 + q\mathbb{Z}_p$. This makes sense, because if we extend this projection to all of $\mathbb{Z}_p$ by mapping non-units to 0, and then restrict back to $\mathbb{Z}$, we get the Dirichlet character ω. The apparently confusing notation turns out, then, not to be so bad, because all the different maps denoted by ω are closely related.

Problem 190 When $p = 2$, some of the above needs to be modified. What changes are needed?

Both Sage and GP can compute the Teichmüller character. In GP, it is just `teichmuller(a)` where a is a p-adic number. In Sage, one first defines the field, as usual:

```
K=Qp(7)
a=K(5)
K.teichmuller(a)
```

To introduce one more bit of notation, one often uses $\langle x \rangle$ to denote the projection of x on $U_p = 1 + q\mathbb{Z}_p$, so that when $p \neq 2$ the direct product decomposition looks like

$$x = \omega(x)\langle x \rangle.$$

When $p = 2$, we can write $x = \pm\langle x \rangle$ with $\langle x \rangle \in U_2 = 1 + 4\mathbb{Z}_2$.

The next problem gives a different way of obtaining ω.

Problem 191 Show that if $x \in \mathbb{Z}_p$, then we have

$$\omega(x) = \lim_{n \to \infty} x^{p^n}.$$

(Hint: one idea is to start with the expression of x as a product of $\omega(x)$ and $\langle x \rangle$.)

These results allow us to write any $x \in \mathbb{Q}_p$ as a product: first, factor out a power of p; the remaining factor is in $\mathbb{Z}_p^\times$, and so we can write it as a product of a root of unity and a 1-unit. So:

Corollary 5.8.3 *If $p \neq 2$, any nonzero x in $\mathbb{Q}_p$ can be written uniquely in the form*

$$x = \omega(y)\langle y \rangle p^n$$

where $n \in \mathbb{Z}$, $y \in \mathbb{Z}_p^\times$, ω is the Teichmüller character, and $\langle y \rangle \in 1 + p\mathbb{Z}_p$.
When $p = 2$, any nonzero x in $\mathbb{Q}_2$ can be written uniquely as

$$x = \pm\langle y \rangle 2^n$$

with $\langle y \rangle \in 1 + 4\mathbb{Z}_2$.

This allows us to extend the p-adic logarithm to all of $\mathbb{Q}_p$: if we assume that the logarithm of a product is the sum of the logarithms of the factors, then we should have

$$\log_p(x) = \log_p(\omega(y)\langle y\rangle p^n) = \log_p(\omega(y)) + \log_p(\langle y\rangle) + n\log_p(p).$$

The logarithm is already defined on the 1-units, so we know $\log_p(\langle y\rangle)$. Since $\omega(y)$ is a root of unity, we should define $\log_p(\omega(y)) = 0$. So to extend the logarithm to all of $\mathbb{Q}_p$ we just need to decide what $\log_p(p)$ is. The most common choice is to set $\log(p) = 0$. This extended logarithm is called the Iwasawa logarithm.

The p-adic logarithm and exponential functions are implemented in both Sage and GP. The functions are called log and exp. Here is GP:

```
gp > a=8+O(7^20)
%1 = 1 + 7 + O(7^20)
gp > log(a)
%2 = 7 + 3*7^2 + 7^3 + 6*7^4 + 5*7^5 + 2*7^6 + 7^7 + 5*7^8
        + 4*7^9 + 4*7^10 + 2*7^11 + 5*7^12 + 7^13 + 5*7^14
        + 6*7^15 + 2*7^16 + 2*7^17 + 2*7^18 + 7^19 + O(7^20)
gp > exp(log(a))
%3 = 1 + 7 + O(7^20)
```

Or in Sage:

```
sage: K=Qp(7)
sage: a=K(8)
sage: log(a)
7 + 3*7^2 + 7^3 + 6*7^4 + 5*7^5 + 2*7^6 + 7^7 + 5*7^8
  + 4*7^9 + 4*7^10 + 2*7^11 + 5*7^12 + 7^13 + 5*7^14
  + 6*7^15 + 2*7^16 + 2*7^17 + 2*7^18 + 7^19 + O(7^20)
```

You can also use a.log(). Good to know they agree!

The logarithm in GP is actually the Iwasawa logarithm, so you get an answer for any input:

```
gp > log(7+O(7^20))
%4 = O(7^19)
gp > b=9+O(7^20)
%5 = 2 + 7 + O(7^20)
gp > log(b)
%6 = 2*7 + 7^2 + 5*7^3 + 5*7^4 + 2*7^5 + 3*7^6 + 5*7^7
        + 6*7^10 + 2*7^11 + 7^12 + 4*7^13 + 4*7^14
        + 3*7^15 + 7^16 + 6*7^17 + 2*7^18 + O(7^20)
gp > exp(log(b))
```

```
%7 = 1 + 2*7 + 3*7^2 + 6*7^3 + 7^4 + 2*7^5 + 6*7^6 + 7^7
       + 2*7^8 + 2*7^9 + 7^10 + 5*7^12 + 2*7^13 + 7^14
       + 7^15 + 2*7^16 + 4*7^17 + 3*7^18 + 3*7^19 + O(7^20)
```

Of course, there is no chance that $\exp(\log(x)) = x$ when we use the extended logarithm.

Sage is more generous: it lets you choose which extended logarithm you want:

```
sage: K=Qp(7)
sage: a=K(7)
sage: a.log(p_branch=1)
  1 + O(7^20)
sage: a.log(p_branch=0)
  O(7^20)
sage: b=K(35)
sage: b.log(p_branch=0)
  7 + 7^2 + 2*7^3 + 6*7^4 + 7^5 + 5*7^7 + 5*7^8 + 7^9
    + 6*7^10 + 2*7^11 + 7^12 + 6*7^13 + 5*7^15 + 3*7^16
    + 2*7^17 + 6*7^18 + 4*7^19 + O(7^20)
```

If you try computing the logarithm of a non-unit Sage without specifying p_branch, you get an error. Notice that the extended p-adic logarithm function is *not* given by a power series. For more on the extended logarithm and information on how to extend the exponential function as well see [53, V.4].

5.9   The Binomial Series

We want to conclude our exploration of the p-adic elementary functions by considering binomial series and the functions they define. In $\mathbb{R}$, we know that the function $x \mapsto (1+x)^\alpha$ can be expanded as a power series which converges for $|x| < 1$:

$$(1 + X)^\alpha = \mathbf{B}(\alpha, X) = \sum_{n=0}^\infty \binom{\alpha}{n} X^n,$$

where

$$\binom{\alpha}{n} = \frac{\alpha(\alpha - 1)\ldots(\alpha - n + 1)}{n!}.$$

We want to use this series to define the p-adic version of this function. (Of course, as is the case over $\mathbb{R}$, this is only new when α is not a positive integer, but it will work in that case also.) In the p-adic context, the convergence properties of the series will depend on the choice of the p-adic number α. We only consider the case when $\alpha \in \mathbb{Z}_p$ is a p-adic *integer*. The case when $\alpha \in \mathbb{Q}_p$ but is not in $\mathbb{Z}_p$ is actually easier, and we leave it as an exercise for the reader.

So take a p-adic integer α, and consider the binomial series

$$\mathbf{B}(\alpha, X) = (1 + X)^{\alpha} = \sum_{n=0}^{\infty} \binom{\alpha}{n} X^n.$$

The first thing is to check that the coefficients are p-adic integers.

Lemma 5.9.1 *If $\alpha \in \mathbb{Z}_p$ and $n \geq 0$, then $\binom{\alpha}{n} \in \mathbb{Z}_p$.*

PROOF: For each n, consider the polynomial

$$P_n(X) = \frac{X(X-1)\dots(X-n+1)}{n!} \in \mathbb{Q}[X].$$

Just as any polynomial does, $P_n(X)$ defines a continuous function from $\mathbb{Q}_p$ to $\mathbb{Q}_p$. Now, we know that the binomial coefficient $\binom{m}{n}$ of two *positive integers* $m, n \in \mathbb{Z}_+$ is in $\mathbb{Z}$. Hence, for $\alpha \in \mathbb{Z}_+$, we have

$$P_n(\alpha) = \binom{\alpha}{n} \in \mathbb{Z}.$$

In other words, the continuous function P_n maps the set $\mathbb{Z}_+$ of positive integers to $\mathbb{Z}$. By continuity, it must map the closure of $\mathbb{Z}_+$ in $\mathbb{Z}_p$ to the closure of $\mathbb{Z}$. But remember that any element in $\mathbb{Z}_p$ is the limit of a sequence of positive integers (the partial sums of its p-adic expansion). Hence the closure of $\mathbb{Z}_+$ is all of $\mathbb{Z}_p$, and we conclude that P_n maps $\mathbb{Z}_p$ to $\mathbb{Z}_p$, which is what we want to prove.   □

Corollary 5.9.2 *If $\alpha \in \mathbb{Z}_p$ and $|x| < 1$, the series*

$$\mathbf{B}(\alpha, x) = \sum_{n=0}^{\infty} \binom{\alpha}{n} x^n$$

converges.

PROOF: Clear.   □

Problem 192 The Corollary makes no claim that the radius of convergence is in fact equal to 1, nor that the series diverges when $|x| = 1$. What are the facts?

Problem 193 Investigate to what extent Sage and GP implement the computation of a^b when $a, b \in \mathbb{Q}_p$. If so, do they use the binomial function?

As for the logarithm and exponential, it follows from an equality of formal power series that for $\alpha = a/b \in \mathbb{Z}_{(p)}$ and $|x| < 1$ we have

$$\left(\mathbf{B}\left(\frac{a}{b}, x \right) \right)^b = (1 + x)^a,$$

so that it makes sense to write

$$\mathbf{B}\left(\frac{a}{b}, x\right) = (1+x)^{a/b}.$$

This suggests that we should *define*, for any $\alpha \in \mathbb{Z}_p$ and any $x \in p\mathbb{Z}_p$,

$$(1+x)^{\alpha} := \mathbf{B}(\alpha, x).$$

One should be careful, however, to distinguish the p-adic function $\mathbf{B}(a/b, x)$ from its real analogue, *even when x is rational and $1 + x$ is a b-th power in* $\mathbb{Q}$. The following neat example is taken from [42, IV.1].

EXAMPLE: (Following Koblitz.) Let $p = 7$, $\alpha = 1/2$, and $x = 7/9$, so that $x \in 7\mathbb{Z}_7$ and $1 + x = 16/9$ is a rational square. In $\mathbb{R}$, we have

$$(1+x)^{1/2} = \frac{4}{3}.$$

In $\mathbb{Q}_7$, on the other hand, we have $|x| = 1/7$, so that, for $n \geq 1$,

$$\left|\binom{1/2}{n} x^n\right| \leq |x|^n = \frac{1}{7^n} < 1.$$

This implies that

$$(1+x)^{1/2} = 1 + \sum_{n \geq 1} \binom{1/2}{n} x^n \in 1 + 7\mathbb{Z}_7,$$

or, in terms of absolute values, that

$$\left|(1+x)^{1/2} - 1\right| < 1.$$

But

$$\left|\frac{4}{3} - 1\right| = \left|\frac{1}{3}\right| = 1,$$

so that we cannot have $\mathbf{B}(\frac{1}{2}, \frac{7}{9}) = \frac{4}{3}$. In fact, what happens is that in $\mathbb{Q}_7$ we have

$$(1 + 7/9)^{1/2} = \mathbf{B}\left(\frac{1}{2}, \frac{7}{9}\right) = -\frac{4}{3} = 1 - \frac{7}{3} \in 1 + 7\mathbb{Z}_7.$$

This shows that the same series $\sum a_n$ with $a_n \in \mathbb{Q}$ can converge in both $\mathbb{R}$ and some $\mathbb{Q}_p$, but have different limits (even different rational limits), since the topologies are completely different. Of course, the ratio between two different n-th roots will be an n-th root of unity.

Despite the risks, we *will* write $(1 + x)^{\alpha}$ instead of $\mathbf{B}(\alpha, x)$, and let the context decide in which field we are working. The point is to keep in mind that the meaning of the symbol depends on the underlying field.

Problem 194 Study the convergence properties of the binomial series when α is not a p-adic integer.

Problem 195 Show that the value of $\mathbf{B}(\alpha, x)$ *does not* depend on the field we are working in when $x \in \mathbb{Q}$ and $\alpha \in \mathbb{Z}$ is an integer.

The next exercise is [42, IV.1, Exercise 11]. It attempts to decide exactly when $(1 + x)^{1/2}$ is equal to the positive square root.

Problem 196 (Koblitz) Choose $x \in \mathbb{Q}$ such that $1 + x$ is a square in $\mathbb{Q}$; say $\sqrt{1 + x} = a/b$ with a and b positive and relatively prime. Let S be the set of primes (including the infinite prime, if applicable) for which the binomial series $\mathbf{B}(1/2, x)$ converges in $\mathbb{Q}_p$. (The limit will have to be a square root of $1 + x$, hence will equal either a/b or $-a/b$.) Prove that:

i) If p is an odd prime, then $p \in S$ if and only if $p|(a + b)$ or $p|(a - b)$, and in $\mathbb{Q}_p$ we will have $\mathbf{B}(1/2, x) = -a/b$ in the first case, $\mathbf{B}(1/2, x) = a/b$ in the second.

ii) We will have $2 \in S$ if and only if a and b are both odd; the limit in $\mathbb{Q}_2$ will be a/b if $a \equiv b \pmod 4$, and $-a/b$ if $a \equiv -b \pmod 4$.

iii) We will have $\infty \in S$ if and only if $0 < a/b < \sqrt{2}$, and the sum in $\mathbb{R}$ will always be a/b.

iv) There is no x for which the set S is empty, and S will have only one element if and only if $x \in \{8, \frac{16}{9}, 3, \frac{5}{4}\}$.

v) Except for the x mentioned in the previous item, there always exist primes $p, q \in S$ such that the sum in $\mathbb{Q}_p$ is different from the sum in $\mathbb{Q}_q$.

For other interesting results along these lines, tracking what happens when we look at the same series in various different $\mathbb{Q}_p$, see the article [12].

5.10   Interpolation

The idea of interpolating a known function to obtain a related p-adic function has become very important in number theory, where the standard targets for this method have been the zeta and L-functions. The point of this section is to give a first example of the interpolation problem. Our example is very simple, and it illustrates only some of the many ideas that have arisen in the literature. We refer the reader to the standard references.[9]

In the previous section, we considered the binomial series, and used it to define a p-adic function $x \mapsto x^\alpha$ for $x \in 1 + p\mathbb{Z}_p$ and $\alpha \in \mathbb{Z}_p$. What we would like to do in this section is to invert the situation, and think of x^α as *a function of* α. We can interpret this as an interpolation problem, in the following way.

---

[9]The idea of working out this example is due to Koblitz, who goes through a similar discussion in [42].

Suppose $n \in \mathbb{Z}_p$ is any p-adic integer, and α is an integer. Then it certainly makes sense to compute n^α. Thus, we can consider the function

$$f(\alpha) = n^\alpha,$$

which is well-defined for $\alpha \in \mathbb{Z}$. What we would like to do is to extend this function to the widest possible range of p-adic values of α. Since $\mathbb{Z}$ is dense in $\mathbb{Z}_p$, such an extension, if continuous, is unique, because two continuous functions that coincide on a dense subset are identical. Indeed, one can even work with smaller subsets of $\mathbb{Z}$: the set of positive integers, or the set of negative integers, or any other set of integers which is dense in $\mathbb{Z}_p$. The problem of finding such an extension is called the problem of finding a p-adic interpolation of the function $f(\alpha) = n^\alpha$.

The first thing to say about p-adic interpolation is that in a certain sense the whole thing is trivial. This is because we know perfectly well when it is that a function defined on a dense subset of $\mathbb{Z}_p$ has a continuous extension to all of $\mathbb{Z}_p$. The crucial notion here is *uniform continuity*. The reason it is relevant to the interpolation problem is a well-known theorem which we leave as an exercise:

Problem 197 Show that any continuous function defined on a compact set is automatically uniformly continuous and bounded.

Problem 198 Can you give an example of a function $\mathbb{Z} \longrightarrow \mathbb{Z}_p$ which is continuous but not uniformly continuous? (This may be a little hard.)

Now suppose our $f(\alpha)$ could indeed be extended to $\mathbb{Z}_p$. Then, since $\mathbb{Z}_p$ is compact, the extension would have to be bounded and uniformly continuous. Hence (restricting back), so would $f(\alpha)$. It turns out that in fact these two conditions are sufficient.

Proposition 5.10.1 *Let S be a dense subset of $\mathbb{Z}_p$, and let $f : S \longrightarrow \mathbb{Q}_p$ be a function. Then there exists a continuous extension $\tilde{f} : \mathbb{Z}_p \longrightarrow \mathbb{Q}_p$ of f to $\mathbb{Z}_p$ if and only if f is bounded and uniformly continuous. If it exists, this extension is unique.*

PROOF: We know that the condition is necessary, and that the extension is unique if it exists, by the discussion above. The difficulty is to prove the sufficiency, i.e., to show that uniform continuity and boundedness are enough to guarantee the existence of the extension.

The key is the continuity. If $x \in \mathbb{Z}_p$, there exists a sequence

$$\alpha_1, \alpha_2, \ldots, \alpha_k, \ldots$$

of elements of S which tends to x (because S is dense). If $\tilde{f}$ exists, then we will have

$$\tilde{f}(x) = \lim_{k \to \infty} \tilde{f}(\alpha_k) = \lim_{k \to \infty} f(\alpha_k).$$

This shows the way to proceed.

First of all, since the sequence (α_k) tends to x, it is a Cauchy sequence, so that

$$\lim_{k\to\infty} |\alpha_{k+1} - \alpha_k| = 0.$$

Since f is uniformly continuous and bounded, it follows that

$$\lim_{k\to\infty} |f(\alpha_{k+1}) - f(\alpha_k)|) = 0$$

(check!), so that the $f(\alpha_k)$ form a Cauchy sequence, hence have a limit in $\mathbb{Q}_p$. Now we can *define* $\tilde{f}$ by the condition we know it has to satisfy:

$$\tilde{f}(x) := \lim_{k\to\infty} f(\alpha_k)$$

for any sequence (α_k) converging to x. This gives the extension. □

There are a whole bunch of things to check, and the reader should:

Problem 199 Check that the image of a Cauchy sequence (α_k) by a bounded and uniformly continuous function f is again a Cauchy sequence.

Problem 200 Check that the function $\tilde{f}$ defined above does not depend on the choice of the sequences (α_k).

Problem 201 Check that the function $\tilde{f}$ defined above is indeed a continuous function on $\mathbb{Z}_p$.

One less obvious fact is that one can replace $\mathbb{Z}_p$ in the proposition by any compact subset of $\mathbb{Q}_p$:

Problem 202 Check that the proposition remains true if we replace $\mathbb{Z}_p$ by any compact subset of $\mathbb{Q}_p$, such as $\mathbb{Z}_p^\times$, $1 + p\mathbb{Z}_p$, or $p^m\mathbb{Z}_p$. (Hint: the point is that only the compactness was used.)

This result may seem to completely settle the issue, but that is far from being the case, for several important reasons. For one thing, one often wants to know more about $\tilde{f}$ than its bare existence. For example, can it be written as a power series? Does it extend to a set larger than $\mathbb{Z}_p$? Can we give a good method to compute (better: to approximate) it? Another point is that we can exploit the "if and only if" in the proposition: if what we want to prove is the uniform continuity, then finding an interpolation will prove just that! Finally, thinking in terms of interpolation often gives us useful new ideas, as we shall see below when we get to the nitty-gritty of our example.

Before we go on to the example, however, it may be useful to unwind what uniform continuity really means in our case. We will take $f(\alpha)$ to be a function defined on a dense subset S of $\mathbb{Z}_p$, with values in $\mathbb{Q}_p$. Then being "close" in S amounts to being congruent modulo a high power of p, and being close in $\mathbb{Q}_p$ is the same. Hence, f will be uniformly continuous if it satisfies the following congruence condition:

Given $m \in \mathbb{Z}$, there exists an $N \in \mathbb{Z}$ such that

$$\alpha \equiv \beta \pmod{p^N} \implies f(\alpha) \equiv f(\beta) \pmod{p^m}.$$

Thus, uniform continuity has a simple translation in terms of congruence properties. This turns out to be quite important.

Now we return to the exponential function $\alpha \mapsto n^\alpha$. This is defined, at first, for $\alpha \in \mathbb{Z}$ and $n \in \mathbb{Z}_p$, and we would like to extend it to all $\alpha \in \mathbb{Z}_p$. The answer, as it happens, depends quite seriously on n.

First of all, suppose n is a 1-unit, that is, $n \in 1 + p\mathbb{Z}_p$. Then we can use the binomial series to get our interpolation:

Corollary 5.10.2 *For any $n \in 1 + p\mathbb{Z}_p$ there exists a continuous function $f_n : \mathbb{Z}_p \longrightarrow \mathbb{Q}_p$ such that for any $\alpha \in \mathbb{Z}$ we have $f_n(\alpha) = n^\alpha$.*

PROOF: We can just define $f_n(\alpha) = \mathbf{B}(\alpha, n - 1)$, which converges because we are assuming $n \in 1 + p\mathbb{Z}_p$. Checking continuity, however, is not all that easy (remember that we want continuity in α, rather than in n, so it is not just a matter of saying that power series are continuous functions). We leave the verification to the reader as a challenging problem.   □

Problem 203 Show that $\mathbf{B}(\alpha, x)$ is continuous as a function of α.

One might also prove Corollary 5.10.2 by a more direct route, showing that if $n \in 1 + p\mathbb{Z}_p$ then $\alpha \mapsto n^\alpha$ is bounded and uniformly continuous. Boundedness is easy: any integral power of n will be in $\mathbb{Z}_p$ (and even in $1 + p\mathbb{Z}_p$), because n is a 1-unit (there are negative powers in this game too!). As for uniform continuity, that is also not hard to show; notice, first that

$$(1 + pk)^{p^m} \equiv 1 \pmod{p^{m+1}},$$

so that if $\beta = \alpha + ip^m$ we get

$$n^\beta = n^\alpha \cdot (n^{p^m})^i \equiv n^\alpha \pmod{p^{m+1}},$$

which is what we want. This establishes the *existence* of f_n. Proving that $f_n(\alpha) = \mathbf{B}(\alpha, n - 1)$ requires showing that the latter is continuous.

Problem 204 Another method to obtain the interpolation of $\alpha \mapsto n^\alpha$ would be to define

$$n^\alpha = \exp_p(\alpha \log_p(n)).$$

Does this work? (Notice that it is much easier to check the continuity in this case.)

So we have extended n^α to $\alpha \in \mathbb{Z}_p$ when $n \in 1 + p\mathbb{Z}_p$. We would like, however, to consider more general p-adic integers. Unfortunately, that turns out to be quite tricky. To begin with, suppose p divides n. Then, as the integer α becomes bigger, n^α becomes p-adically closer and closer to zero.

This messes everything up. For example, take $n = p$, and look at the sequence $\alpha_k = 1 + p^k$. Then

$$\lim_{k \to \infty} \alpha_k = 1, \qquad \text{but} \qquad \lim_{k \to \infty} p^{\alpha_k} = 0 \neq p^1,$$

so that the map $\alpha \mapsto p^\alpha$ is not even continuous.

We might have a better chance if we tried to work only with p-adic units. When $p = 2$, this gives nothing new, since $\mathbb{Z}_2^\times = 1 + 2\mathbb{Z}_2$. For odd primes p, however, we know that $1 + p\mathbb{Z}_p$ is a subgroup of index $p - 1$ in $\mathbb{Z}_p^\times$, so that going from $n \in 1 + p\mathbb{Z}_p$ to $n \in \mathbb{Z}_p^\times$ would be progress. Even this, however, turns out to be a little tricky, basically because of the presence of the roots of unity.

Let $p \neq 2$; for $n \in \mathbb{Z}_p^\times$, we will try to interpolate the function $\alpha \mapsto n^\alpha$. Since we have already done this for $n \in 1 + p\mathbb{Z}_p$, the easiest way to do this is to use the known relation between $\mathbb{Z}_p^\times$ and its subgroup $1 + p\mathbb{Z}_p$. Recall that we showed that there is a direct product decomposition

$$\mathbb{Z}_p^\times = V \times U_1 \cong \mathbb{F}_p^\times \times (1 + p\mathbb{Z}_p),$$

and that for $x \in \mathbb{Z}_p^\times$ this decomposition gives $x = \omega(x)\langle x \rangle$ with $\omega(x) \in V$ and $\langle x \rangle \in 1 + p\mathbb{Z}_p$. Then, for any integer α, we have

$$n^\alpha = \omega(n)^\alpha \langle n \rangle^\alpha.$$

The first thing to note is that $\omega(n)$ is a $(p-1)$-st root of unity, and hence if $\alpha \equiv \alpha_0 \pmod{p - 1}$ we can re-write the formula as

$$n^\alpha = \omega(n)^{\alpha_0} \langle n \rangle^\alpha.$$

Now, since $\langle n \rangle \in 1 + p\mathbb{Z}_p$, we already know how to interpolate its part of the function, i.e., we know how to interpolate the function $\alpha \mapsto \langle n \rangle^\alpha$. But this is almost enough to solve the problem, since we've reduced everything to this known interpolation together with the choice of α_0. In fact, the best way to think of this is to do a complete turnaround, and change the function to be interpolated!

Rather than considering the function $\alpha \mapsto n^\alpha$ for all integers α, consider it only for those integers congruent to a fixed α_0 modulo $(p - 1)$. There are of course $p - 1$ different functions of this kind, each corresponding to a choice of α_0. The kicker, of course, is that the set of integers α which are congruent to a fixed α_0 is itself *dense* in $\mathbb{Z}_p$, so that it makes sense to ask for an interpolation from this set to all of $\mathbb{Z}_p$. And this, by the discussion above, is easily done: consider the p-adic function $f_{\alpha_0} : \mathbb{Z}_p \longrightarrow \mathbb{Z}_p$ given by

$$f_{\alpha_0}(\alpha) = \omega(n)^{\alpha_0} \langle n \rangle^\alpha.$$

This, first of all, makes sense, by the discussion above, since we do know how to compute the α-th power of the 1-unit $\langle n \rangle$. Next, it does coincide with the function $\alpha \mapsto n^\alpha$ whenever α is an integer satisfying $\alpha \equiv \alpha_0 \pmod{p-1}$. So it does give a (somewhat skewed) solution to our interpolation problem, which we state as a theorem:

Proposition 5.10.3 *Let $n \in \mathbb{Z}_p^\times$ and $\alpha_0 \in \{0, 1, \ldots, p-1\}$, and let*

$$A_{\alpha_0} = \{\alpha \in \mathbb{Z} : p \nmid \alpha \text{ and } \alpha \equiv \alpha_0 \pmod{p-1}\} \subset \mathbb{Z}.$$

Then

$$f_{\alpha_0}(\alpha) = \omega(n)^{\alpha_0} \langle n \rangle^\alpha$$

defines a function $f_{\alpha_0} : \mathbb{Z}_p \longrightarrow \mathbb{Z}_p$ such that

$$f_{\alpha_0}(\alpha) = n^\alpha \qquad \text{whenever} \quad \alpha \in A_{\alpha_0}.$$

Notice that all the different f_{α_0} coincide if $n \in 1 + p\mathbb{Z}_p$, so that this is a genuine extension of our first interpolation. What happens, though, if we compute f_{α_0} on the wrong sort of $\alpha \in \mathbb{Z}$? Well, we get something like

$$\begin{aligned}
f_{\alpha_0}(\alpha) &= \omega(n)^{\alpha_0} \langle n \rangle^\alpha \\
&= \omega(n)^{\alpha_0 - \alpha} \omega(n)^\alpha \langle n \rangle^\alpha \\
&= \omega(n)^{\alpha_0 - \alpha} n^\alpha.
\end{aligned}$$

In words, f_{α_0} actually interpolates a function that is slightly different from our original function: rather than giving n^α, it gives a "twisted" version which ends up being equal to a root of unity times n^α. For the special α's that belong to A_{α_0}, the root of unity disappears, and we get our original function. So we're close, but we haven't really done exactly what we set out to do. This is in fact as good a result as one might hope for, as the example in the next problem shows.

Problem 205 Show that the function $\mathbb{Z} \longrightarrow \mathbb{Z}$ given by $\alpha \mapsto (-1)^\alpha$ can only be interpolated to a function $\mathbb{Z}_p \longrightarrow \mathbb{Z}_p$ when $p = 2$ (in which case -1 is a 1-unit). For $p = 3$, the proposition above claims that there exist *two* functions f_0 and f_1 which "together" give an interpolation. Describe the two functions f_0 and f_1. (Hint: they are not very interesting.)

Some readers may find this situation a bit unsatisfactory: rather than one interpolating function, we have ended up with a whole bunch, each of which gives an interpolation for a restriction of the original function to a smaller set. One way of jazzing this up a bit is the following. The collection of all the f_{α_0} together define a function

$$\mathcal{F} : \mathbb{Z}_p \times \mathbb{Z}/(p-1)\mathbb{Z} \longrightarrow \mathbb{Z}_p$$

given by $\mathcal{F}(\alpha, \alpha_0) = f_{\alpha_0}(\alpha)$. Now, one has the "diagonal inclusion"

$$\mathbb{Z} \hookrightarrow \mathbb{Z}_p \times \mathbb{Z}/(p-1)\mathbb{Z},$$

given by $\alpha \mapsto (\alpha, \alpha)$. (The first α to be thought of as an element of $\mathbb{Z}_p$, the second as an integer modulo $p - 1$.) In other words, if $\alpha \in \mathbb{Z}$, its image under the inclusion is the pair (α, α_0), where $\alpha_0 \equiv \alpha \pmod{p - 1}$. Thus, if we restrict $\mathcal{F}$ to the image of $\mathbb{Z}$ we get

$$\alpha \mapsto \mathcal{F}(\alpha, \alpha_0) = f_{\alpha_0}(\alpha) = \omega(n)^{\alpha_0} \langle n \rangle^{\alpha} = n^{\alpha}.$$

This means that we *can* think of $\mathcal{F}$ as giving an interpolation of the function $\alpha \mapsto n^{\alpha}$, provided we think of $\mathbb{Z}$ as included in this larger set.

Interpolation problems of this kind are very important in the applications of p-adic analysis to number theory, and several of the features of our toy example persist in the more interesting ones. First, one often has to "remove the p-part," which in our case was accomplished by restricting the base n to be a p-adic unit. Second, the interpolation often requires us to consider "twisted" versions of the original function. In our case, these were the several f_{α_0} functions: restricting one of the f_{α_0} to $\mathbb{Z}$ does not give the function $\alpha \mapsto n^{\alpha}$, but rather

$$\alpha \mapsto \omega(n)^{\alpha_0 - \alpha} n^{\alpha}.$$

This kind of modification, when something is multiplied by a root of unity, is often referred to as "twisting." The upshot: one cannot interpolate the function $\alpha \mapsto n^{\alpha}$, but one can interpolate appropriate twists of that function. This phenomenon is quite common.

An obvious question should be mentioned here: what is the point? Why should one want to interpolate "classical" functions in this fashion?

The question is hard to answer in elementary terms, without delving into the specific interpolation problems that mathematicians have been interested in. But we can give some idea of what is going on by saying that many classical functions have interesting "special values," that is, their values at certain magical points have a special significance. For example, the values of the Riemann zeta function $\zeta(s)$ at positive even integers involve the Bernoulli numbers, which hide within themselves quite a lot of information about the arithmetic of cyclotomic fields. (Its pole at $s = 1$ also carries this kind of information.)

Now suppose one can interpolate these special values with a p-adic function. This gives us a p-adic function which shares with its classical analogue the same (or similar, if a twist creeps in) special values. Well, this means that one can get information on those values by looking at either function... and the p-adic function is often easier to handle. This yields a basic strategy that has been applied over and over in modern number theory, with very interesting results. The reader may want to browse through the articles in [16] to get some idea of what the goals of this particular enterprise are. To begin to study the enterprise itself, one might start with the treatment in [42].

6   Vector Spaces and Field Extensions

Up to now, we have kept our attention focused on the field $\mathbb{Q}$ and its p-adic completions. We have already felt, however, the need to consider other fields (for example, when we dealt with the zeros of a function defined by a power series). In fact, just as we have emphasized the natural analogy between the p-adic fields $\mathbb{Q}_p$ and the field $\mathbb{R}$ of real numbers, it is a very natural thing to do to look for an extension of $\mathbb{Q}_p$ that is analogous to the complex numbers. In other words, we would like to look for ways to extend $\mathbb{Q}_p$ in order to obtain a field that is not only complete (so that we can do analysis), but also algebraically closed (so that all non-constant polynomials have roots). This turns out to be more subtle (and therefore more interesting) than one might expect. It turns out, first of all, that to get an algebraically closed field one must make a very large extension of $\mathbb{Q}_p$. This extension turns out not to be complete any more, so there is no other recourse but to go through the completion process again, and this finally yields the field we wanted. This is very different from the classical case, where going from $\mathbb{R}$ to an algebraically closed field is just a small step (just add i), and the resulting field (the complex numbers) is already complete. The goal of this chapter is to tell the p-adic version of this story in its entirety.

In order to get there, we begin by considering vector spaces over $\mathbb{Q}_p$ and the norms one might define on them. This is a step in the right direction, since any field containing $\mathbb{Q}_p$ will also be a $\mathbb{Q}_p$-vector space. We then go on to considering the fields themselves. This will necessarily involve some knowledge of abstract algebra; as usual, we have tried to make the facts we use explicit, in order to make it easier to look up the material we need in the standard texts. We start with finite field extensions, and only after we have understood them well do we try to go on to an algebraic closure.

The reader should note that we have taken one of two possible points of view in addressing our subject. We will be investigating extensions of the p-adic fields $\mathbb{Q}_p$. It would be just as interesting to consider extensions of $\mathbb{Q}$ itself, and to attempt to construct a theory of absolute values on such fields. This leads to an interesting theory, which we have decided not to address at all (because it requires more knowledge of Galois theory than we wish to assume, and because it properly belongs in an introduction to algebraic number theory). This means that we must of necessity fail to mention certain topics, such as the extension to bigger fields of the product formula, of Ostrowski's theorem, or of the local-global principle. Instead, we

© Springer Nature Switzerland AG 2020
F. Q. Gouvêa, *p-adic Numbers*, Universitext, https://doi.org/10.1007/978-3-030-47295-5_6

take a "strictly local" perspective: we are living in the p-adic world from the start. There is a good discussion of the global (or semi-local) aspect in [14] and in many introductions to algebraic number theory.

Once we have absolute values on extensions of $\mathbb{Q}_p$, we will be in a position to extend to such fields much of what was done in Chapters 4 and 5. Rather than do so in full detail, we will often be content with "this clearly extends;" the reader for whom the "clearly" is not clear should go back and check.[1] We will also need to prove a few results about these fields that will allow us to understand what goes on when one puts them all together to get an algebraic closure.

6.1   Normed Vector Spaces over Complete Valued Fields

The algebraic part of the theory of vector spaces over $\mathbb{Q}_p$ is, of course, identical to the theory of vector spaces over any other field. This is simply because that part of the theory does not depend on the specific field at all: it only requires the knowledge of the basic field properties. Therefore, we won't bother to discuss the basics about vector spaces, subspaces, bases, dimension, and so on.

What we would like to focus on, then, is the point where the vector spaces acquire a metric. This is usually done by putting a *norm* on the vector space. For example, in the classical case, we can metrize $\mathbb{R}^2$ using the norm

$$\|(x,y)\| = \sqrt{x^2 + y^2},$$

and similarly for all the $\mathbb{R}^n$. Of course, there isn't just one choice of norm. For example, the following two choices of norms on $\mathbb{R}^2$ are also popular:

$$\|(x,y)\|_1 = |x| + |y|$$

and

$$\|(x,y)\|_{\text{sup}} = \max\{|x|, |y|\}$$

(the subscript 1 here has nothing to do with the subscripts on the p-adic norms; there should be no serious confusion involved).

We want to build up an analogous theory for norms on vector spaces over $\mathbb{Q}_p$. We begin, as we did in Chapter 2, by considering a general theory of normed vector spaces over valued fields, because it is no more difficult than doing things over $\mathbb{Q}_p$. As we did then, we will restrict to $\mathbb{Q}_p$ whenever that makes things easier.

We begin with a field $\Bbbk$, which we assume has an absolute value $|\ |$ on it. (We do not make any assumption about whether the absolute value is archimedean, but we *do* assume it is non-trivial, because otherwise things are pretty silly.) In order to get an interesting theory, we assume that $\Bbbk$ is

---

[1]Instructors should note that this may mean that more time than usual may need to be spent on this chapter!

complete with respect to its absolute value. We will also assume for simplicity that $\Bbbk$ is of *characteristic zero*,[2] so that it contains $\mathbb{Q}$. The reader should keep both $\mathbb{R}$ and $\mathbb{Q}_p$ in mind as examples.

Let V be a vector space over $\Bbbk$. At first we make no further assumptions on V, but later we will want to concentrate on the case where V is finite-dimensional.

Definition 6.1.1 *Let $\Bbbk$ be a complete valued field of characteristic zero with a nontrivial absolute value $|\ |$. A* norm *on a $\Bbbk$-vector space V is a function*

$$\|\ \| : V \longrightarrow \mathbb{R}_+$$

satisfying the following conditions:

   i) $\|\mathbf{v}\| = 0$ if and only if $\mathbf{v} = 0$,

   ii) for any two vectors $\mathbf{v}, \mathbf{w} \in V$, we have $\|\mathbf{v} + \mathbf{w}\| \leq \|\mathbf{v}\| + \|\mathbf{w}\|$,

   iii) for any $\mathbf{v} \in V$ and any $\lambda \in \Bbbk$, we have $\|\lambda \mathbf{v}\| = |\lambda|\,\|\mathbf{v}\|$.

A $\Bbbk$-vector space V which has a norm $\|\ \|$ is called a normed vector space over $\Bbbk$.

In other words, a norm is just a way to measure the size of vectors, and the conditions merely require that it behave as we would expect such a notion of length to behave. One is tempted, of course, to introduce the notion of non-archimedean norms, but it is less clear that it is a good idea. For example, consider the norm on $V = \mathbb{Q}_p \times \mathbb{Q}_p$ given by

$$\|(x, y)\| = \sqrt{|x|_p^2 + |y|_p^2}.$$

One easily checks that this is indeed a norm, but that it does not satisfy the naïve analogue of the non-archimedean inequality (by which we mean that something like

$$\|(x + x', y + y')\| \leq \max\{\|(x, y)\|, \|(x', y')\|\},$$

does not hold). But this norm is still "non-archimedean" in the sense that given two vectors it may not be possible to find an integer multiple of one which is bigger than the other (check this!). In fact, this suggests that normed vector spaces over non-archimedean complete fields are automatically "non-archimedean" in any reasonable sense, so that there is nothing to define.

Given a norm, we can easily define a metric (i.e., a way of measuring distance) on V by saying that the distance between two vectors is (what else?) the size of their difference:

---

[2]The reader will recall, I hope, that the characteristic of a field is the smallest number of ones that need to be added together to get zero, when this is possible, and is zero when it is not possible. For example, the characteristic of $\mathbb{Q}$ is zero, and the characteristic of $\mathbb{F}_p$ is p. It is an easy exercise to prove that if the characteristic is non-zero, then it must be a prime number.

Definition 6.1.2 *Let V be a normed vector space with norm $\| \; \|$. We define a metric on V by putting, for any $\mathbf{v}, \mathbf{w} \in V$,*

$$d(\mathbf{v}, \mathbf{w}) = \|\mathbf{v} - \mathbf{w}\|.$$

Problem 206 Show that the metric thus defined is indeed a metric, that is, it has the properties listed in Problem 50.

Once we have a metric, we have, as in Chapter 2, a topology, so that we can talk about open and closed balls, open sets, and convergence. (We urge the reader who is hesitant about this to re-read the appropriate section of Chapter 2.)

Problem 207 Let V be a normed vector space. The point of this problem is to check that the metric $d(\mathbf{v}, \mathbf{w})$ (or, equivalently, the norm it is derived from) relates well to the operations in V:

i) Fix $\mathbf{v}_0, \mathbf{w}_0 \in V$. Show that for any $\varepsilon > 0$ there exists a $\delta > 0$ such that, whenever $d(\mathbf{v}, \mathbf{v}_0) < \delta$ and $d(\mathbf{w}, \mathbf{w}_0) < \delta$, we have $d(\mathbf{v} + \mathbf{w}, \mathbf{v}_0 + \mathbf{w}_0) < \varepsilon$. In other words, addition is a continuous function.

ii) Fix $\mathbf{v}_0 \in V$ and $\lambda_0 \in \Bbbk$. Show that for any $\varepsilon > 0$ there exists a $\delta > 0$ such that, whenever $d(\mathbf{v}, \mathbf{v}_0) < \delta$ (distance in V) and $d(\lambda, \lambda_0) < \delta$ (distance in $\Bbbk$), we have $d(\lambda\mathbf{v}, \lambda_0\mathbf{v}_0) < \varepsilon$. In other words, multiplication of a vector by an element of $\Bbbk$ is a continuous function.

This shows that the metric $d(\mathbf{v}, \mathbf{w})$ makes V a *topological vector space* over the topological field $\Bbbk$. (Compare Problem 51.)

Let's consider some examples. For these, we assume V is finite-dimensional, and we fix a basis $\{\mathbf{v}_1, \mathbf{v}_2, \ldots, \mathbf{v}_n\}$. Any vector in V can then be written (uniquely) in the form $\mathbf{v} = a_1\mathbf{v}_1 + a_2\mathbf{v}_2 + \cdots + a_n\mathbf{v}_n$ with $a_i \in \Bbbk$, and we exploit this to obtain norms on V from the absolute value on $\Bbbk$:

i) We can define a norm by putting

$$\|a_1\mathbf{v}_1 + a_2\mathbf{v}_2 + \cdots + a_n\mathbf{v}_n\|_{\sup} = \max_{1 \leq i \leq n} |a_i|.$$

This is called the sup-norm on V with respect to our choice of basis.

ii) We can also define, for each real number $r \geq 1$, the r-norm

$$\|a_1\mathbf{v}_1 + a_2\mathbf{v}_2 + \cdots + a_n\mathbf{v}_n\|_r = \left(|a_1|^r + |a_2|^r + \cdots + |a_n|^r \right)^{1/r}.$$

These are analogous to norms on spaces of functions that are often used in analysis. Notice that they do depend on the choice of basis.

If $\Bbbk = \mathbb{R}$, $V = \mathbb{R}^2$, $\mathbf{v}_1 = (1, 0)$, $\mathbf{v}_2 = (0, 1)$, and we take $r = 1$ or $r = 2$, we get the examples mentioned in the introduction to this section. We leave it to the reader to check that these are indeed norms.

Problem 208 Check that the sup-norm and the r-norms are indeed norms.

Problem 209 Let $\Bbbk = \mathbb{R}$, $V = \mathbb{R}^2$, and use the canonical basis $\{(1,0),(0,1)\}$. Sketch the closed ball of radius 1 with respect to (a) the sup-norm, (b) the r-norms for $r = 1$, 2, 3.

Problem 210 Show, with an example, that the norms we have defined depend quite seriously on the choice of basis. (Hint: this is very easy; just use the simplest vector space you can think of.)

Problem 211 Let $V = \mathbb{Q}_p \times \mathbb{Q}_p$, and define $\|(x,y)\| = |x+y|$. Does this define a norm?

We need to define a notion of equivalence for norms, just as we defined equivalence of absolute values.

Definition 6.1.3 *We say two norms* $\| \; \|_1$ *and* $\| \; \|_2$ *on a* $\Bbbk$-*vector space* V *are equivalent if there exist positive real numbers* C *and* D *such that, for every vector* $\mathbf{v} \in V$, *we have*

$$\|\mathbf{v}\|_1 \leq C\|\mathbf{v}\|_2 \qquad and \qquad \|\mathbf{v}\|_2 \leq D\|\mathbf{v}\|_1.$$

To get a good feeling for this notion, the reader is invited to work through a few elementary facts about it:

Problem 212 Show that two norms on V are equivalent if and only if they define the same topology on V (i.e., a set is open with respect to one norm if and only if it is open with respect to the other).

Problem 213 Sometimes it's useful to state the condition for equivalence in another way. Suppose $\| \; \|_1$ and $\| \; \|_2$ are equivalent. Show that any open ball around 0 with respect to norm $\| \; \|_1$ contains an open ball around 0 with respect to $\| \; \|_2$ and is contained in an open ball around 0 with respect to $\| \; \|_2$. Show that this condition is equivalent to the two inequalities.

Problem 214 Show that if two norms are equivalent, then they have the same Cauchy sequences; in other words, a sequence is Cauchy with respect to one of them if and only if it is Cauchy with respect to the other.

One should note that the fact that equivalent norms give the same Cauchy sequences is a priori *stronger* than the fact that they induce the same topology. That the two things end up being the same in our case is directly linked to the fact that our metric comes from a norm on a vector space, and that extra structure yields extra information.

Problem 215 Show that the norms on $\mathbb{R}^2$ that we mentioned above are equivalent. (Hint: your sketches from Problem 209 might prove helpful.)

Problem 216 Let $V = \mathbb{Q}_p \times \mathbb{Q}_p$, and define the norms

$$\|(a, b)\|_{\mathsf{sup}} = \max\{|a|, |b|\}$$

and

$$\|(a, b)\|_1 = |a| + |b|.$$

Prove that these norms are equivalent.

Once we have a metric, we can ask about completeness, just as in Chapter 3. Recall that we say V is complete with respect to a norm $\|\ \|$ if every Cauchy sequence in V (with respect to $\|\ \|$) converges. (Note that this depends only on the equivalence class of the norm, which is as we want it.) It is well known that $\mathbb{R}^2$, for example, is complete with respect to all of the norms mentioned above. Here is another example where one can show completeness for a whole bunch of spaces and norms in one blow:

Proposition 6.1.4 *Let V be a finite-dimensional vector space over a complete valued field $\Bbbk$. Choose a basis $\{v_1, v_2, \ldots, v_m\}$ for V, and let $\|\ \|$ be the sup-norm with respect to this basis. Then V is complete. Specifically, a sequence (w_n) with*

$$w_n = a_{1n}v_1 + a_{2n}v_2 + \cdots + a_{mn}v_m$$

is Cauchy in V if and only if the sequences of basis coefficients (a_{1n}), (a_{2n}), $\ldots$, (a_{mn}) are Cauchy sequences in $\Bbbk$, and the limit is obtained by taking the limits of the coefficients:

$$\lim_{n \to \infty} w_n = (\lim_{n \to \infty} a_{1n})v_1 + (\lim_{n \to \infty} a_{2n})v_2 + \cdots + (\lim_{n \to \infty} a_{mn})v_m.$$

PROOF: Since the norm is simply given by the largest of the basis coefficients, saying that $\|w_{n_1} - w_{n_2}\|$ tends to zero just amounts to saying that *all* the differences $a_{in_1} - a_{in_2}$ do. That is enough to prove everything we've claimed is true. $\square$

Finally, here are some problems that suggest some avenues for further exploration:

Problem 217 Let V and W be normed vector spaces, and write $\|\ \|_v$ and $\|\ \|_w$ for their norms. Let $f : V \longrightarrow W$ be a linear transformation. Show that the following are equivalent:

 i) f is continuous at $0 \in V$;

 ii) $\sup\limits_{\|v\|_v \leq 1} \|f(v)\|_w$ is finite;

 iii) there exists an M such that we have $\|f(v)\|_w \leq M\|v\|_v$ for all $v \in V$;

 iv) f is continuous at all $v \in V$.

Problem 218 (Hard) Let V be the space of sequences (a_n) with $a_n \in \mathbb{Q}_p$ and $\lim a_n = 0$. Define a norm on V by $\|(a_n)\| = \sup_n |a_n|$.

i) Is V complete with respect to this norm?

ii) Consider the subspace $W \subset V$ defined by the condition that $\sum |a_n|$ converges (in $\mathbb{R}$, of course). Is W a closed subspace of V? On W, we have two norms: the norm induced by the norm on V, and the 1-norm given by $\|(a_n)\|_1 = \sum |a_n|$. Are these norms equivalent?

Problem 219 Let V be the space of all polynomials with coefficients in $\mathbb{Q}_p$. Choose a positive real number $c \in \mathbb{R}$ and define, for $f(X) = a_n X^n + \cdots + a_1 X + a_0$,

$$\|f(X)\|_c = \max_{0 \leq i \leq n} |a_i| c^i.$$

i) Show that this is a norm on V.

ii) Is V complete with respect to this norm?

iii) We know how to multiply polynomials. Is it true that the norm we just defined is multiplicative, i.e., that $\|f(X)g(X)\|_c = \|f(X)\|_c \|g(X)\|_c$?

iv) Explain why this norm is interesting.

v) Now suppose we vary c; we get a whole family of norms. Are they equivalent?

6.2 Finite-dimensional Normed Vector Spaces

The problems at the end of the previous section already hint that there is a fundamental difference between finite- and infinite-dimensional spaces when it comes to the theory of norms. This is indeed the case, and in this section we prove the fundamental theorem about finite-dimensional normed vector spaces over complete fields. What this theorem says is that, after Proposition 6.1.4, we already know all that there is to know about the finite-dimensional case. This is because it turns out that *any* norm on such a vector space is equivalent to the sup-norm (with respect to any given basis); in particular, all the sup-norms are equivalent. The proof of this result, which we give next, is often given only for locally compact complete fields. Since we want to apply it to infinite extensions of $\mathbb{Q}_p$, we need to know if for all complete valued fields. The proof we give follows Cassels in [14, Ch. 7, Lemma 2.1].

Theorem 6.2.1 *Let V be a finite-dimensional vector space over a complete valued field $\Bbbk$. Then any two norms on V are equivalent. Moreover, V is complete with respect to the metric induced by any norm.*

This is a tricky theorem to prove, so we do this in several parts. As we go, we will use the assumptions that V is finite-dimensional and that $\Bbbk$ is complete very strongly. As the examples in the previous section show, the theorem is false for infinite-dimensional normed spaces.

Take V to be a finite-dimensional vector space over $\Bbbk$ (which is assumed to be complete, of course). Fix a basis $\{v_1, v_2, \ldots, v_n\}$ of V, and let $\| \ \|_0$ be the sup-norm with respect to this basis. Finally, let $\| \ \|_1$ be any other norm on V. We want to prove that $\| \ \|_1$ is equivalent to $\| \ \|_0$, which means that we want to show that there are positive real numbers C and D such that, for every $\boldsymbol{v} \in V$, we have

$$\|\boldsymbol{v}\|_1 \le C\|\boldsymbol{v}\|_0 \qquad \text{and} \qquad \|\boldsymbol{v}\|_0 \le D\|\boldsymbol{v}\|_1.$$

The first inequality is not very hard to obtain:

Proposition 6.2.2 *Let*

$$C = n \cdot \max_{1 \le i \le n} \|\boldsymbol{v}_i\|_1.$$

Then we have, for any $\boldsymbol{v} \in V$,

$$\|\boldsymbol{v}\|_1 \le C\|\boldsymbol{v}\|_0.$$

PROOF: Take $\boldsymbol{v} \in V$, and write it in terms of the basis as

$$\boldsymbol{v} = a_1\boldsymbol{v}_1 + a_2\boldsymbol{v}_2 + \cdots + a_n\boldsymbol{v}_n.$$

Then $\|\boldsymbol{v}\|_0 = \max |a_i|$. Now just follow the path of least resistance:

$$\|\boldsymbol{v}\|_1 = \|a_1\boldsymbol{v}_1 + a_2\boldsymbol{v}_2 + \cdots + a_n\boldsymbol{v}_n\|_1$$

$$\le \|a_1\boldsymbol{v}_1\|_1 + \|a_2\boldsymbol{v}_2\|_1 + \cdots + \|a_n\boldsymbol{v}_n\|_1$$

$$= |a_1|\|\boldsymbol{v}_1\|_1 + |a_2|\|\boldsymbol{v}_2\|_1 + \cdots + |a_n|\|\boldsymbol{v}_n\|_1$$

$$\le n \max |a_i| \max \|\boldsymbol{v}_i\|_1 = C \max |a_i| = C\|\boldsymbol{v}\|_0,$$

which is exactly what we want. $\qquad\square$

The converse inequality takes a lot more proving. We will do it by induction on the dimension of V.

Proposition 6.2.3 *There exists a positive real number D such that, for every $\boldsymbol{v} \in V$, we have $\|\boldsymbol{v}\|_0 \le D\|\boldsymbol{v}\|_1$. In particular, V is complete with respect to $\| \ \|_1$.*

PROOF: (Take a deep breath. Here goes.) Notice, first of all, that once the inequality is proved, it follows that $\| \ \|_1$ is equivalent to the sup-norm, and we already know that V is complete with respect to $\| \ \|_0$, so that V will also be complete with respect to $\| \ \|_1$. In other words, once we have proved the first statement of the proposition, we will have proved the second statement too.

We will prove the inequality by induction on the dimension of V, noting first that it is trivially true for spaces of dimension 1.[3] Thus, we only need

---

[3]Prove it!

to prove the induction step: assume that the proposition is true for spaces of dimension $n - 1$, and show that it is then also true for spaces of dimension n.

Let V, then, be a space of dimension n. As above, we fix a basis

$$\{v_1, v_2, \ldots, v_n\}.$$

We want to show that there exists a number D such that

$$\|w\|_0 \leq D\|w\|_1 \qquad \text{for all } w \in V.$$

Well, suppose not. In that case, the quotient $\|w\|_1/\|w\|_0$ must get arbitrarily close to zero as w ranges through the non-zero vectors in V (because otherwise we can let $E > 0$ be a number such that the quotient is always bigger than E, and then taking $D = 1/E$ will do the trick). This means that, given any integer m we can find a vector w_m such that

$$\|w_m\|_1 < \frac{1}{m} \|w_m\|_0.$$

We want to argue that the w_m can be chosen in a particular (rather peculiar) way. Note, first, that the sup-norm $\|w_m\|_0$ is equal to the largest of the n basis coefficients. Since there are finitely many basis vectors and infinitely many ms, there must be some index i such that there are infinitely many ms for which $\|w_m\|_0$ is equal to the i-th basis coefficient. (Got it? Read it again.) After permuting the basis vectors, we can assume that $i = n$, i.e., that there are infinitely many ms such that $\|w_m\|_0 =$ the absolute value of the n-th basis coefficient. Let $m_1 < m_2 < \cdots < m_k < \ldots$ be the sequence of those ms, arranged in increasing order; we will now restrict ourselves to the corresponding sequence of vectors

$$w_{m_1}, w_{m_2}, \ldots, w_{m_k}, \ldots$$

Recall that these satisfy the inequality

$$\|w_{m_k}\|_1 < \frac{1}{m_k} \|w_{m_k}\|_0,$$

and that we've also arranged things so that if we set β_k equal to the n-th basis coefficient of w_{m_k}, then $\|w_{m_k}\|_0 = |\beta_k|$.

Now consider the vectors $\beta_k^{-1} w_{m_k}$. These have two nice properties.

i) For each of these vectors the n-th basis coefficient is 1, so that we can write

$$\beta_k^{-1} w_{m_k} = u_k + v_n,$$

with u_k belonging to the subspace $W \subset V$ spanned by the vectors $v_1, v_2, \ldots, v_{n-1}$. (We take this equation as the definition of the u_k.)

ii) We have

$$\|\mathbf{u}_k + \mathbf{v}_n\|_1 = \|\beta_k^{-1}\mathbf{w}_{m_k}\|_1 = |\beta_k|^{-1}\|\mathbf{w}_{m_k}\|_1 = \frac{\|\mathbf{w}_{m_k}\|_1}{\|\mathbf{w}_{m_k}\|_0} < \frac{1}{m_k}$$

where m_k is an infinite increasing sequence of integers.

It follows that we have constructed a sequence of vectors $\mathbf{u}_k$, all of which lie in the $(n-1)$-dimensional subspace W, and such that the norms $\|\mathbf{u}_k + \mathbf{v}_n\|_1$ (where, remember, $\mathbf{v}_n$ is the n-th vector in our chosen basis for V) tend to zero as $k \to \infty$.

Now clearly, the $\mathbf{u}_k$ form a Cauchy sequence in W, since

$$\|\mathbf{u}_{k+l} - \mathbf{u}_k\|_1 \le \|\mathbf{u}_{k+l} + \mathbf{v}_n\|_1 + \|\mathbf{u}_k + \mathbf{v}_n\|_1 < \frac{1}{m_{k+l}} + \frac{1}{m_k}.$$

By induction, we know that W is complete, so that there must be a vector $\mathbf{u} \in W$ such that $\mathbf{u}_k \to \mathbf{u}$. But then we must have

$$\|\mathbf{u} + \mathbf{v}_n\|_1 = \lim \|\mathbf{u}_k + \mathbf{v}_n\|_1 = 0,$$

which means that $\mathbf{u} = -\mathbf{v}_n$, which is a contradiction, since $\mathbf{v}_n \notin W$ by the definition of the subspace W. This contradiction shows that D must exist, and therefore proves the theorem. □

That is quite a long haul, but worth it, since it says that, as long as our vector spaces are finite-dimensional, the theory is essentially quite simple, and we might as well work with the sup-norm all the time.

There is one extra property of finite-dimensional normed spaces that is worth pointing out. This has to do with *local compactness*, which we discussed above when we showed that the p-adic fields $\mathbb{Q}_p$ were locally compact, as are $\mathbb{R}$ and $\mathbb{C}$ (see Chapter 4). In the vector space context, we have the following:

Proposition 6.2.4 *Let $\Bbbk$ be a locally compact complete valued field, and let V be a finite-dimensional (and therefore complete) normed vector space over $\Bbbk$. Then V is locally compact.*

PROOF: To show that V is locally compact, we need to find a neighborhood of the zero vector which is compact. The neighborhood we will choose will be the closed unit ball B around zero, so that

$$B = \{\mathbf{v} \in V : \|\mathbf{v}\| \le 1\}.$$

Using the main theorem, we see that we can take any norm on V (being locally compact is a topological property, and all the norms are equivalent). We choose the sup-norm with respect to some fixed basis $\{\mathbf{v}_1, \mathbf{v}_2, \ldots \mathbf{v}_n\}$. Then a vector $\mathbf{v} = a_1\mathbf{v}_1 + a_2\mathbf{v}_2 + \cdots + a_n\mathbf{v}_n$ belongs to B if and only if each a_i belongs to the closed unit ball in $\Bbbk$. This is promising, since we know that the closed unit ball in $\Bbbk$ is compact (because we are assuming that $\Bbbk$ is locally compact).

In Chapter 4, we saw that to prove that a set is compact it is enough to show that it is complete and that it is totally bounded (which means: for every positive number ε there is a finite covering of the set by balls of radius ε). The first part is done already: B is a closed subset of a complete space, and therefore is complete.

To show that B is totally bounded, we use the fact that the unit ball in $\Bbbk$ is totally bounded. Given an ε, cover the unit ball in $\Bbbk$ with a finite number, say N, of balls of radius ε. Let $c_1, c_2, \ldots c_N$ be the centers of those balls. Then consider the N^n vectors in V each of whose basis coefficients is one of the c_i. Around each of these vectors, take a ball of radius ε. We claim that these balls cover B, that is, that any vector in B belongs to at least one of them.

To see why, take a vector $\mathbf{v} = a_1\mathbf{v}_1 + a_2\mathbf{v}_2 + \cdots + a_n\mathbf{v}_n \in B$. Since this means that the coefficients a_j are in the unit ball in $\Bbbk$, we know that each a_j is within less than ε of one of the centers c_i; call this one c_{i_j}. Then $\mathbf{v}$ belongs to the ball of radius ε (remember, with respect to the sup-norm) around the vector $c_{i_1}\mathbf{v}_1 + c_{i_2}\mathbf{v}_2 + \cdots + c_{i_n}\mathbf{v}_n$, which proves what we wanted.   □

Problem 220 Draw a picture to explain the proof we just gave. It should clarify things immensely.

The converse of this proposition is also true: any locally compact normed vector space over $\Bbbk$ is of necessity finite-dimensional. This is harder to prove, however, so we leave it to the reader to puzzle it out or look it up.

In contrast to the finite-dimensional case, the theory of infinite-dimensional normed vector spaces is quite rich and complex. It is the starting point of the field called "functional analysis," which has a long and distinguished history. Given our point of view, of course, we would mostly be interested in *non-archimedean* functional analysis, which is a much younger, but still very interesting, subject. We refer the interested reader to the references. One can start with (the relevant sections of) [3], [56], and [53]. For more advanced material, see [46] , [64], [57], [50], [23].

6.3   Extending the p-adic Absolute Value

We now go on to what we are really interested in, which is considering extensions of the field $\mathbb{Q}_p$. These are simply fields K containing $\mathbb{Q}_p$. For example, if 2 is not a square in $\mathbb{Q}_p$, we might want to consider the extension $K = \mathbb{Q}_p(\sqrt{2})$. More generally, we might want to obtain K by adjoining a root of some irreducible polynomial, or even to consider a field like $\mathbb{Q}_p(X)$ (rational functions with coefficients in $\mathbb{Q}_p$). For much of this section, we will restrict ourselves to *finite* extensions (the definition follows just below).

Our main goal will be to construct a p-adic absolute value on any finite extension K of $\mathbb{Q}_p$. Of course, we will require that it agree with the p-adic absolute value for $x \in \mathbb{Q}_p$. Under that assumption, there turns out to be a *unique* way to extend the p-adic absolute value to K. The uniqueness will allow us to put together all of these p-adic absolute values to get an absolute value on the algebraic closure of $\mathbb{Q}_p$. Finally, we will show that this algebraic closure is *not* a finite extension of $\mathbb{Q}_p$, in contrast to the case of $\mathbb{R}$, whose algebraic $\mathbb{C} = \mathbb{R}(i)$ is an extension of degree two.

So let K be a field containing $\mathbb{Q}_p$. This means, among other things, that K is a vector space over $\mathbb{Q}_p$, and we say that K is a *finite* extension of $\mathbb{Q}_p$ if its dimension as a $\mathbb{Q}_p$-vector space is finite. We will write $[K : \mathbb{Q}_p] = \dim_{\mathbb{Q}_p} K$, and call this number the *degree* of K over $\mathbb{Q}_p$. We want to consider absolute values on K, but to keep things interesting we will require that these absolute values extend the p-adic absolute value on $\mathbb{Q}_p$. In other words, we are looking for a function $|\ | : K \longrightarrow \mathbb{R}_+$ which is an absolute value, and hence satisfies the usual properties:

  i) $|x| = 0$ if and only if $x = 0$,

  ii) $|xy| = |x|\,|y|$ for any $x,\, y \in K$,

  iii) $|x + y| \le |x| + |y|$ for any $x,\, y \in K$,

and that also satisfies the extra condition that

  iv) $|\lambda| = |\lambda|_p$ whenever $\lambda \in \mathbb{Q}_p$.

There are several things to note. First, any such function will be a norm on K as a $\mathbb{Q}_p$-vector space (restrict x to $\mathbb{Q}_p$ in the second property, and we have the defining properties of a norm). Second, the absolute value $|\ |$ will have to be non-archimedean, since this depends only on the absolute values of the elements of $\mathbb{Z}$, which are in $\mathbb{Q}_p$ (see Theorem 2.2.4).

We begin by showing that *if* such an absolute value exists, it must have certain properties. Later, we will use these properties to obtain a construction which shows that the extension we are looking for *does* exist.[4]

The first thing is easy:

Proposition 6.3.1 *Let K be a finite extension of $\mathbb{Q}_p$. If there exists an absolute value $|\ |$ on K extending the p-adic absolute value on $\mathbb{Q}_p$, then*

  i) K is complete with respect to $|\ |$, and

  ii) we can take the limit of a sequence in K by taking the limits of the coefficients with respect to any given basis $\{x_1, x_2, \ldots, x_n\}$ of K as a $\mathbb{Q}_p$-vector space.

---

[4]This is another standard mathematician's ruse: study the properties an object must have if it exists, and this may lead to a proof that it does exist. St. Anselm would understand.

In particular, the topology on K induced by $|\ |$ is simply the unique topology on K as a normed $\mathbb{Q}_p$-vector space, and therefore is independent of the particular choice of absolute value.

PROOF: Obvious, of course, because all norms on a finite-dimensional vector space are equivalent. The statement about convergence just says that they are equivalent to the sup-norm with respect to any given basis. □

Problem 221 Let $p = 5$. Check that 2 is not a square in $\mathbb{Q}_5$. Let $K = \mathbb{Q}_5(\sqrt{2})$. Give an example of a norm on K which is not an absolute value. Can you arrange things in your example so that the norm gives the same as the 5-adic absolute value when computed on elements of $\mathbb{Q}_5$?

A very important fact follows from the proposition.

Corollary 6.3.2 *There is at most one absolute value on K extending the p-adic absolute value on $\mathbb{Q}_p$.*

PROOF: Suppose $|\ |$ and $\|\ \|$ are two absolute values on K which extend the p-adic absolute value. We first show that they are equivalent[5] (as absolute values), and then we show that they are identical.

To show that $|\ |$ and $\|\ \|$ are equivalent, we need to show that for any $x \in K$, we have

$$|x| < 1 \iff \|x\| < 1.$$

To see this, remember that $|x| < 1$ if and only if $x^n \to 0$ with respect to the topology defined by $|\ |$, and similarly that $\|x\| < 1$ if and only if $x^n \to 0$ with respect to the topology defined by $\|\ \|$. But we already know that $|\ |$ and $\|\ \|$ are equivalent as norms on the vector space K, and hence define the same topology. Therefore, we have convergence with respect to one absolute value exactly when we have convergence with respect to the other, and this proves our claim. (Notice how seriously the field structure, rather than just the vector space structure, comes into that argument.)

This shows that $|\ |$ and $\|\ \|$ are equivalent absolute values on K; according to Proposition 3.1.3, this means that there is a positive real number α such that we have $|x| = \|x\|^\alpha$ for *every* $x \in K$. But $|x|$ and $\|x\|$ must be *equal* whenever $x \in \mathbb{Q}_p$, since both absolute values extend the p-adic absolute value; computing both at $x = p$ shows that we must have $\alpha = 1$, i.e., the two absolute values are the same. □

We know, then, that there can be at most one extension of the p-adic absolute value to K, and that K will be complete with respect to that extension. None of this establishes, however, that such an extension exists.[6] To show the existence of the absolute value, we will need to give a construction.

---

[5]The definition is at the beginning of Chapter 3; see especially Proposition 3.1.3.

[6]We do know that there are many vector space norms on K, and we can arrange for these to have the right value on elements of $\mathbb{Q}_p$, but it is not at all clear that any of these norms will be an absolute value, i.e., will work well with the multiplication in K.

One consequence of the uniqueness, however, should be noted (and will be used when we construct the absolute value). It is simply this: suppose that we have two extensions K and L, one containing the other, so that, say, $\mathbb{Q}_p \subset L \subset K$, and suppose that we have found absolute values $|\ |_L$ on L and $|\ |_K$ on K, both extending the p-adic absolute value on $\mathbb{Q}_p$. The restriction of $|\ |_K$ to elements of L is an absolute value on L which extends the p-adic absolute value; by uniqueness, it must be the same as $|\ |_L$. In other words, if $x \in L \subset K$, then

$$|x|_L = |x|_K.$$

In words, *the absolute value of x does not depend on the context.* We will use this, when defining $|x|$, in two different ways: at times we will want to work in $\mathbb{Q}_p(x)$, the smallest extension of $\mathbb{Q}_p$ containing x; at other times, we will want to work in a bigger field that may have nicer properties.

In order to be able to give the construction, we need to recall a few facts from the theory of field extensions. We assume that the reader has met these concepts before, and hence only sketch out the basic facts; for more details, see any standard text on abstract algebra or the survey in [33, Ch. 6].

So let K and F be fields, and assume that $F \subset K$ and that $[K : F]$ is finite, so that K/F is a finite field extension. Recall that we are always assuming that our fields have characteristic zero.

Let $\mathbf{C}$ be any algebraically closed field containing F (or, to be fancy, *fix an inclusion of F into such a field $\mathbf{C}$, and identify F with its image under the inclusion*). Suppose K/F is a field extension. We want to consider field homomorphisms $\sigma : K \hookrightarrow \mathbf{C}$ which induce the identity (or, if we're being fancy, our fixed inclusion) on F. (Recall that any such homomorphism is necessarily injective.)[7] We will say the field extension K/F is *normal* if all such $\sigma : K \hookrightarrow \mathbf{C}$ have the same image. Another way to say this is to identify K with one of its images, and then say that K/F is normal if every σ maps K to itself. If K/F is normal, then we can think of σ as an automorphism $K \longrightarrow K$ which induces the identity on F. To make a picture, any such σ fits into a diagram like this:

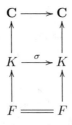

where the vertical arrows are inclusions.

---

[7]Recall that a field homomorphism is a mapping that (i) sends 1 to 1, and (ii) works well with the field operations, so that $\sigma(x + y) = \sigma(x) + \sigma(y)$ and $\sigma(xy) = \sigma(x)\sigma(y)$. It is a nice exercise, especially recommended to the reader who is unsure of his footing at this point, to show that such a function must always be injective.

When K/F is normal, it is clear that the choice of $\mathbf{C}$ doesn't much matter, since any σ maps K to itself anyway. We call a map $\sigma : K \longrightarrow K$ which induces the identity on F an *automorphism of the extension* K/F. It is known that when K/F is finite and normal (and of characteristic zero)[8] the automorphisms of K/F form a finite group[9] whose order is equal to the degree $[K : F]$ (this group is called the *Galois group* of the field extension). For a summary of these results (and a lot more details), see [33, §6.6.7].

The following problems give a few examples.

Problem 222 Let $F = \mathbb{Q}$ and $K = \mathbb{Q}(i)$, where, as usual, $i^2 = -1$. Show that this extension is normal. In fact, show that any extension that is obtained by adding to F the square root of some element will be normal. (Hint: there are only two possibilities for σ.)

Problem 223 Let $F = \mathbb{Q}$ and $K = \mathbb{Q}(\sqrt[3]{2})$. Show that K/F is not a normal extension by taking $\mathbf{C} = \mathbb{C}$, considering K as a subfield of $\mathbb{R}$ (and hence of $\mathbb{C}$) in the obvious way, and noting that any $\sigma : K \longrightarrow \mathbb{C}$ must map $\sqrt[3]{2}$ to a cube root of 2. What are the choices?

Problem 224 Let $F = \mathbb{Q}$ and $K = \mathbb{Q}(\sqrt[3]{2}, \zeta)$, where

$$\zeta = \frac{-1 + i\sqrt{3}}{2}$$

is a cube root of 1. Show that K/F is a normal extension. Show also that K is the smallest normal extension of $\mathbb{Q}$ containing $\sqrt[3]{2}$.

Normal extensions are very nice, and it is comforting (and useful) to know that the process suggested above in the case of $\mathbb{Q}(\sqrt[3]{2})$ and $\mathbb{Q}(\sqrt[3]{2}, \zeta)$ works in general: given any finite extension K/F, there exists a finite *normal* extension of F containing K. The smallest such is called the *normal closure* of K/F. This will be useful in what follows, because it means that to construct the absolute value of an element $x \in K$ we might as well assume that K is a normal extension (otherwise just replace K by its normal closure, since the absolute value does not depend on the context).

The crucial fact that we will need is that there exists a function

$$\mathbf{N}_{K/F} : K \longrightarrow F,$$

which is called the *norm from K to F*. (It is a bit unfortunate that this "norm" has the same name as the vector space "norm," but both terms have been standard for such a long time that there is no chance of ever changing them. Watch out for the context to avoid confusion.) This will be useful because it gives a natural way to "go down" from elements of the bigger field K to elements of F.

The norm function can be defined in several ways, each useful in certain contexts; here are three:

---

[8]In characteristic p, an extra condition, called "separability," is needed.
[9]It is easy to prove that they form a group.

i) Take $\alpha \in K$, think of K as a finite-dimensional F-vector space, and consider the F-linear map from K to K given by multiplication by α. Since this is linear, it corresponds to a matrix. Then we define $\mathbf{N}_{K/F}(\alpha)$ to be the determinant of this matrix.

ii) Take $\alpha \in K$, and consider the subextension $F(\alpha)$, i.e., the smallest field containing both F and α (this is clearly a subfield of K). Set $r = [K : F(\alpha)]$ to be the degree of K as an extension of $F(\alpha)$. Let

$$f(X) = X^n + a_{n-1}X^{n-1} + \cdots + a_1 X + a_0 \in F[X]$$

be the minimal polynomial of α over F, that is, the lowest degree monic polynomial with coefficients in F such that $f(\alpha) = 0$. Then we define $\mathbf{N}_{K/F}(\alpha) = (-1)^{nr} a_0^r$.

iii) Suppose the extension K/F is *normal*. Then we can define $\mathbf{N}_{K/F}(\alpha)$ to be the product of all the $\sigma(\alpha)$, where σ runs through the (finite) set of all the automorphisms of K/F.

Before we discuss why these definitions are equivalent, we note some useful facts. First, if $\alpha \in F$ (rather than in the bigger field K), then $\mathbf{N}_{K/F}(\alpha) = \alpha^n$, where $n = [K : F]$ is the degree of the extension. (This is essentially obvious from any of the definitions—check!) Next, norms are multiplicative. This is probably easiest to see from the first definition, since determinants are multiplicative, but it's pretty obvious from the last one, too. (Less so for the middle definition—can you give a direct proof using that version?) In any case, we will need to know that

$$\mathbf{N}_{K/F}(\alpha\beta) = \mathbf{N}_{K/F}(\alpha)\mathbf{N}_{K/F}(\beta)$$

for any α, $\beta \in K$. Notice, by contrast, that the norm of a sum has no clear relation to the norms of the summands.

The equivalence of these definitions is not hard to prove; we suggest that the reader who has not seen it proved work through the next few exercises.

Problem 225 Prove the equivalence of the first two definitions when $K = F(\alpha)$ by considering the basis of K which consists of $\{1, \alpha, \alpha^2, \ldots, \alpha^{n-1}\}$.

Problem 226 Now suppose $K \neq F(\alpha)$. Let $n = [F(\alpha) : F]$ and let $r = [F : K(\alpha)]$. Then we can find a basis for F over K of the form $\alpha^i b_j$, where $i = 0, 1, \ldots, n-1$ and $b_1, b_2, \ldots b_r$ are a basis of K over $F(\alpha)$. Using this basis, show that the first and second definitions agree in this case as well.

Problem 227 Suppose K/F is normal and that $K = F(\alpha)$. Show that the images $\sigma(\alpha)$ as σ runs through the automorphisms of K/F are exactly the roots of the polynomial $f(X)$. (It's easy to see that any $\sigma(\alpha)$ is a root—just compute $\sigma(f(\alpha))$—but it's less clear that for each root there is a unique σ for which $\sigma(\alpha)$ is equal to that root.) Conclude that the second and third definitions are equivalent in this case.

Problem 228 Finish off the proof that all three definitions give the same answer. (One loose end to consider is the case where K/F is normal, but K is not equal to $F(\alpha)$. What then?)

Problem 229 Suppose K/F is *not* normal. Can you give a version of the third definition that makes sense?

Problem 230 Show that if we have three fields $F \subset L \subset K$, then, for any $\alpha \in K$, we have

$$\mathbf{N}_{L/F}\big(\mathbf{N}_{K/L}(\alpha)\big) = \mathbf{N}_{K/F}(\alpha).$$

After all that theory, we need some concrete examples to keep us afloat. Let's take a really easy one, and put $F = \mathbb{Q}_5$, $K = \mathbb{Q}_5(\sqrt{2})$. Take a generic element $a + b\sqrt{2} \in K$; let's compute its norm using all three definitions:

i) A basis for K over $\mathbb{Q}_5$ is $\{1, \sqrt{2}\}$. The linear map "multiplication by $a + b\sqrt{2}$" maps 1 to $a + b\sqrt{2}$ and $\sqrt{2}$ to $2b + a\sqrt{2}$, so its matrix with respect to our basis is

$$\begin{bmatrix} a & 2b \\ b & a \end{bmatrix},$$

which has determinant $a^2 - 2b^2$. Therefore, $\mathbf{N}_{K/F}(a + b\sqrt{2}) = a^2 - 2b^2$.

ii) We will have $r = 1$ unless $b = 0$, in which case $r = 2$. If $b = 0$, we have $\alpha = a$, whose minimal polynomial is $X - a$, and the norm is then $(-1)^2(-a)^2 = a^2$. If $b \neq 0$, we must work out the minimal polynomial; it must be of degree two. Since $(a + b\sqrt{2})^2 = a^2 + 2b^2 + 2ab\sqrt{2}$, we will get zero by combining as follows:

$$(a + b\sqrt{2})^2 - 2a(a + b\sqrt{2}) + (a^2 - 2b^2) = 0.$$

(Can you see how that was found?) Hence, the minimal polynomial is

$$X^2 - 2aX + (a^2 - 2b^2),$$

and the norm is $a^2 - 2b^2$. Thus, whether b is zero or not, we have $\mathbf{N}_{K/F}(a + b\sqrt{2}) = a^2 - 2b^2$.

iii) Finally, we have two automorphisms: the identity, and

$$\sigma : a + b\sqrt{2} \mapsto a - b\sqrt{2}.$$

The product of the images of $a + b\sqrt{2}$ is

$$(a + b\sqrt{2})(a - b\sqrt{2}) = a^2 - 2b^2,$$

so that once again we have $\mathbf{N}_{K/F}(a + b\sqrt{2}) = a^2 - 2b^2$.

The general case is nowhere near as easy, of course. Here are a few more relatively simple examples:

Problem 231 Do the same for

i) a general quadratic extension $\mathbb{Q}_p(\sqrt{n})$,

ii) some specific elements of the extension $\mathbb{Q}(\sqrt[3]{2}, \zeta)$ with $\zeta = (-1+i\sqrt{3})/2$ (notice that this is an extension of degree 6; working with the general element wouldn't be too pleasant).

To see why the norm is going to play a central role, notice the following. Suppose $K/\mathbb{Q}_p$ is a normal extension, and let σ be an automorphism. Let $| \ |$ be an absolute value on K. Then the function $x \mapsto |\sigma(x)|$ is also an absolute value on K (check!), and also gives the p-adic absolute value over $\mathbb{Q}_p$, since σ induces the identity on $\mathbb{Q}_p$. But we have shown that there is *only one* such absolute value! Thus, we must have $|\sigma(x)| = |x|$ for any $x \in K$. Multiplying over all the σ's (and remembering that there are exactly $n = [K : \mathbb{Q}_p]$ of them) we get that

$$\left| \prod_\sigma \sigma(x) \right| = |x|^n.$$

Now, since the product is equal to the norm, this translates to

$$|x|^n = |\mathbf{N}_{K/\mathbb{Q}_p}(x)|,$$

or, taking the root,

$$|x| = \sqrt[n]{|\mathbf{N}_{K/\mathbb{Q}_p}(x)|}.$$

But this last gives a formula which we can compute just from the knowledge of the p-adic absolute value, since the norm is an element of $\mathbb{Q}_p$!

So far, that only works for normal extensions, but note the following:

Lemma 6.3.3 *Let L and K be finite extensions of $\mathbb{Q}_p$ which form a tower:* $\mathbb{Q}_p \subset L \subset K$. *Let $x \in L$. Set $m = [L : \mathbb{Q}_p]$ and $n = [K : \mathbb{Q}_p]$. Then*

$$\sqrt[m]{|\mathbf{N}_{L/\mathbb{Q}_p}(x)|_p} = \sqrt[n]{|\mathbf{N}_{K/\mathbb{Q}_p}(x)|_p}.$$

PROOF: We have

$$\mathbf{N}_{K/F}(x) = \mathbf{N}_{L/\mathbb{Q}_p}(\mathbf{N}_{K/L}(x)),$$

and $\mathbf{N}_{K/L}(x) = x^{[K:L]}$. Remembering that $[K : \mathbb{Q}_p] = [K : L][L : \mathbb{Q}_p]$ and plugging everything into the formulas gives the equality. $\square$

This is very nice, since it says that the value of $\sqrt[n]{|\mathbf{N}_{K/\mathbb{Q}_p}(x)|_p}$ is the same for any field K containing x (as above, $n = [K : \mathbb{Q}_p]$). In particular, this shows that it must be equal to the absolute value of x also when the extension is not normal (pass to the normal closure!). In other words, we have proved the following:

Proposition 6.3.4 *If there is an absolute value on K extending the p-adic absolute value, then it must be given by the formula*

$$|x| = \sqrt[n]{|\mathbf{N}_{K/\mathbb{Q}_p}(x)|_p},$$

where $n = [K : \mathbb{Q}_p]$ is the degree of the extension.

Notice that the fact that the value of our formula does not depend on the choice of field containing x matches exactly the fact that the same is true of the absolute value (if it exists). This is encouraging, since we want to prove that the formula does define an absolute value. We are now, finally, in a position to prove that it does. As we will see, the hard part is showing the non-archimedean inequality, for which we will use Lemma 2.2.3.

Theorem 6.3.5 *Let $K/\mathbb{Q}_p$ be a finite extension of degree n. The function $|\ | : K \longrightarrow \mathbb{R}_+$ defined by*

$$|x| = \sqrt[n]{|\mathbf{N}_{K/\mathbb{Q}_p}(x)|_p}$$

is a non-archimedean absolute value on K which extends the p-adic absolute value on $\mathbb{Q}_p$.

PROOF: Several things are immediate. First, $|x| = 0$ will only happen if $\mathbf{N}_{K/\mathbb{Q}_p}(x) = 0$, which (using the first definition of the norm) will only happen if multiplication by x is not invertible; since K is a field, that only happens if $x = 0$. Next, since $\mathbf{N}_{K/\mathbb{Q}_p}(xy) = \mathbf{N}_{K/\mathbb{Q}_p}(x)\mathbf{N}_{K/\mathbb{Q}_p}(y)$, we will certainly have $|xy| = |x|\,|y|$. Finally, if $x \in \mathbb{Q}_p$ then $\mathbf{N}_{K/\mathbb{Q}_p}(x) = x^n$, so that (since absolute values are positive real numbers) $|x| = \sqrt[n]{|x|_p^n} = |x|_p$.

We will finish the proof by showing that condition (*iii*) from Lemma 2.2.3 holds. Looking at the definition, we see that $|x| \leq 1$ will happen exactly when $|\mathbf{N}_{K/\mathbb{Q}_p}(x)|_p \leq 1$. Hence, what we need to show is that

$$|\mathbf{N}_{K/\mathbb{Q}_p}(x)|_p \leq 1 \Longrightarrow |\mathbf{N}_{K/\mathbb{Q}_p}(x - 1)|_p \leq 1,$$

or, in more algebraic terms, that

$$\mathbf{N}_{K/\mathbb{Q}_p}(x) \in \mathbb{Z}_p \Longrightarrow \mathbf{N}_{K/\mathbb{Q}_p}(x - 1) \in \mathbb{Z}_p.$$

We will do this by using the definition of the norm in terms of the minimal polynomial.

By Lemma 6.3.3, we may assume that $K = \mathbb{Q}_p(x)$ is the smallest field containing x (and note that we will always have $\mathbb{Q}_p(x) = \mathbb{Q}_p(x - 1)$, since any field containing x will also contain $x - 1$ and vice versa). Let

$$f(X) = X^n + a_{n-1}X^{n-1} + \cdots + a_1 X + a_0$$

be the minimal polynomial for x. Since the minimal polynomial for $x - 1$ must have the same degree and has $x - 1$ as a root, it must be

$$f(X + 1) = (X + 1)^n + a_{n-1}(X + 1)^{n-1} + \cdots + a_1(X + 1) + a_0$$
$$= X^n + (a_{n-1} + n)X^{n-1} \cdots + (1 + a_{n-1} + \cdots + a_1 + a_0).$$

(We only need the last coefficient, of course: each $(X + 1)^k$ has constant term 1, and it is multiplied by a_k.) Thus, using the second definition for the norm, we have

$$\mathbf{N}_{K/\mathbb{Q}_p}(x) = (-1)^n a_0$$

and

$$\mathbf{N}_{K/\mathbb{Q}_p}(x - 1) = (-1)^n (1 + a_{n-1} + \cdots + a_1 + a_0).$$

What we want to prove will follow, then, from the assertion that if

$$f(X) = X^n + a_{n-1}X^{n-1} + \cdots + a_1 X + a_0$$

is an irreducible polynomial and $a_0 \in \mathbb{Z}_p$, then we have

$$1 + a_{n-1} + \cdots + a_1 + a_0 \in \mathbb{Z}_p.$$

In fact, we will prove something that is even better.

Lemma 6.3.6 *If* $f(X) = X^n + a_{n-1}X^{n-1} + \cdots + a_1 X + a_0$ *is a monic irreducible polynomial with coefficients in* $\mathbb{Q}_p$ *and* $a_0 \in \mathbb{Z}_p$, *then all of the coefficients* $a_{n-1}, \ldots, a_1, a_0$ *belong to* $\mathbb{Z}_p$.

PROOF OF THE LEMMA: This is the crux of the matter, and we follow the proof given by Neukirch in [25, Ch. 6, §4]. We will use Hensel's Lemma for polynomials, Theorem 4.7.2, to show that if some of the coefficients are not in $\mathbb{Z}_p$ then $f(X)$ will be reducible.

So let $f(X) = X^n + a_{n-1}X^{n-1} + \cdots + a_1 X + a_0$, and assume that $a_0 \in \mathbb{Z}_p$ but some $a_j \notin \mathbb{Z}_p$. Choose m to be the smallest exponent such that $p^m a_i \in \mathbb{Z}_p$ for every i, and "clear denominators" by multiplying the whole polynomial by p^m. Set

$$g(X) = p^m f(X) = b_n X^n + b_{n-1}X^{n-1} + \cdots + b_1 X + b_0,$$

so that $b_i = p^m a_i$. Since $f(X)$ is monic, $b_n = p^m$ is divisible by p; since $a_0 \in \mathbb{Z}_p$ (our main hypothesis), $b_0 = p^m a_0$ is also divisible by p; by our choice of m, all the b_i are in $\mathbb{Z}_p$, and at least one is not divisible by p. Let k be the smallest i such that b_i is not divisible by p. Then $0 < k < n$, $b_0, b_1, \ldots, b_{k-1}$ are all divisible by p and b_k is not, so we have a factorization

$$g(X) \equiv (b_n X^{n-k} + \cdots + b_k) X^k \pmod{p}.$$

The two factors are clearly relatively prime modulo p. By the second form of Hensel's Lemma, it follows that $g(X) = p^m f(X)$ is reducible, and therefore so is $f(X)$ itself. This contradiction proves the lemma, and therefore also the theorem. □

An alternative proof that the extended absolute value satisfies the ultra-metric inequality is given in [53, II.3.4]. Our proof is very algebraic, using polynomials and Hensel's Lemma; Robert's uses, instead, the fact that K is locally compact.

So now we have the extension we needed. In other words, given any finite extension K of $\mathbb{Q}_p$, we have shown that there exists a unique absolute value on K which extends the p-adic absolute value on $\mathbb{Q}_p$; we call it, of course, the p-adic absolute value on K. We know that K is complete with respect to this absolute value.

To complete this section, we go on to consider an algebraic closure of $\mathbb{Q}_p$. This is a field $\overline{\mathbb{Q}}_p$ which contains all the roots of all the polynomials with coefficients in $\mathbb{Q}_p$. To construct it, we just take the union of all the finite extensions of $\mathbb{Q}_p$ (and then we prove that this is an algebraically closed field).

We claim that we have already constructed an absolute value on the algebraic closure. The point is this: given any $x \in \overline{\mathbb{Q}}_p$, the extension $\mathbb{Q}_p(x)$ is finite (its degree is the degree of the minimal polynomial of x over $\mathbb{Q}_p$). Since x then lives in the finite extension $\mathbb{Q}_p(x)$, we can define $|x|$ by using the unique extension of the p-adic absolute value to $\mathbb{Q}_p(x)$. But we already know that this absolute value does not depend on the field we take it in; in other words, it just depends on x itself (as the root of some polynomial over $\mathbb{Q}_p$). Thus, it makes sense to say it is the absolute value of the element $x \in \overline{\mathbb{Q}}_p$. This shows that we have actually defined a function

$$| \, | : \overline{\mathbb{Q}}_p \longrightarrow \mathbb{R}_+$$

which extends the p-adic absolute value, and it is easy to see that this function is an absolute value. Our construction, then, shows that there is a unique p-adic absolute value on $\overline{\mathbb{Q}}_p$.

Problem 232 Prove that the function we have defined is an absolute value, i.e., that it satisfies the three conditions listed in the beginning of this section.

It is not clear (in fact, it is not true) that $\overline{\mathbb{Q}}_p$ is complete with respect to this absolute value, because $\overline{\mathbb{Q}}_p$ is an *infinite* extension of $\mathbb{Q}_p$. Proving this will take knowing a lot more about the absolute value on $\overline{\mathbb{Q}}_p$. For now, we will content ourselves to showing that $\overline{\mathbb{Q}}_p$ is indeed an infinite extension of $\mathbb{Q}_p$. To do this, it is enough to show that there are irreducible polynomials of arbitrarily large degree over $\mathbb{Q}_p$. Since the root of an irreducible polynomial of degree n generates an extension of degree n, this means that $\overline{\mathbb{Q}}_p$ contains extensions of degree n for every n, and hence is not a finite extension. We conclude this section by showing that this is in fact the case. We first need to prove "Gauss's Lemma" in our context.

Lemma 6.3.7 (Gauss's Lemma in $\mathbb{Q}_p$) *Suppose that $f(X) \in \mathbb{Z}_p[X]$ factors (in a non-trivial way) in $\mathbb{Q}_p[X]$, so that*

$$f(X) = g(X)h(X)$$

with $g(X), h(X) \in \mathbb{Q}_p[X]$ and non-constant. Then there exist non-constant polynomials $g_0(X), h_0(X) \in \mathbb{Z}_p[X]$ such that $f(X) = g_0(X)h_0(X)$.

PROOF: If $k(X) = a_n X^n + \cdots + a_1 X + a_0 \in \mathbb{Q}_p[X]$ is any polynomial, define

$$w(k(X)) = \min_{0 \le i \le n} v_p(a_i).$$

This is a kind of p-adic valuation on polynomials, since $w(k(X))$ is the largest power of p that divides *all* the coefficients of $k(X)$. It is easy to see that for $a \in \mathbb{Q}_p$ we have $w(a \, k(X)) = v_p(a) + w(k(x))$. Also, it is clear that $k(X) \in \mathbb{Z}_p[X]$ if and only if $w(k(X)) \ge 0$. We will use the "valuation" w to prove the lemma.

Step 1: If the lemma is true for the case when $w(f(X)) = 0$, then it is true in general.

Proof of Step 1: Since $f(X) \in \mathbb{Z}_p[X]$, we know that $w(f(X)) \ge 0$. If $w(f(X)) > 0$ then let a be a coefficient with the smallest valuation (i.e., one that is least divisible by p). By the definition of w we have $w(f(X)) = v_p(a)$; since we know $f(X) \in \mathbb{Z}_p[X]$, we know that $a \in \mathbb{Z}_p$. Then it is clear that $w(a^{-1} f(X)) = 0$; set $\tilde{f}(X) = a^{-1} f(X)$ (and in particular $\tilde{f}(X) \in \mathbb{Z}_p[X]$). If we set, say, $\tilde{g}(X) = a^{-1} g(X)$, we now have $\tilde{f}(X) = \tilde{g}(X)h(X)$ and $w(\tilde{f}(X)) = 0$.

Since we are assuming that the theorem is true when $w(\tilde{f}(X)) = 0$, we can decompose $\tilde{f}(X)$ as a product of two polynomials in $\mathbb{Z}_p[X]$, say, $\tilde{f}(X) = G_0(X) H_0(X)$. Then we have

$$f(X) = a\tilde{f}(X) = aG_0(X) H_0(X),$$

and, since we know $a \in \mathbb{Z}_p$, this decomposition is in $\mathbb{Z}_p[x]$: just absorb the a into one of the factors by putting $g_0(X) = aG_0(X)$ and $h_0(X) = H_0(X)$, and we get $f(X) = g_0(X) h_0(X)$ as desired.

This proves Step 1. In other words, *we may assume, without loss, that $w(f(X)) = 0$*, i.e., that at least one coefficient of $f(X)$ is a p-adic unit.

Step 2: The lemma is indeed true when $w(f(X)) = 0$.

Proof of Step 2: Assume, then, that $w(f(X)) = 0$. Using the same reasoning as above, we can find $b \in \mathbb{Q}_p$ such that $w(b \, g(X)) = 0$ and $c \in \mathbb{Q}_p$ such that $w(c \, h(X)) = 0$. If we write $g_1(X) = b \, g(X)$ and $h_1(X) = c \, h(X)$, then we can write

$$f_1(X) = bc \, f(X) = g_1(X)h_1(X).$$

Write $\bar{k}(X) \in \mathbb{F}_p[X]$ for the reduction modulo p of a polynomial $k(X) \in \mathbb{Z}_p[X]$. We have set things up so that $\bar{g}_1(X)$ and $\bar{h}_1(X)$ are both non-zero; it follows that $\bar{f}_1(X)$ is also non-zero, and hence that $w(f_1(X)) = w(bc\,f(X)) = 0$. Since we had already arranged things so that $w(f(X)) = 0$, it follows that $v_p(bc) = 0$, so that bc is a p-adic unit. Then we have

$$f(X) = (bc)^{-1}f_1(X) = (bc)^{-1}g_1(X) \cdot h_1(X).$$

Taking $g_0(X) = (bc)^{-1}g_1(X)$ and $h_0(X) = h_1(X)$ then gives the desired factorization. □

When everything is assumed monic, the lemma is even easier:

Problem 233 Suppose that $f(X) \in \mathbb{Z}_p[X]$ is monic and factors as a product $f(X) = g(X)\,h(X)$, with $g(X)$ and $h(X) \in \mathbb{Q}_p[X]$ and monic. Show that then $g(X)$ and $h(X)$ must be in $\mathbb{Z}_p[X]$. (Hint: the main difference between this and the lemma is that we are assuming that the factors are monic.)

In particular, we get the following:

Corollary 6.3.8 *Let $f(X) \in \mathbb{Z}_p[X]$ be a monic polynomial whose reduction modulo p is irreducible in $\mathbb{F}_p[X]$. Then $f(X)$ is irreducible over $\mathbb{Q}_p$.*

PROOF: If $f(X)$ factors over $\mathbb{Q}_p$, then it factors over $\mathbb{Z}_p$ by Gauss's Lemma; reducing the factorization modulo p gives a factorization over $\mathbb{F}_p$, which cannot exist. □

Problem 234 That was pretty quick; fill in the details. For example, how do we know that the factorization modulo p is non-trivial?

Problem 235 Is the assumption that $f(X)$ is monic really necessary?

Notice that this corollary has a kind of converse in the "second form" of Hensel's Lemma (Theorem 4.7.2), which says that, under certain conditions, factorizations over $\mathbb{F}_p$ lift to factorizations over $\mathbb{Z}_p$.

It is well known that there are many irreducible polynomials in $\mathbb{F}_p[X]$. In fact, for every n one can show[10] that there is an irreducible polynomial of degree n in $\mathbb{F}_p[X]$ whose roots generate the unique extension of degree n of $\mathbb{F}_p$. Choosing any lift of such a polynomial to a monic polynomial in $\mathbb{Z}_p[X]$ gives an irreducible polynomial of degree n in $\mathbb{Q}_p[X]$. Adjoining a root of this polynomial then gives an extension of $\mathbb{Q}_p$ of degree n, which in some sense "comes from" the extension of $\mathbb{F}_p$. So we have proved that

---

[10]This is one of the few facts about finite fields that we will need to make use of in this chapter. Most of them are easily proved—see any introductory book on abstract algebra for the details. What we are using here is the fact that, for each $n \geq 1$, the finite field $\mathbb{F}_p$ has a unique extension of degree n (up to isomorphism). This extension is a field with p^n elements. It is usually denoted by $\mathbb{F}_{p^n}$, and, since it is a separable extension, there exists a polynomial $f(X)$ such that $\mathbb{F}_{p^n}$ is obtained by adjoining a root of $f(X)$ to $\mathbb{F}_p$.

Corollary 6.3.9 *For each integer $n \geq 1$ there is an extension of $\mathbb{Q}_p$ which has degree exactly n and which "comes from" the unique extension of degree n of the finite field $\mathbb{F}_p$.*

In particular,

Corollary 6.3.10 *The algebraic closure $\overline{\mathbb{Q}}_p$ is an infinite extension of $\mathbb{Q}_p$.*

We should note the contrast, at this point, between $\mathbb{R}$ and $\mathbb{Q}_p$. The algebraic closure of $\mathbb{R}$ is $\mathbb{C}$, which is an extension of degree two, and is therefore complete with respect to the ∞-adic absolute value. This is a point, then, at which the p-adic and the classical theories diverge quite sharply.

Before we consider the algebraic closure in more detail, we need a better grasp of the properties of finite extensions of $\mathbb{Q}_p$. That is the point of the next section. Before we delve in, however, we prove one final result about polynomials that gives us still more finite extensions of $\mathbb{Q}_p$.

Proposition 6.3.11 (Eisenstein Irreducibility Criterion) *Let*

$$f(X) = a_n X^n + \cdots + a_1 X + a_0 \in \mathbb{Z}_p[X]$$

be a polynomial satisfying the conditions

i) $|a_n| = 1$,

ii) $|a_i| < 1$ for $0 \leq i < n$, and

iii) $|a_0| = 1/p$.

Then $f(X)$ is irreducible over $\mathbb{Q}_p$.

PROOF: Suppose $f(X)$ is reducible in $\mathbb{Q}_p[X]$. By the Lemma, it is then reducible over $\mathbb{Z}_p$, i.e., there exist $g(X)$, $h(X) \in \mathbb{Z}_p[X]$ such that

$$f(X) = g(X)\, h(X).$$

Write

$$g(X) = b_r X^r + \cdots + b_1 X + b_0$$

and

$$h(X) = c_s X^s + \cdots + c_1 X + c_0,$$

with $r + s = n$; since $|a_n| = 1$ and $a_n = b_r c_s$, we must have $|b_r| = |c_s| = 1$. As above, using bars to denote reduction modulo p, we have $\bar{f}(X) = \bar{g}(X)\,\bar{h}(X)$. On the other hand, the hypotheses imply that $\bar{f}(X) = \bar{a}_n X^n$. Then we must have $\bar{g}(X) = \bar{b}_r X^r$ and $\bar{h}(X) = \bar{c}_s X^s$. In particular, both b_0 and c_0 must be divisible by p. But then $a_0 = b_0 c_0$ will be divisible by p^2, so that $|a_0| \leq 1/p^2$, contradicting our third assumption. This shows that $f(X)$ must be irreducible. $\qquad\square$

The reader will note that the irreducible polynomials furnished by the Eisenstein criterion (we might call them *Eisenstein polynomials*) are certainly reducible modulo p (*very* reducible: modulo p, they look essentially like a power of X). In other words, the irreducible polynomials provided by this criterion are very different from the ones we found before. So what we have here is *another* infinite family of finite extensions of $\mathbb{Q}_p$.

Problem 236 Given that Eisenstein polynomials factor modulo p, why can't we use Hensel's Lemma to factor them in $\mathbb{Z}_p$?

To conclude this section, here are a few more problems about polynomials:

Problem 237 Is the function w defined in the proof of Gauss's Lemma above a valuation? (Hint: the difficult bit is to show that $w(f(X)g(X)) = w(f(X))+w(g(X))$. Notice that the proof of the Lemma would be greatly simplified if we could use this identity.)

Problem 238 (This needs Galois theory.) In the situation of Corollary 6.3.9, show that the extension of $\mathbb{Q}_p$ is normal, and that its Galois group is isomorphic to the Galois group of the corresponding extension of $\mathbb{F}_p$.

Problem 239 Use Lemma 6.3.7 above to show "Gauss's Lemma," which says that if a polynomial $f(X) \in \mathbb{Z}[X]$ factors over $\mathbb{Q}$, then it factors over $\mathbb{Z}$. (This may be taken as another example of how to use "local"—i.e., p-adic—methods to prove "global" results.)

Problem 240 Can the Eisenstein criterion also be turned into a "global" result? In other words, does it give us a way to determine irreducibility over $\mathbb{Q}$?

Problem 241 Does Gauss's Lemma extend to polynomials in several variables? If so, does Problem 239 also extend to that case?

6.4 Finite Extensions of $\mathbb{Q}_p$

The point of this section is to gather information about finite extensions of $\mathbb{Q}_p$. On one level, what we want to say is that much of the structure we have found in $\mathbb{Q}_p$ extends without effort. Our main interest, however, is to see what information this gives us about finite extensions of $\mathbb{Q}_p$.

To help us understand, we will keep a few standard examples in mind as we go along; at each step, we will consider (usually in a problem) how the result that has just been proved looks in the particular case of our examples.

Here are the three examples:

i) Let $p = 5$; we have checked that 2 is not a square in $\mathbb{Q}_5$, so we let $F_1 = \mathbb{Q}_5(\sqrt{2})$. This is an extension of degree 2, with basis $\{1, \sqrt{2}\}$.

ii) Again, let $p = 5$. It is clear that 5 itself is not a square in $\mathbb{Q}_5$. We let $F_2 = \mathbb{Q}_5(\sqrt{5})$. This is also an extension of degree 2, with basis $\{1, \sqrt{5}\}$.

iii) Our third example is more complicated. We let $p = 3$. We adjoin to $\mathbb{Q}_3$ a cube root of unity and a square root of 2: $F_3 = \mathbb{Q}_3(\zeta, \sqrt{2})$, where $\zeta^3 = 1$ but $\zeta \neq 1$. F_3 is an extension of $\mathbb{Q}_3$ of degree 4; both $\mathbb{Q}_3(\zeta)$ and $\mathbb{Q}_3(\sqrt{2})$ are subextensions of degree 2.

Throughout this section, K will be a finite extension of degree n of $\mathbb{Q}_p$, and we will write $|\ | = |\ |_p$ for the p-adic absolute value (extended to K as above). We already know that the absolute value makes K a locally compact topological field, that K is complete with respect to its absolute value, and that the absolute value on K is given by the formula

$$|x| = \sqrt[n]{|\mathbf{N}_{K/\mathbb{Q}_p}(x)|_p}.$$

Our next step is to show that this absolute value is discrete. Recall that in $\mathbb{Q}_p$, the absolute value of any non-zero element was always of the form p^v, with v an integer; in fact, this is what allowed us to define the p-adic valuation v_p. Looking at the formula for the absolute value on K, we immediately see that the absolute value of any non-zero $x \in K$ is of the form p^v, where $v \in \frac{1}{n}\mathbb{Z}$, since it is the n-th root of the absolute value of some element of $\mathbb{Q}_p$. This spurs us on to define:

Definition 6.4.1 *Let K be a finite extension of $\mathbb{Q}_p$, and let $|\ |$ be the p-adic absolute value on K. For any $x \in K$, $x \neq 0$, we define the p-adic valuation $v_p(x)$ to be the unique rational number satisfying*

$$|x| = p^{-v_p(x)}.$$

We extend the definition formally by setting $v_p(0) = +\infty$.

It is easy to see that v_p is a valuation, in the sense we defined:

i) $v_p(x + y) \geq \min\{v_p(x), v_p(y)\}$, and

ii) $v_p(xy) = v_p(x) + v_p(y)$.

As before, we use the standard conventions about how to interpret these equations when one of x, y, or $x + y$ is zero.

It is useful to notice that since we know exactly how to compute the p-adic absolute value of an element of K, we also know how to compute v_p. Here is the formula: for any $x \in K^{\times}$,

$$v_p(x) = \frac{1}{n} v_p\big(\mathbf{N}_{K/\mathbb{Q}_p}(x)\big).$$

This reduces computing v_p to computing norms.

Problem 242 Let $x = 1 + 3\sqrt{2} \in F_1$. Compute $v_5(x)$. Do the same for $x = \sqrt{2}$, $x = 1 + 5\sqrt{2}$, and $x = 5\sqrt{2}$. (Hint: the easiest way is probably to consider the images under automorphisms to compute the norm, and then use the basic formula.)

Problem 243 Let $x = 4 + \sqrt{5} \in F_2$. Compute $v_5(x)$. Do the same for $x = \sqrt{5}$, $x = 5 + \sqrt{5}$, $x = 10 - 3\sqrt{5}$, and $x = 1 + \sqrt{5}$.

Problem 244 Let $x = \sqrt{2} \in F_3$. Compute $v_3(x)$. Do the same for $x = \zeta$, $x = 1 - \zeta$ (be careful!), $x = 10 - 3\sqrt{2}$, and $x = 2 + 3\zeta$. (This problem is a little harder than the previous two problems.)

We know that the image of v_p is contained in $\frac{1}{n}\mathbb{Z}$ (in fact, that is obvious from the formula above). But we do not yet know exactly what it is. The next result tells us what *kind* of subset of $\mathbb{Q}$ it is.

Proposition 6.4.2 *The p-adic valuation v_p is a homomorphism from the multiplicative group $K^\times$ to the additive group $\mathbb{Q}$. Its image is of the form $\frac{1}{e}\mathbb{Z}$, where e is a divisor of $n = [K : \mathbb{Q}_p]$.*

PROOF: That v_p is a homomorphism is just property (ii) above; its image is therefore an additive subgroup of $\mathbb{Q}$. We already know that the image is contained in $\frac{1}{n}\mathbb{Z}$. We also know that the image contains all of $\mathbb{Z}$, since the image of v_p on $\mathbb{Q}_p^\times$ does. Choose $x \in K$ with $v_p(x) = \frac{d}{e}$ with d and e relatively prime so that the denominator e is the largest possible. (This makes sense because it is clear that e must be a divisor of n, so that the range of possible denominators is bounded.) Now, since d and e are relatively prime, there must be a multiple of d which is congruent to 1 modulo e, i.e., we can find integers r and s such that $rd = 1 + se$. But then

$$r\frac{d}{e} = \frac{1 + se}{e} = \frac{1}{e} + s$$

is in the image; since $s \in \mathbb{Z}$ is in the image, it follows that $1/e$ is in the image. Since e was chosen to be the largest possible denominator in the image, it follows that the image must be exactly $\frac{1}{e}\mathbb{Z}$, and we are done.   □

The image of the p-adic valuation on a field K is called the *value group* of K. The number e is an invariant of the field extension $K/\mathbb{Q}_p$, and therefore we give it a name:

Definition 6.4.3 *Let $K/\mathbb{Q}_p$ be a finite extension, and let $e = e(K/\mathbb{Q}_p)$ be the unique positive integer (dividing $n = [K : \mathbb{Q}_p]$) defined by*

$$v_p(K^\times) = \frac{1}{e}\mathbb{Z}.$$

We call e the ramification index of K over $\mathbb{Q}_p$. We say the extension $K/\mathbb{Q}_p$ is unramified if $e = 1$. We say the extension is ramified if $e > 1$, and totally ramified if $e = n$. Finally, we write $f = f(K/\mathbb{Q}_p) = n/e$ and call it the residual degree of K over $\mathbb{Q}_p$.

The notations e and f are traditional for these two numbers. Notice that at this point f has simply been defined as the "other factor" of n, which does not explain the name "residual degree." We will soon give it a more interesting interpretation that justifies the choice of nomenclature. Before we do that, however, we need to explore the structure of K a little further.

Problem 245 Compute e for the fields F_1, F_2, and F_3. (Hint: we made sure to have one example each of unramified, totally ramified, and ramified-but-not-totally extensions.)

In $\mathbb{Q}_p$, the number p played a special role, due to the fact that it was an element of smallest positive valuation, $v_p(p) = 1$. This meant that any element $x \in \mathbb{Z}_p$ with $v_p(x) > 0$ was divisible by p, and in fact, we could interpret $v_p(x)$ as a multiplicity: any $x \in \mathbb{Q}_p$ can be written as $x = p^{v_p(x)}u$, where u is a p-adic unit, i.e., satisfies $v_p(u) = 0$. To do something similar in K, we need an element whose valuation is exactly $1/e$. But since $1/e$ is in the value group, such elements exist.

Definition 6.4.4 *Let $K/\mathbb{Q}_p$ be a finite extension, and let $e = e(K/\mathbb{Q}_p)$. We say an element $\pi \in K$ is a* uniformizer *if $v_p(\pi) = 1/e$.*

Notice that there are many uniformizers, just as there are many elements of $\mathbb{Z}_p$ whose valuation is exactly 1. In what follows, we will *choose* a uniformizer π, and fix it throughout the discussion. We should remark that in the *unramified* case, we have $e = 1$, and we can (and usually will) simply take $\pi = p$.

Problem 246 Find uniformizers for F_1, F_2, and F_3. (Only F_3 takes some thought.)

Having set this up, we can describe the algebraic structure of K. First of all, recall that we defined the valuation ring

$$\mathcal{O} = \mathcal{O}_K = \{x \in K : |x| \leq 1\} = \{x \in K : v_p(x) \geq 0\}$$

and its maximal ideal

$$\mathfrak{p} = \mathfrak{p}_K = \{x \in K : |x| < 1\} = \{x \in K : v_p(x) > 0\}.$$

$\mathcal{O}_K$, as we saw in Chapter 2, is a local ring, and the residue field is the quotient

$$\mathsf{k} = \mathcal{O}_K/\mathfrak{p}_K.$$

The basic facts about these rings are easy to describe.

Proposition 6.4.5 *Let notations be as above, and fix a uniformizer π in K. Then:*

    i) The ideal $\mathfrak{p}_K \subset \mathcal{O}_K$ is principal, and π is a generator.

ii) *Any element* $x \in K$ *can be written in the form* $x = u\pi^{ev_p(x)}$, *where* $u \in \mathcal{O}_K^\times$ *is a unit, and therefore satisfies* $v_p(u) = 0$. *In particular,* $K = \mathcal{O}_K[\frac{1}{\pi}]$.

iii) *The residue field* k *is a finite extension of* $\mathbb{F}_p$ *whose degree is less than or equal to the degree* $[K : \mathbb{Q}_p]$. *In particular, the number of elements in* k *is a power of* p. *(The exact number of elements will be determined below.)*

iv) *Any element of* $\mathcal{O}_K$ *is the root of a monic polynomial with coefficients in* $\mathbb{Z}_p$.

v) *Conversely, if* $x \in K$ *is the root of a monic polynomial with coefficients in* $\mathbb{Z}_p$, *then* $x \in \mathcal{O}_K$.

vi) $\mathcal{O}_K$ *is a compact topological ring. The sets* $\pi^m \mathcal{O}_K$, $m \in \mathbb{Z}$, *form a fundamental system of neighborhoods of zero in* K, *which is a totally disconnected, Hausdorff, locally compact topological space.*

vii) *Let* ℓ *be the number of elements of the residue field* k *and let* $A = \{c_1, c_2, \ldots, c_\ell\} \subset \mathcal{O}_K$ *be a fixed set of representatives for the elements of* k, *i.e., for the cosets of* $\mathfrak{p}_K$ *in* $\mathcal{O}_K$. *Then any* $x \in K$ *has a unique representation as a p-adic expansion*

$$x = \sum_{i=-m}^{\infty} a_i \pi^i = a_{-m}\pi^{-m} + \cdots + a_0 + a_1\pi + a_2\pi^2 + \ldots,$$

where each $a_i \in A$. *In other words, every element of* K *has a unique expansion in powers of* π *with coefficients chosen from the "digits"* $c_1, c_2, \ldots, c_\ell$.

Problem 247 Prove the proposition. (Some hints: (i) is a matter of computing v_p; (ii) is pretty much the same. For (iii), the crucial observation is that if a set of elements of $\mathcal{O}$ is linearly dependent over $\mathbb{Q}_p$, then the set of their reductions modulo π is linearly dependent over $\mathbb{F}_p$. Item (iv) was pretty much proved when we constructed the absolute value, and (v) is immediate from that construction. The rest is identical to what we did for $\mathbb{Q}_p$.)

Problem 248 Work out $\mathcal{O}_K$ and k for each of our running examples.

Given that k is a finite extension of $\mathbb{F}_p$, its degree is another natural invariant of the extension $K/\mathbb{Q}_p$. It turns out, however, to be a number we have already introduced, namely the residual degree.

Theorem 6.4.6 *Still using the notations above, let* $f = f(K/\mathbb{Q}_p)$ *be the residual degree of* K *over* $\mathbb{Q}_p$ *(see Definition 6.4.3). Then* $[k : \mathbb{F}_p] = f$, *so that* f *is the degree of the residue field extension* $k/\mathbb{F}_p$. *In particular,* $k = \mathbb{F}_{p^f}$ *is a finite field with* p^f *elements.*

PROOF: This is a big one. Ready? Here we go.

Let $m = [\mathsf{k} : \mathbb{F}_p]$, and let $e = e(K/\mathbb{Q}_p)$ be the ramification index. What the theorem says is that $e \cdot m = n = [K : \mathbb{Q}_p]$. First of all, choose elements $\alpha_1, \alpha_2, \ldots, \alpha_m \in \mathcal{O}_K$ such that their images $\bar{\alpha}_1, \bar{\alpha}_2, \ldots \bar{\alpha}_m \in \mathsf{k}$ are a basis of k over $\mathbb{F}_p$. (In particular, they must be non-zero, so that the α's are actually in $\mathcal{O}_K^\times$.) As we noted above, the α's are clearly linearly independent over $\mathbb{Q}_p$ (given a dependence relation, scale so that the coefficients are integral and at least one is a unit, then reduce modulo π; this gives a dependence relation over $\mathbb{F}_p$). To prove the theorem, we show how to complete this set to a basis of K over $\mathbb{Q}_p$.

The idea is to use the uniformizer π. Consider the elements

$$\alpha_1, \alpha_2, \ldots, \alpha_m,$$
$$\pi\alpha_1, \pi\alpha_2, \ldots, \pi\alpha_m,$$
$$\pi^2\alpha_1, \pi^2\alpha_2, \ldots, \pi^2\alpha_m,$$
$$\ldots,$$
$$\pi^{e-1}\alpha_1, \pi^{e-1}\alpha_2, \ldots, \pi^{e-1}\alpha_m.$$

We claim these form a basis of K over $\mathbb{Q}_p$. Note that, if so, the theorem follows, since we then have $n = e \cdot m$.

Proving our claim requires several steps. First of all, if every element of $\mathcal{O}_K$ is a $\mathbb{Q}_p$-linear combination of the $\pi^i\alpha_j$, then so is every element of K, since for any $x \in K$ we can find a power of p such that $p^r x \in \mathcal{O}_K$, find the expansion of this element, then divide by p^r.

Now consider $x \in \mathcal{O}_K$. We will show that x is a $\mathbb{Z}_p$-linear combination of the elements listed above. First, reducing modulo π, we can write $\bar{x}$ as a combination of the $\bar{\alpha}_j$; in other words, we have

$$x = x_{0,1}\alpha_1 + x_{0,2}\alpha_2 + \cdots + x_{0,m}\alpha_m + \text{a multiple of } \pi,$$

with $x_{0,j} \in \mathbb{Z}_p$. Now repeat the same reasoning to the multiple of π to get

$$x = x_{0,1}\alpha_1 + x_{0,2}\alpha_2 + \cdots + x_{0,m}\alpha_m$$
$$+ x_{1,1}\pi\alpha_1 + x_{1,2}\pi\alpha_2 + \cdots + x_{1,m}\pi\alpha_m$$
$$+ \text{a multiple of } \pi^2.$$

Repeating this e times, and noticing that π^e and p differ by a unit (because they have the same valuation!), we see that our x can be written as

$$x = x_{0,1}\alpha_1 + x_{0,2}\alpha_2 + \cdots + x_{0,m}\alpha_m$$
$$+ x_{1,1}\pi\alpha_1 + x_{1,2}\pi\alpha_2 + \cdots + x_{1,m}\pi\alpha_m$$
$$+ \ldots$$
$$+ x_{e-1,1}\pi^{e-1}\alpha_1 + x_{e-1,2}\pi^{e-1}\alpha_2 + \cdots + x_{e-1,m}\pi^{e-1}\alpha_m$$
$$+ px',$$

where all the coefficients $x_{i,j}$ are in $\mathbb{Z}_p$ and $x' \in \mathcal{O}_K$. Now apply the same reasoning to x'. This will give new coefficients $x_{i,j} + px'_{i,j}$, for which the equality holds modulo p^2. Continuing in this fashion produces, for each (i, j), a convergent series in $\mathbb{Z}_p$:

$$x_{i.j} + px'_{i,j} + p^2 x''_{i,j} + \dots$$

Let $y_{i,j} \in \mathbb{Z}_p$ be the sum. Then we get

$$x = \sum_{i=0}^{e-1} \sum_{j=1}^{m} y_{i,j} \pi^i \alpha_j,$$

which shows x is a $\mathbb{Z}_p$-linear combination of the $\pi^i \alpha_j$.

To show that the $\pi^i \alpha_j$ are independent, assume for a contradiction[11] that we have a nontrivial linear dependence relation

$$\sum_{i,j} x_{i,j} \pi^i \alpha_j = 0$$

with $x_{i,j} \in \mathbb{Q}_p$. Since not all the coefficients are zero, we can scale the entire relation to make sure that the $x_{i,j}$ are all in $\mathbb{Z}_p$ and that at least one is not divisible by p. We will find a contradiction by showing that in fact all of the $x_{i,j}$ are divisible by p.

Reducing this equation modulo π gives a dependence relation for the $\bar{\alpha}_j$ over $\mathbb{F}_p$; this must be trivial because the $\bar{\alpha}_j$ are a basis. Hence all of the coefficients $x_{0,j}$ must reduce to zero, i.e., must be divisible by p. If $e = 1$ this contradicts our assumption that at least one coefficient is a unit.

If $e > 1$, all we know so far is that all the $x_{0,j}$ are divisible by p, and hence by π. This makes the whole relation divisible by π; divide through. Notice that $x_{0,j}/\pi$ will still be divisible by π, since its valuation is at least $1 - \frac{1}{e} > 0$. Now reduce modulo π again. We know that most of the equation is still divisible by π, and using the same reasoning as before, we can conclude that the $x_{1,j}$ must all be divisible by p. If $e = 2$ we have a contradiction again.

Continuing in this fashion, we get that *all* the $x_{i,j}$ are divisible by p, which contradicts our initial assumption. It follows that no such linear dependence relation exists, and we are finally done.   □

After that long proof, it is well to remind ourselves of what we have obtained. We have shown that the degree $n = [K : \mathbb{Q}_p]$ of a finite extension of $\mathbb{Q}_p$ breaks up as a product $n = e \cdot f$, where e measures the change of the image of the p-adic valuation v_p and $f = [\mathsf{k} : \mathbb{F}_p]$ measures the change in the residue field.

Problem 249 This was a messy proof. It's probably wise to work out the precise details.

---

[11] This proof has a similar flavor to the standard proof that $\sqrt{2}$ is irrational: $a^2 = 2b^2$ forces both a and b to be even.

6.5   Classifying Extensions of $\mathbb{Q}_p$

We now know some important invariants of finite extensions $K/\mathbb{Q}_p$: the *degree* n, the *ramification index* e, and the *residual degree* f. These are related by $n = ef$. In this section, we consider the special cases $e = n$ and $e = 1$.

We begin by giving a partial description of the totally ramified finite extensions of $\mathbb{Q}_p$. It is a standard result in field theory that any extension of a field of characteristic zero (such as $\mathbb{Q}_p$) is generated by adjoining the root of an irreducible polynomial. In the case of totally ramified extensions, we can say exactly what kind of polynomial.

Proposition 6.5.1 *Let $K/\mathbb{Q}_p$ be a totally ramified finite extension of $\mathbb{Q}_p$, so that $e(K/\mathbb{Q}_p) = n = [K : \mathbb{Q}_p]$. Then $K = \mathbb{Q}_p(\pi)$, where π, as above, is a uniformizer. Furthermore, π is a root of a polynomial*

$$f(X) = X^n + a_{n-1}X^{n-1} + \cdots + a_1 X + a_0 \in \mathbb{Q}_p[X]$$

that satisfies the conditions of the Eisenstein criterion, i.e., $p|a_i$ for $0 \leq i < n$ and $p^2 \nmid a_0$.

PROOF: Let π be a uniformizer, so that $v_p(\pi) = 1/n$, or, equivalently, $|\pi| = p^{-1/n}$. Take $f(X)$ to be the minimal polynomial for π over $\mathbb{Q}_p$. Recall that we can compute the absolute value of π in terms of its norm. It goes like this: if the degree of $f(X)$ is s (which must be a divisor of n) and its last coefficient is a_0, we set $r = n/s$, and then the norm of π is $(-1)^n a_0^r$. Once we know the norm, we can compute the absolute value; this gives the equation

$$p^{-1/n} = |\pi| = \sqrt[n]{|a_0^r|} = \sqrt[s]{|a_0|}.$$

Now, since a_0 is in $\mathbb{Q}_p$, its absolute value is an integral power of p. Looking at the equation, we see that we must have $s = n$ (so that $f(X)$ is of degree n) and $|a_0| = p^{-1}$.

The fact that the degree is n shows that $K = \mathbb{Q}_p(\pi)$, and $|a_0| = p^{-1}$ is exactly what we claimed about this coefficient of $f(X)$. It remains to show our claim about the other coefficients. For this, let $\pi_1 = \pi, \pi_2, \ldots, \pi_n$ be the roots of $f(X)$. Note, first, that all of the roots have the same minimal polynomial, hence the same norm, hence the same absolute value. In particular, we have $|\pi_i| < 1$ for every i. Now, the coefficients of $f(X)$ are sums of products of the roots (write $f(X)$ as the product of the $(X - \pi_i)$ and expand); it follows that we must have $|a_i| < 1$ for $1 < i < n$, and we are done.   □

Problem 250 In the case of F_2, which is totally ramified, what is the Eisenstein polynomial?

Problem 251 Let $p = 3$ and let $K = \mathbb{Q}_3(\zeta)$ be the field obtained by adjoining a cube root of unity. Check that this is a totally ramified extension of degree 2, and find the Eisenstein polynomial given by the proposition.

This is quite a remarkable result, but it only gives a partial description of the totally ramified extensions of $\mathbb{Q}_p$, because we don't have a way of deciding whether two different Eisenstein polynomials give the same extension (or at least not yet; it can be done using Theorem 6.8.2). It turns out that for each n there are only *finitely many* different totally ramified extensions of degree n.

Sage can create totally ramified extensions of $\mathbb{Q}_p$ if you give it an Eisenstein polynomial. First we create the base field and the polynomial ring.

```
sage: K=Qp(7)
sage: K
 7-adic Field with capped relative precision 20
sage: S.<x>=ZZ[]
sage: S
 Univariate Polynomial Ring in x over Integer Ring
```

Now we can create an Eisenstein polynomial and tell Sage to make the field extension.

```
sage: f=x^3-49*x^2+35*x+7
sage: F.<w>=K.ext(f)
sage: F
 7-adic Eisenstein Extension Field in w defined
 by x^3 - 49*x^2 + 35*x + 7
```

The field F now allows the usual computations to happen, with the element w being the uniformizer.

```
sage: F(12)
 5 + 6*w^3 + 5*w^4 + 3*w^5 + 5*w^8 + 3*w^9 + 3*w^10 + 3*w^12
 + 6*w^13 + w^14 + 2*w^15 + 6*w^16 + 5*w^17 + 4*w^18 + 4*w^19
 + 6*w^20 + 5*w^21 + 4*w^22 + 4*w^23 + 4*w^24 + 2*w^26
 + 6*w^27 + 2*w^28 + 4*w^29 + 6*w^30 + 2*w^31 + 3*w^32
 + 5*w^33 + 3*w^34 + 6*w^35 + 3*w^36 + 6*w^37 + w^38 + 5*w^41
 + 4*w^42 + 4*w^43 + 3*w^44 + 2*w^45 + 5*w^46 + 6*w^47
 + w^48 + 3*w^51 + 6*w^52 + w^53 + 4*w^54 + 2*w^55 + 5*w^56
 + 3*w^57 + 3*w^58 + O(w^60)
```

Notice $K = \mathbb{Q}_7$ had precision $O(7^{20})$; the extension of degree 3 $F = \mathbb{Q}_7(w)$ has the same precision, but in terms of w that is $O(w^{60})$.

We would say $v_7(w) = 1/3$, but Sage does it differently:

```
sage: F(w).valuation()
 1
sage: F(7).valuation()
 3
```

Sage prefers to make w have valuation 1, which means the valuation of 7 must be 3 when we think of it as an element of F. Our normalization has the advantage of being field-independent.

It is natural to look for a similar result in the other extreme case, i.e., for unramified extensions. It turns out that those are even simpler, but to be able to prove that we will have to have one more tool in our kit. That tool is Hensel's Lemma.

Theorem 6.5.2 (Hensel's Lemma) *Let K be a finite extension of $\mathbb{Q}_p$, and let π be a uniformizer. Let $F(X) = a_0 + a_1 X + a_2 X^2 + \cdots + a_n X^n$ be a polynomial whose coefficients are in $\mathcal{O}_K$. Suppose that there exists an $\alpha_1 \in \mathcal{O}_K$ such that*

$$F(\alpha_1) \equiv 0 \pmod{\pi}$$

and

$$F'(\alpha_1) \not\equiv 0 \pmod{\pi},$$

where $F'(X)$ is the (formal) derivative of $F(X)$. Then there exists an $\alpha \in \mathcal{O}_K$ such that $\alpha \equiv \alpha_1 \pmod{\pi}$ and $F(\alpha) = 0$.

Recall that π is a generator of the maximal ideal $\mathfrak{p}_K$, so that we can also write the conditions as congruences modulo $\mathfrak{p}_K$, or in terms of absolute values. The proof is identical to the one we gave in Chapter 4.

Problem 252 Prove Theorem 6.5.2.

Problem 253 Formulate and prove a version of Theorem 4.7.2 (Hensel's Lemma for polynomials) that works over K.

Problem 254 Formulate and prove a version of Problem 120 (the stronger form of Hensel's Lemma to which we occasionally needed to resort) that works over K.

The crucial observation, for all three problems, is that there is really nothing to do: *exactly* the same proofs work.

As before, we can use Hensel's Lemma to obtain roots of unity in K. The point is that the non-zero elements of the residue field k (which, remember, has p^f elements) form a cyclic group[12] with $p^f - 1$ elements. This means that for each m dividing $p^f - 1$, there are exactly m roots of $F_m(X) = X^m - 1$ in $k^\times$. Choosing any lift of these to $\mathcal{O}_K^\times$ gives us m non-congruent "approximate roots." This sets us up for Hensel's Lemma, since the derivative is $F_m'(X) = mX^{m-1}$, which will be non-zero (m is a divisor of $p^f - 1$, hence not divisible by p, and our approximate roots are units). Hensel's Lemma gives us m non-congruent (and therefore m different) m-th roots of unity in $\mathcal{O}_K^\times$. Since

---

[12]That the non-zero elements of any field form a group is sort of obvious. That the non-zero elements of a finite field form a *cyclic* group follows from the fact that any *finite* subgroup of a field is cyclic, which the reader may have met in her abstract algebra course, and which otherwise is a very nice and challenging exercise.

this is true for any m dividing $p^f - 1$, it means that K contains the full cyclic group of $(p^f - 1)$-st roots of unity. In other words:

Corollary 6.5.3 *Let $K/\mathbb{Q}_p$ be a finite extension, and let $f = f(K/\mathbb{Q}_p)$. Then $\mathbb{O}_K^\times$ contains the cyclic group of $(p^f - 1)$-st roots of unity.*

Problem 255 Describe what roots of unity are given by this corollary in each of the fields F_1, F_2, and F_3. In each case, can you decide whether there are any other roots of unity?

Of course, if K contains the $(p^f - 1)$-st roots of unity, then it also contains the m-th roots of unity for any m dividing $p^f - 1$. We can also turn this around: given an m which is not divisible by p, one can always find an f such that $p^f \equiv 1 \pmod{m}$ (the group of invertible elements of $\mathbb{Z}/m\mathbb{Z}$ is, after all, finite), which means that m divides $p^f - 1$. So by taking fields with larger and larger f, we get all the prime-to-p-th roots of unity.

Except for p-power roots of unity, this description is complete. First, if $f = f(K/\mathbb{Q}_p)$ and K contains any other roots of unity, that is, m-th roots of unity for some m which is relatively prime to $p^f - 1$, then they must be 1-units, since their reduction modulo π must be equal to 1.

Problem 256 (This problem just asks you to verify carefully what we have just asserted.) Suppose $x \in K$ satisfies $x^m = 1$.

i) Show that $x \in \mathbb{O}_K^\times$, i.e., that x is a unit in K.

ii) Show that if m is relatively prime to $p^f - 1$, then $x \equiv 1 \pmod{\pi}$, so that $x \in 1 + \mathfrak{p}_K$.

Next, a 1-unit can be an m-th root of unity only if m is a power of p. We show this by a direct argument. First we make the following useful remark:

Lemma 6.5.4 *If $x \equiv 1 \pmod{\pi}$, then $x^p \equiv 1 \pmod{\pi^2}$, and, more generally, $x^{p^r} \equiv 1 \pmod{\pi^{r+1}}$.*

PROOF: An easy exercise on the binomial theorem. Notice that unless $e = 1$ we can in fact do much better than is stated. □

Now it's easy: if ζ is a 1-unit, and $\zeta^m = 1$ for some m prime to p, then we begin with

$$\zeta \equiv 1 \pmod{\pi}.$$

Now choose any r such that $p^r \equiv 1 \pmod{m}$ (this certainly exists, as we observed above). Then, taking p^r-th powers, we get

$$\zeta = \zeta^{p^r} \equiv 1 \pmod{\pi^{r+1}}.$$

Iterating (or just replacing r by a multiple), we see that in fact ζ is congruent to 1 modulo an arbitrarily large power of π. It follows that $\zeta = 1$. (Otherwise, what would be the valuation of $\zeta - 1$?)

Problem 257 Prove the lemma.

Problem 258 Push the argument above a little harder to show that if m is prime to p then two different m-th roots of 1 will never be congruent modulo π.

Problem 259 We outline an alternative way to show that 1-units cannot be m-th roots of unity if m is prime to p. Suppose ζ is a 1-unit, and $\zeta^m = 1$ for some m prime to p. Taking a power of ζ, we get an ℓ-th root of unity ζ_1, where ℓ is a prime not equal to p. Let $x_1 = 1 - \zeta_1 \in \mathfrak{p}_K$. Then we have

$$(1 - x_1)^\ell - 1 = 0.$$

Expand the left-hand side, and rearrange to get a contradiction.

One interesting way to read the last few paragraphs is to see that they tell us something about the structure of the 1-units, i.e., the elements of $U_1 = 1 + \pi \mathcal{O}_K$. This is clearly a group, since

$$(1 + \pi x)(1 + \pi y) = 1 + \pi x + \pi y + \pi^2 xy,$$

and

$$(1 + \pi x)^{-1} = 1 - \pi x + (\pi x)^2 - (\pi x)^3 + \dots,$$

which clearly converges and belongs to U_1. Similarly, each of the sets $U_n = 1 + \pi^n \mathcal{O}_K$ are subgroups.

Problem 260 Show that for any n the quotient U_n/U_{n+1} is a p-group (i.e., its order is a power of p). (Hint: you need to show that it is a finite abelian group, and that the order of any element is a power of p.)

The upshot is that we have obtained an almost complete description of the roots of unity in K: if we set $f = f(K/\mathbb{Q}_p)$, then K contains $p^f - 1$ non-congruent $(p^f - 1)$-st roots of unity, and possibly some p-power roots of unity. These last will be 1-units.

We are now ready to go back to what started us on this roots-of-unity excursus, that is, to describe the unramified extensions of $\mathbb{Q}_p$.

Proposition 6.5.5 *For each f there is exactly one unramified extension of degree f. It can be obtained by adjoining to $\mathbb{Q}_p$ a primitive $(p^f - 1)$-st root of unity.*

PROOF: It's best to clarify from the start what we mean by "exactly one." We don't just mean all such extensions are isomorphic, but also that any such extension is a normal extension. One way to express this is to say that if we fix an algebraic closure $\overline{\mathbb{Q}}_p$ then there is a unique subfield $K \subset \overline{\mathbb{Q}}_p$ which is unramified and has degree f over $\mathbb{Q}_p$.

Let $\bar{\alpha}$ be a generator of the cyclic group of non-zero elements of $\mathbb{F}_{p^f}$. Then $\mathbb{F}_{p^f} = \mathbb{F}_p(\bar{\alpha})$ is an extension of degree f (check the usual references on abstract algebra for the details). Let

$$\bar{g}(X) = X^f + \bar{a}_{f-1}X^{f-1} + \cdots + \bar{a}_1 X + \bar{a}_0$$

be the minimal polynomial for $\bar{\alpha}$ over $\mathbb{F}_p$. Lifting $\bar{g}(X)$ to $g(X) \in \mathbb{Z}_p[X]$ any way we like, we get an irreducible polynomial over $\mathbb{Q}_p$. If α is a root of $g(X)$, then $K = \mathbb{Q}_p(\alpha)$ is an extension of degree f. The residue field k of K clearly contains a root of $\bar{g}(X)$ (the reduction of α modulo $\mathfrak{p}_K$), hence we must have $[\mathsf{k} : \mathbb{F}_p] \geq f$; since, on the other hand, the degree of the residue field is at most equal to the degree of $K/\mathbb{Q}_p$, we have $[\mathsf{k} : \mathbb{F}_p] \leq [K : \mathbb{Q}_p] = f$, it follows that $[\mathsf{k} : \mathbb{F}_p] = f = [K : \mathbb{Q}_p]$, so that $K/\mathbb{Q}_p$ is unramified. We also see that $\mathsf{k} = \mathbb{F}_{p^f}$.

This shows that there always *exists* an unramified extension of degree f (it is, in fact, the extension we considered at the end of the previous section). We still need to show the uniqueness. To do that, we will show that any extension $K/\mathbb{Q}_p$ which is unramified and of degree f will have to be equal to the extension obtained by adjoining a primitive $(p^f - 1)$-st root of unity.

By the corollary above, we already know that K must contain all the $(p^f - 1)$-th roots of unity. Hence, to show the equality we want, all we need to do is show that the smallest field extension of $\mathbb{Q}_p$ which contains the $(p^f - 1)$-st roots of unity is already of degree f, and hence must be all of K.

So choose β to be a primitive $(p^f - 1)$-th root of unity in K. Then we have

$$\mathbb{Q}_p \subset \mathbb{Q}_p(\beta) \subset K.$$

Now, the powers of β are exactly all the $(p^f - 1)$-th roots of unity, and we know, from Problem 258, that they are all distinct modulo π. This means that $\bar{\beta}$ is a $(p^f - 1)$-th root of unity, so that the residue field of the extension $\mathbb{Q}_p(\beta)/\mathbb{Q}_p$ contains $\mathbb{F}_{p^f} = \mathsf{k}$. Since the degree of the residue field extension is certainly less than or equal to the degree of the extension of $\mathbb{Q}_p$, it follows that the degree of $\mathbb{Q}_p(\beta)$ over $\mathbb{Q}_p$ is at least f. Since $K/\mathbb{Q}_p$ is of degree f, it follows that $K = \mathbb{Q}_p(\beta)$.

Finally, the roots of the minimal polynomial for β over $\mathbb{Q}_p$ are exactly the primitive $(p^f - 1)$-th roots of unity, all of which are powers of β, which shows $K = \mathbb{Q}_p(\beta)$ is a normal extension. $\qquad\square$

Problem 261 For $p = 5$, consider the extensions $F_1 = \mathbb{Q}_5(\sqrt{2})$ and $K = \mathbb{Q}_5(\sqrt{3})$. Show that they are both unramified and of degree 2. Conclude that they are equal. How can this be? The theorem also says that either extension is the same as the one obtained by adjoining a primitive 24-th root of unity. Can you find a few terms of the 5-adic expansion of a primitive 24-th root of unity? (Remember that we can take $\pi = 5$ as uniformizer, because we know the extension is unramified, so that the first problem is to choose a convenient set of "digits.")

Problem 262 Find the largest subfield of F_3 which is an unramified extension of $\mathbb{Q}_3$.

Problem 263 Let $K = \mathbb{Q}_3(\sqrt[3]{2})$ be the extension of $\mathbb{Q}_3$ obtained by adjoining a cube root of 2. Show that this extension is totally ramified.

One way to understand what we have just proved is this: we have shown that for each extension $\mathsf{k}/\mathbb{F}_p$ there is a unique unramified extension $K/\mathbb{Q}_p$ whose residue field is k. This is very cool: given k, which is an object of characteristic p, we can get a unique K, which is of characteristic zero. It is easy to start in characteristic 0 and produce something of characteristic p by modding out. But moving in the opposite direction seems harder.

Of course, our description of K depends on having at hand both $\mathbb{Q}_p$ and roots of unity. Might there be a way to construct K (or, equivalently, the valuation ring $\mathcal{O}_K$) *directly* from the finite field k? This was one of the motivations for the creation of *Witt vectors*, which provide an explicit way to take a field k of characteristic p and create a ring $W(\mathsf{k})$ of characteristic zero. If k is a finite field, $W(\mathsf{k})$ is exactly the valuation ring $\mathcal{O}_K$ of the unique unramified extension corresponding to k. But the construction works for *any* field of characteristic p, and even more generally than that. The definition of $W(\mathsf{k})$ is complicated, however, so we will not give the details; a good source is [61, Ch. II, §6].

The unique unramified extension K_f of degree f has several interesting properties. If $m = p^f - 1$, we know that K_f is generated by the m-th roots of unity. So if ζ is any primitive m-th root of unity, we have $K_f = \mathbb{Q}_p(\zeta)$. Let $F(X)$ be the minimal polynomial for ζ; the roots of $F(X)$ are exactly all the primitive m-th roots of unity, which we know are also powers of ζ, so any field that contains a primitive m-th root of unity contains K_f. But we can do a bit better, because we also know that no two primitive m-th roots of unity are congruent mod p, so $F'(\zeta) \not\equiv 0 \pmod{p}$. That puts us in a Hensel's Lemma situation, which allows us to prove a neat result.

Corollary 6.5.6 *Let K be a finite extension of $\mathbb{Q}_p$ and let $m = p^f - 1$. If there exists $c \in K$ and a primitive m-th root of unity in $\overline{\mathbb{Q}}_p$ such that $c \equiv \zeta \pmod{p}$, then K contains the degree f unramified extension of $\mathbb{Q}_p$.*

PROOF: Let $F(X)$ be the minimal polynomial for ζ. Then, since $c \equiv \zeta \pmod{p}$, we have $F(c) \equiv 0 \pmod{p}$ and $F'(c) \not\equiv 0 \pmod{p}$. By Hensel's Lemma, it follows that K contains a root of $F(X)$, i.e., a primitive m-th root of unity. Hence K contains the unramified extension of degree f. □

Moving from "contains something congruent to ζ" to "contains ζ" feels like magic, but it's typical of what Hensel's Lemma can do. It might be fun to consider to what extent this corollary can be generalized. See Theorem 6.8.2 for a related result.

Sage has built-in functions to create unramified extensions. The command K.<u>=Qq(125) tells Sage that K is the unramified extension of $\mathbb{Q}_5$ corresponding to the finite field with 5^3 elements, whose generator over $\mathbb{Q}_5$ is to be named u. The command R.<u>=Zq(125) creates the ring $W(\mathbb{F}_{125})$.

```
sage: K.<u>=Qq(125)
sage: K
 5-adic Unramified Extension Field in u defined by
 x^3 + 3*x + 3
sage: a=K(1+5*u^5)
sage: a
 1 + (2*u^2 + 4*u + 4)*5 + (4*u^2 + u + 1)*5^2 + 4*u^2*5^3
   + 4*u^2*5^4 + 4*u^2*5^5 + 4*u^2*5^6 + 4*u^2*5^7
   + 4*u^2*5^8 + 4*u^2*5^9 + 4*u^2*5^10 + 4*u^2*5^11
   + 4*u^2*5^12 + 4*u^2*5^13 + 4*u^2*5^14 + 4*u^2*5^15
   + 4*u^2*5^16 + 4*u^2*5^17 + 4*u^2*5^18 + 4*u^2*5^19
   + O(5^20)
```

Notice that the digits are the standard lifts of elements of $\mathbb{F}_{125}$, namely polynomials $a + bu + c^2$ where $a, b, c \in \{0, 1, 2, 3, 4\}$.

The two main results of this section, describing the totally ramified and the unramified extensions of $\mathbb{Q}_p$, together yield a rather good description of arbitrary extensions. We won't go into it here in detail; basically, one shows that any extension is obtained by first taking an unramified extension, and then taking a totally ramified extension of the resulting field. (Of course, this requires knowing what it means for an extension L/K to be totally ramified; we have only defined this when $K = \mathbb{Q}_p$.) This is proved in, for example, [34, Ch. 14] or [42, §III.3]. Using that and Krasner's Lemma (Theorem 6.8.2) one can then show that there are finitely many extensions of degree n, and indeed find them all.

In fact, GP contains a function `padicfields(p,n)` that returns a list of generating polynomials for all $K/\mathbb{Q}_p$ of degree n. There is a variant `padicfields(p,n,1)` that gives even more information in the form

```
[polynomial, e, f, d, c]
```

We get a polynomial whose root generates K, the invariants e and f, an integer d so that the discriminant[13] of K is p^d, and the number c of distinct conjugates[14] of K in a fixed algebraic closure. For example,

```
gp > padicfields(3,4,1)
%1 = [[x^4 + 13*x^3 + 64*x^2 + 61*x + 40, 1, 4, 0, 1],
      [x^4 + 2*x^3 + 11*x^2 + 10*x + 4, 2, 2, 2, 1],
      [x^4 + 2*x^3 + 2*x^2 + 7*x + 16, 2, 2, 2, 1],
      [x^4 + 3, 4, 1, 3, 2],
      [x^4 + 6, 4, 1, 3, 2]]
```

So the unique unramified extension of degree 4 is defined over $\mathbb{Q}_3$ by the polynomial $x^4 + 13x^3 + 64x^2 + 61x + 40$. There are two extensions[15] with $e = f = 2$ and there are two totally ramified extensions.

---

[13] No, we haven't defined what this is.
[14] This is just the number of embeddings $K \hookrightarrow \overline{\mathbb{Q}}_p$. If K is normal then $c = 1$.
[15] One of these is our F_3; which?

The following problems move in the direction of getting a more precise description of totally ramified extensions.

Problem 264 Let $K/\mathbb{Q}_p$ be a totally ramified extension of degree e which satisfies the extra condition that p does not divide e (such extensions are called *tamely ramified*). Show that K can be obtained by adjoining to $\mathbb{Q}_p$ a root of a polynomial of the form $X^e - pu$, where $u \in \mathbb{Z}_p^\times$ is a p-adic unit.

Problem 265 Let K be a finite extension of $\mathbb{Q}_p$. Is there an analogue of the Eisenstein Criterion for polynomials with coefficients in K? If so, state it and prove it.

Problem 266 What would it mean for an extension L/K to be totally ramified? Would the analogue of Proposition 6.5.1 still be true?

Since there is a unique unramified extension of each degree, one nice thing we can do is to consider the union of all these extensions. (Equivalently, we could apply the Witt vector construction to the algebraic closure of $\mathbb{F}_p$.) The result will be an infinite extension of $\mathbb{Q}_p$, and will contain all the unramified extensions of $\mathbb{Q}_p$. It is called *the maximal unramified extension of $\mathbb{Q}_p$*, sometimes denoted by $\mathbb{Q}_p^{\mathrm{unr}}$.

Problem 267 Let m be an integer which is not divisible by p. Show that the maximal unramified extension of $\mathbb{Q}_p$ contains the m-th roots of unity. Conclude that we can describe $\mathbb{Q}_p^{\mathrm{unr}}$ as being obtained by adjoining to $\mathbb{Q}_p$ all the prime-to-p-th roots of unity.

The following two problems ask you to obtain important information on $\mathbb{Q}_p^{\mathrm{unr}}$ and $\overline{\mathbb{Q}}_p$. Make sure you either solve them or check the answers in the back of the book.

Problem 268 The p-adic absolute value and the p-adic valuation v_p makes sense, of course, on $\mathbb{Q}_p^{\mathrm{unr}}$. What is the image of v_p? What is the residue field?

Problem 269 The last problem also makes perfect sense if we replace $\mathbb{Q}_p^{\mathrm{unr}}$ by an algebraic closure $\overline{\mathbb{Q}}_p$ of $\mathbb{Q}_p$. Do the answers change in that case?

6.6 Analysis

Just as in the case of $\mathbb{Q}_p$, once we have a field with an absolute value we can do elementary analysis. In fact, all we need to point out is that *most of what we did in Chapter 5 extends without any difficulty*, because we were careful, when we proved our results, never (well, hardly ever) to use anything that specifically requires that the field was $\mathbb{Q}_p$ rather than an extension. More specifically, any argument that only uses estimates on absolute values will be valid in general. The only issue we have to keep in mind is ramification: any argument that depends on using the fact that p is a uniformizer for $\mathbb{Z}_p$ will need to be changed to use a uniformizer π; if $v_p(\pi) = 1/e$, we will have a larger range of possible absolute values.

In other words, we already know a lot of things, which we list. Let K be a finite extension of $\mathbb{Q}_p$ and let $|\ |$ be the unique extension of the p-adic absolute value to K.

$i)$ A sequence (a_n) in K is Cauchy if and only if

$$\lim_{n \to \infty} |a_{n+1} - a_n| = 0.$$

$ii)$ If a sequence (a_n) converges to a non-zero limit a, then we have $|a_n| = |a|$ for all sufficiently large n.

$iii)$ A series $\sum a_n$ in K converges if and only if its general term tends to zero.

$iv)$ Proposition 5.1.4 holds for double series in K.

$v)$ A power series $f(X) = \sum a_n X^n$ with coefficients $a_n \in K$ defines a continuous function on an open ball of radius $\rho = 1/\limsup \sqrt[n]{|a_n|}$; the function extends to the closed ball of radius ρ if $|a_n|\rho^n \to 0$ as $n \to \infty$.

$vi)$ Proposition 5.4.2, Theorem 5.4.3, and Problem 159 are true for power series with coefficients in K.

$vii)$ Functions defined by power series are differentiable, and their derivatives are defined by the formal derivative of the original series.

$viii)$ If $f(X) = \sum a_n X^n$ and $g(X) = \sum b_n X^n$ are power series with coefficients in K, x_m is a convergent sequence contained in the intersection of the disks of convergence of f and g, and we have $f(x_m) = g(x_m)$ for all m, then $a_n = b_n$ for all n.

$ix)$ Strassman's Theorem holds without any change beyond replacing $\mathbb{Q}_p$ by K and $\mathbb{Z}_p$ by $\mathcal{O}_K$.

$x)$ The corollaries to Strassman's Theorem therefore also extend.

$xi)$ The usual power series defines a p-adic logarithm function

$$\log_p : U_1 \longrightarrow K,$$

where

$$U_1 = \{x \in \mathcal{O}_K : |x - 1| < 1\} = 1 + \pi \mathcal{O}_K.$$

This function satisfies the functional equation

$$\log_p(xy) = \log_p(x) + \log_p(y) \quad \text{for any } x, y \in U_1.$$

xii) The usual power series defines an exponential function

$$\exp_p : D \longrightarrow K,$$

where

$$D = \{x \in \mathcal{O}_K : |x| < p^{-1/(p-1)}\}.$$

This function satisfies the functional equation

$$\exp_p(x + y) = \exp_p(x)\exp_p(y) \quad \text{for any } x, y \in D.$$

(Notice that when e is big there will certainly be elements in $\mathcal{O}_K$ whose absolute values are less than 1 but not less than $p^{-1/(p-1)}$, so that the restriction in the domain is more serious for finite extensions than it was for $\mathbb{Q}_p$ itself.)

xiii) If $x \in D$, then $\exp_p(x) \in 1 + D$ and we have

$$\log_p(\exp_p(x)) = x.$$

xiv) If $x \in 1 + D$, then $\log_p(x) \in D$ and we have

$$\exp_p(\log_p(x)) = x.$$

xv) The p-adic logarithm gives a homomorphism from the multiplicative group $U_1 = 1 + \pi\mathcal{O}_K$ to a bounded additive subgroup of K. (But, in contrast to what happens in $\mathbb{Q}_p$, the image is usually not contained in $\pi\mathcal{O}_K$.)

xvi) The p-adic logarithm gives an isometric isomorphism from the multiplicative group $1 + D$ to the additive group D, which is itself isomorphic to the additive group of $\mathcal{O}_K$.

xvii) For each $\alpha \in \mathbb{Z}_p$, the binomial series $(1+x)^\alpha = \mathbf{B}(\alpha, x)$ converges whenever $|x| < 1$ (i.e., for $x \in \pi\mathcal{O}_K = \mathfrak{p}_K$). (We need to keep the condition $\alpha \in \mathbb{Z}_p$ because we used the fact that $\mathbb{Z}$ is dense in $\mathbb{Z}_p$ to conclude that the binomial coefficients were p-adic integers. $\mathbb{Z}$ is certainly not dense in $\mathcal{O}_K$.) In other words, u^α is well-defined whenever $u \in 1 + \mathfrak{p}_K$ is a 1-unit in $\mathcal{O}_K$ and $\alpha \in \mathbb{Z}_p$ is a p-adic integer.

PROOF: Most of these are just a matter of re-reading the proofs, so we will leave those as exercises. The only significant change is (xv). Recall that in $\mathbb{Q}_p$ we proved that $\log_p(U_1) \subset p\mathbb{Z}_p$ when $p \neq 2$ (because $U_1 = 1 + D$ in this case) and that $\log_2(U_1) \subset 4\mathbb{Z}_2$. What we claim now is that $\log(U_1) \subset \pi^n\mathcal{O}_K$ for some $n \in \mathbb{Z}$, but it is possible that $n < 0$.

Suppose $x \in K$. If $|x| < 1$, then $x \in \pi\mathcal{O}_K$, so $|x| \leq |\pi|$. We know

$$\log_p(1 + x) = \sum_{n=1}^{\infty} \frac{(-1)^{n+1}x^n}{n}.$$

Now

$$\left| \frac{(-1)^{n+1} x^n}{n} \right| \leq \left| \frac{\pi^n}{n} \right|.$$

You will show in the next exercise that

$$\lim_{n \to \infty} \left| \frac{\pi^n}{n} \right| = 0.$$

Any sequence that converges to zero in $\mathbb{R}$ is bounded, so there exists a $B \in \mathbb{R}$, $B > 0$, such that $\left| \frac{\pi^n}{n} \right| < B$ for all n, which by Corollary 5.1.2 gives $|\log_p(1 + x)| < B$ as well. This shows that the image of the logarithm is bounded. Since it is clearly an additive subgroup, we have proved (xv). □

Problem 270 Show that if $v_p(\pi) = 1/e$ then

$$\lim_{n \to \infty} \left| \frac{\pi^n}{n} \right| = 0.$$

Problem 271 Find an example that shows that the image of the p-adic logarithm is not always contained in $\mathcal{O}_K$.

Problem 272 Satisfy yourself that the other assertions we enumerated are all correct.

This pretty much transports all of the elementary analysis which we developed in Chapter 5 to finite extensions of $\mathbb{Q}_p$. In fact, we will later want to extend it to infinite extensions as well.

6.7   Example: Adjoining a p-th Root of Unity

The discussion in the previous sections was mostly theoretical. It may be helpful to apply it now to a concrete case. We consider, in this section, the field $K = \mathbb{Q}_p(\zeta)$, where ζ is a p-th root of unity and $p \neq 2$. (The case $p = 2$ is a bit trivial.) In other words, ζ satisfies $\zeta^p = 1$ but $\zeta \neq 1$, and is therefore a root of the polynomial

$$\Phi_p(X) = \frac{X^p - 1}{X - 1} = X^{p-1} + X^{p-2} + \cdots + X + 1,$$

which is known as the p-th cyclotomic polynomial. The first thing we need to do, then, is to check that this polynomial is irreducible. For that, we use the Eisenstein criterion:

Lemma 6.7.1 *The polynomial*

$$\Phi_p(X) = X^{p-1} + X^{p-2} + \cdots + X + 1$$

is irreducible over $\mathbb{Q}_p$.

PROOF: The polynomial $\Phi_p(X)$ itself certainly does not satisfy the conditions for the Eisenstein criterion. So we use a little trick.

Let $F(X) = \Phi_p(X + 1)$. It is easy to see that $\Phi_p(X)$ is irreducible if and only if $F(X)$ is. We claim that $F(X)$ does satisfy the conditions in the Eisenstein criterion. To see that, we need to check two things: that all the coefficients except the first are divisible by p, and that the last coefficient is not divisible by p^2.

For the first, recall that, modulo p, taking p-th powers distributes over sums:

$$(a + b)^p \equiv a^p + b^p \pmod{p}.$$

This allows us to compute:

$$F(X) = \Phi_p(X + 1)$$

$$= \frac{(X + 1)^p - 1}{(X + 1) - 1} = \frac{(X + 1)^p - 1}{X}$$

$$\equiv \frac{X^p + 1 - 1}{X} \equiv X^{p-1} \pmod{p},$$

so that, except for the first, all the coefficients of $F(X)$ are divisible by p, as we claimed.

As for the last coefficient, it is equal to $F(0) = \Phi_p(1) = p$, which is certainly not divisible by p^2. The Eisenstein criterion then says that $F(X)$ is irreducible, which proves our assertion. $\qquad\square$

In particular, we can deduce the following things:

- $K = \mathbb{Q}_p(\zeta)$ is an extension of $\mathbb{Q}_p$ of degree $p - 1$ (since that is the degree of the minimal polynomial for ζ).

- Looking at the minimal polynomial, we see that $\mathbf{N}_{K/\mathbb{Q}_p}(\zeta) = 1$, and therefore that $|\zeta| = 1$. (Another way to see this is to note that ζ belongs to $\mathcal{O}_K$, and that so does $\zeta^{-1} = \zeta^{p-1}$, which shows that ζ must be a unit in $\mathcal{O}_K$.)

- The polynomial $F(X) = \Phi_p(X + 1)$ is the minimal polynomial for $\zeta - 1$. Therefore, we have $\mathbf{N}_{K/\mathbb{Q}_p}(\zeta - 1) = p$, and so

$$|\zeta - 1| = p^{-1/(p-1)}.$$

- K is totally ramified, and $\pi = \zeta - 1$ is a uniformizer in K.

- We have $\zeta \equiv 1 \pmod{\pi}$; in other words, ζ is a 1-unit in $\mathcal{O}_K$.

- The fact that ζ is in $\mathcal{O}_K$ shows that any polynomial $a_0 + a_1\zeta + \cdots + a_{p-2}\zeta^{p-2}$, with $a_i \in \mathbb{Z}_p$, is in $\mathcal{O}_K$ (it's clearly unnecessary to consider polynomials in ζ of higher degree, because ζ is a root of $\Phi_p(X)$). In other words, $\mathbb{Z}_p[\zeta] \subset \mathcal{O}_K$. This inclusion is actually an equality—see below.

Since K is totally ramified, we have $e = p-1$, $f = 1$, and the residue field $\mathcal{O}_K/\pi\mathcal{O}_K$ of K is just $\mathbb{F}_p$. That means we can choose the integers $0, 1, \ldots, p-1$ as coset representatives, and it follows that the elements of K can all be written as π-adic expansions of the form

$$a_{-n}\pi^{-n} + a_{-n+1}\pi^{-n+1} + \cdots + a_0 + a_1\pi + \cdots + a_m\pi^m + \ldots$$

where the a_i are integers between 0 and $p-1$. This is very nice, except for a slight problem: suppose we are given the p-adic expansion of an element of $\mathbb{Q}_p$; it is not immediately clear how to obtain its π-adic expansion in a simple way. For example,

Problem 273 What is the π-adic expansion of the integer p?

We said above that it is easy to see that $\mathbb{Z}_p[\zeta] \subset \mathcal{O}_K$, and that in fact we have an equality. To see why, remember that in the proof of Theorem 6.4.6 we showed that any element of $\mathcal{O}_K$ could be written as a $\mathbb{Z}_p$-linear combination of the elements

$$\alpha_1, \alpha_2, \ldots, \alpha_f,$$
$$\pi\alpha_1, \pi\alpha_2, \ldots, \pi\alpha_f,$$
$$\pi^2\alpha_1, \pi^2\alpha_2, \ldots, \pi^2\alpha_f,$$
$$\ldots,$$
$$\pi^{e-1}\alpha_1, \pi^{e-1}\alpha_2, \ldots, \pi^{e-1}\alpha_f$$

where $\alpha_1, \ldots, \alpha_f$ were a set of elements of $\mathcal{O}_K$ reducing to a basis of the residue field k over $\mathbb{F}_p$. In our case, however, $f = 1$, and k is equal to $\mathbb{F}_p$, so we need only one element in the basis: $\alpha_1 = 1$. The result then says that any element of $\mathcal{O}_K$ is a $\mathbb{Z}_p$-linear combination of $1, \pi, \pi^2, \ldots, \pi^{p-2}$ (since $n = e = p-1$). Remembering that we have taken $\pi = \zeta - 1$ and substituting in, this says that any element of $\mathcal{O}_K$ can be written as a polynomial in ζ, and hence that $\mathbb{Z}_p[\zeta] = \mathcal{O}_K$.

To conclude this section, we will point out some interesting things about the field K. First of all, since we have $|(\zeta-1)| < 1$, the series for the logarithm of ζ will converge. Since $\zeta^p = 1$, we must have $p\log_p(\zeta) = \log_p(\zeta^p) = \log_p(1) = 0$, so that $\log_p(\zeta) = 0$. Writing out the series, this says that

$$\sum_{n=1}^{\infty}(-1)^{n+1}\frac{(\zeta-1)^n}{n} = 0,$$

which we can rearrange slightly into

$$\sum_{n=1}^{\infty}\frac{(1-\zeta)^n}{n} = 0,$$

which is a rather remarkable formula. (We've met it before in the case $p = 2$.)

Another interesting result is that there is a $(p-1)$-st root of $-p$ in K. To see why one might want to look for such a thing, remember that $v_p(\pi^e) = 1$, so that π^e differs from p by a unit in $\mathcal{O}_K$. One might argue that the "nicest" choice of π would be one where this unit were the simplest possible unit. What we are about to show is that there is a $\pi_1 \in \mathcal{O}_K$ for which $\pi_1^e = -p$, so that the unit in this case is simply -1. That's pretty nice!

It's also rather tricky: begin by recalling that the norm of $(1 - \zeta)$ is precisely p (look at the minimal polynomial for $\zeta - 1$, which we found above, and notice that since the degree of K is even, $\mathbf{N}_{K/\mathbb{Q}_p}(x) = \mathbf{N}_{K/\mathbb{Q}_p}(-x)$). The norm, remember, can be obtained as the product of the images of our element $1 - \zeta$ under the various automorphisms of K over $\mathbb{Q}_p$. There are $p - 1$ such automorphisms, and they are given by

$$\sigma_i : \zeta \mapsto \zeta^i$$

for $i = 1, 2, \ldots, p-1$. This means that the images of $1 - \zeta$ under the various σ_i are the $1 - \zeta^i$, and the fact that the norm is p gives the equation

$$(1 - \zeta)(1 - \zeta^2) \cdots (1 - \zeta^{p-1}) = p.$$

(We could also get this equality by setting $X = 1$ in the p-th cyclotomic polynomial.) Now, we want to make $(p-1)$-st powers appear, so we do it by brute force, rewriting the equation as

$$(1 - \zeta)^{p-1} \cdot \frac{1 - \zeta^2}{1 - \zeta} \cdots \frac{1 - \zeta^{p-1}}{1 - \zeta} = p.$$

Notice that $(1 - \zeta)^{p-1}$ has the same valuation as p, which suggests that the other factors are units (to be precise: it shows that the *product* of all the other factors is a unit, and this *suggests* that each of the other factors is a unit). To see that this is indeed the case, suppose we can show that the factors are all in $\mathcal{O}_K$. Then their valuations would all be greater than or equal to zero. But the sum of their valuations is the valuation of the product, which is zero. Hence, each of the factors must have valuation zero. In other words, if we can show that all the factors are in $\mathcal{O}_K$, then it will follow that they are all units. But the algebraic identity

$$\frac{1 - \zeta^i}{1 - \zeta} = 1 + \zeta + \cdots + \zeta^{i-1}$$

shows that the factors are indeed in $\mathcal{O}_K$ (since they are polynomials in ζ). Hence, each of the fractions $(1 - \zeta^i)/(1 - \zeta)$ is a unit in $\mathcal{O}_K$.

There is one more thing we can get from the equation

$$\frac{1 - \zeta^i}{1 - \zeta} = 1 + \zeta + \cdots + \zeta^{i-1}.$$

Since $\zeta \equiv 1 \pmod{\pi}$ (because, after all, $\pi = \zeta - 1 \ldots$), and since there are i summands on the right-hand side, we get

$$\frac{1 - \zeta^i}{1 - \zeta} \equiv i \pmod{\pi}.$$

Multiplying all of these gives something congruent modulo π to the product of the integers from 2 to $p - 1$. In other words, we get

$$\frac{1 - \zeta^2}{1 - \zeta} \cdots \cdots \frac{1 - \zeta^{p-1}}{1 - \zeta} \equiv (p - 1)! \pmod{\pi}.$$

Now remember that

$$(p - 1)! \equiv -1 \pmod{p}$$

(this is "Wilson's Theorem" in elementary number theory). Changing sign, and using the previous formula, we see that

$$-\frac{1 - \zeta^2}{1 - \zeta} \cdots \frac{1 - \zeta^{p-1}}{1 - \zeta}$$

is a 1-unit, i.e., is congruent to 1 modulo π. This gives an equation of the form

$$(1 - \zeta)^{p-1} \cdot (\text{a 1-unit}) = -p.$$

What we are after, remember, is to show that $-p$ has a $(p - 1)$-st root in K; from this equation, we will be done if we can show that we can always take a $(p - 1)$-st root of a 1-unit in $\mathcal{O}_K$. But this follows easily from Hensel's Lemma:

Problem 274 Let $u \in 1 + \pi \mathcal{O}_K$, so that u is a 1-unit. Show, using Hensel's Lemma, that the polynomial $X^{p-1} - u$ has a root in $\mathcal{O}_K$.

The upshot: there exists an element $\pi_1 \in \mathcal{O}_K$ such that $\pi_1^{p-1} = -p$. This is interesting in itself, but it also gives an example of the situation described in Problem 264, since we have $e = p-1$ prime to p, and of course $K = \mathbb{Q}_p(\pi_1)$, where π_1 is a root of $X^{p-1} + p$.

Problem 275 This was a long-drawn-out argument. Can you give a simpler proof that K contains a $(p - 1)$-st root of $-p$?

Problem 276 Even nicer than our π_1 would be a uniformizer π_2 such that $\pi_2^{p-1} = p$ (i.e., the unit is just 1). Show that K in general *does not* contain such a π_2. Can it happen, for a specific prime, that such a π_2 does exist? If so, give an example of a prime for which it does exist.

This example shows how powerful an array of tools we have already put together to study $\mathbb{Q}_p$ and its algebraic extensions. The combination of algebraic and analytic techniques is very effective!

One last bit of fun. Consider the problem of finding the roots of the equation $\log_p(x) = 0$ in K (of course, what this really means is that we want to look for roots $x \in 1 + \pi \mathcal{O}_K$, since otherwise $\log_p$ is not defined). This amounts to looking for the zeros of the logarithm series

$$\log(1 + X) = \sum \frac{(-1)^{n+1} X^n}{n}$$

in $\pi \mathcal{O}_K$, and we can do this with Strassman's theorem by changing variables. Write

$$f(X) = \log(1 + \pi X) = \sum \frac{(-1)^{n+1} \pi^n X^n}{n}.$$

Clearly $f(x)$ converges when $x \in \mathcal{O}_K$, and then Strassman's theorem says that the number of roots of $f(X)$ in $\mathcal{O}_K$ (which is the number of roots of $\log_p$ in $1 + \pi \mathcal{O}_K$) is bounded by the integer N defined by the two conditions

$$\left| \frac{\pi^N}{N} \right| = \max_{n \geq 1} \left| \frac{\pi^n}{n} \right| \quad \text{and} \quad \left| \frac{\pi^N}{N} \right| > \left| \frac{\pi^n}{n} \right| \quad \text{if } n > N.$$

So we need to estimate the absolute value

$$\left| \frac{\pi^n}{n} \right|$$

as a function of n. Let's do it with valuations this time: clearly, $v_p(\pi^n) = n/(p-1)$; that's how we chose π to begin with. So

$$v_p\left(\frac{\pi^n}{n} \right) = \frac{n}{p-1} - v_p(n).$$

We need to find the n for which the absolute value is largest; in valuation terms, we want to find the n which makes the valuation smallest. To help us get our bearings, we can tabulate the first few values: see Table 6.1.

The table suggests that the smallest value is $1/(p-1)$, which occurs only when $n = 1$ and when $n = p$. This means $N = p$, so that $\log_p$ has at most p roots in $1 + \pi \mathcal{O}_K$. Since we already know p roots, namely the roots of unity $1, \zeta, \zeta^2, \ldots, \zeta^{p-1}$, we already know all the roots. In particular, this tells us that the roots of unity contained in K are exactly the cyclic group of $p(p-1)$-st roots of unity (the p-th roots we just found, plus the $(p-1)$-st roots that are provided by Hensel's Lemma when $f = 1$).

Problem 277 Prove that our surmise from the table is correct, i.e., that the smallest value for $v_p(\pi^n/n)$ is $1/(p-1)$, and that it occurs last when $n = p$.

Problem 278 Investigate what would change if instead of adjoining a p-th root of unity we adjoin a p^n-th root of unity for some $n > 1$.

n	$v_p(n)$	$v_p\left(\dfrac{\pi^n}{n}\right)$
$n = 1, \ldots, p - 1$	0	$\dfrac{n}{p-1}$
$n = p$	1	$\dfrac{p}{p-1} - 1 = \dfrac{1}{p-1}$
$n = p+1, \ldots, 2p-1$	0	$\dfrac{n}{p-1}$
$n = 2p$	1	$\dfrac{2p}{p-1} - 1 = 1 + \dfrac{2}{p-1}$
$\ldots$	$\ldots$	$\ldots$
$n = p^2$	2	$\dfrac{p^2}{p-1} - 2 = p - 1 + \dfrac{1}{p-1}$

Table 6.1: Computing $v_p(\pi^n/n)$

6.8 On to $\mathbb{C}_p$

We now want to go on to consider the algebraic closure $\overline{\mathbb{Q}}_p$ in earnest. We have already constructed its absolute value, and the next order of business is to show that it is *not* complete with respect to this absolute value. We will then go to its completion, of course, and we will be able to prove that the completion *is* algebraically closed.

To do all that, we need to know a little more about the algebraic closure, and we begin by proving a few useful facts. The first of these is known as "Krasner's Lemma." To be able to state it we need to remind ourselves of what it means for two elements of $\overline{\mathbb{Q}}_p$ to be "conjugate." This is a concept that really comes from Galois theory, but we state it here in a minimalistic fashion.

Definition 6.8.1 *Let K be a subfield of $\overline{\mathbb{Q}}_p$. Two elements a and a' of $\overline{\mathbb{Q}}_p$ are called* conjugate *over K when they are roots of the same monic irreducible polynomial with coefficients in K.*

As we have pointed out above, an equivalent way of saying this is to say that two elements are conjugate when there exists an automorphism $\sigma : \overline{\mathbb{Q}}_p \longrightarrow \overline{\mathbb{Q}}_p$ which induces the identity map on K and sends a to a'. It is clear, from either characterization, that conjugate elements have the same absolute value. (Yes? Good.)

It is worth pointing out that this definition has nothing "p-adic" about it: it works just as well for an arbitrary field K of characteristic zero, provided we replace $\overline{\mathbb{Q}}_p$ by an algebraic closure of K.

What Krasner's Lemma says is that if an element b is "close enough" to a (what this means is defined by the statement of the lemma, in terms of the conjugates of a), then a belongs to the field generated by b. Perhaps we can use the description "b is more complicated than a" to mean that adjoining b gives a field which contains the field generated by a. In that language, the lemma says that b can only be "very close" to a if it is more complicated than a. (Thinking this way makes Krasner's result resemble standard theorems in diophantine approximation.)

This is a somewhat surprising conclusion, since to say that the field generated by a is contained in the field generated by b amounts to saying that a can be written as a polynomial in b. Viewed at from this angle, Krasner's Lemma looks like the prototypical p-adic theorem: it deduces an algebraic fact (a can be written as a polynomial in b) from an analytic fact (b is very close to a).

Here is the precise statement, where we have taken care to be very general in order to be able to use the theorem at a crucial juncture below. The reader should feel free to replace "K" by "$\mathbb{Q}_p$" everywhere in this and the following result if the added generality proves to be a hindrance.

Theorem 6.8.2 (Krasner's Lemma) *Let K be a non-archimedean complete valued field of characteristic zero, and let a and b be elements of the algebraic closure of K. Let $a_1 = a$, a_2, ..., a_n be the conjugates of a over K. Suppose that b is closer to a than any of the conjugates of a, i.e.,*

$$|b - a| < |a - a_i|$$

for $i = 2, 3, \ldots, n$. Then $K(a) \subset K(b)$.

PROOF: This is short and sweet, but uses field theory a bit more seriously than other results we have proved. Let $L = K(b)$ and suppose the theorem is false, that is, that $a \notin L$. Well, then look at $L(a)$ (which, remember, is the smallest extension of L which contains a). Since we are assuming that $a \notin L$, the degree $m = [L(a) : L]$ is bigger than one. Now, there must be m homomorphisms $\sigma : L(a) \longrightarrow \overline{K}$ which are the identity on L (and send a to one of its conjugates, of course). There is at least one such σ for which $\sigma(a) \neq a$ (because if $\sigma(a) = a$ then σ is the identity on $L(a)$, and we're assuming that there's at least one other σ besides the identity); call it σ_0. Since we know, by the uniqueness of the extension of an absolute value, that $|\sigma(x)| = |x|$ for any σ and any $x \in \overline{K}$, we have

$$|\sigma_0(b) - \sigma_0(a)| = |b - a|.$$

But σ_0 fixes L, and b is in L, so $\sigma_0(b) = b$, and the last equality now says

$$|b - \sigma_0(a)| = |b - a|.$$

But then

$$|a - \sigma_0(a)| \leq \max\{|a - b|, |b - \sigma_0(a)|\} = \max\{|a - b|, |b - a|\} = |b - a|,$$

and that's not allowed, since our assumption was that b was closer to a than any of its conjugates, one of which is $\sigma_0(a)$. The contradiction shows that our assumption was wrong, that is, that a *does* belong to L, and that shows that $K(a) \subset K(b)$.   □

Problem 279 Use Krasner's Lemma to give another proof that the field $\mathbb{Q}_p(\zeta_p)$ obtained by adjoining a p-th root of unity contains a $(p-1)$-st root of $-p$.

A really important corollary of Krasner's Lemma tells us that if we have a monic irreducible polynomial $f(X)$, then any polynomial which is close enough to our $f(X)$ shares two main properties of $f(X)$: it is irreducible, and it has a root in the field extension determined by a root of $f(X)$. How close is "close enough" may depend on the specific $f(X)$, of course. Here's the precise statement.

Corollary 6.8.3 *Let K be a non-archimedean complete valued field of characteristic zero. Let*

$$f(X) = X^n + a_{n-1}X^{n-1} + \cdots + a_1 X + a_0 \in K[X]$$

be a monic irreducible polynomial of degree n with coefficients in K, let λ be a root of $f(X)$, and let $L = K(\lambda)$ be the extension of K obtained by adjoining that root. Then there exists a real number $\varepsilon > 0$ such that the following holds:

- *If $g(X) = X^n + b_{n-1}X^{n-1} + \cdots + b_1 X + b_0$ is any monic polynomial of degree n for which we have*

$$|a_i - b_i| < \varepsilon \qquad \text{for all } i = 0, 1, \ldots, n - 1,$$

  then $g(X)$ is irreducible over K and has a root in L.

PROOF: The proof (which is based on the one given in [3]) has two parts. First, we establish that under certain conditions the conclusion holds, and then we show that we can choose ε so that the conditions must hold. We use the notations in the theorem, so λ is a root of $f(X)$ and $L = K(\lambda)$. Since f is irreducible we have $[L : K] = n$.

Let $\lambda_1 = \lambda, \lambda_2, \ldots, \lambda_n$ be the roots of $f(X)$ in $\overline{K}$, and let

$$r = \min_{i \neq j} |\lambda_i - \lambda_j|.$$

Let $g(X)$ be as in the statement: a monic polynomial of degree n. Let $\mu_1, \mu_2, \ldots, \mu_n$ be the roots (listed with multiplicities, of course) of $g(X)$ in $\overline{K}$, so that $g(X) = \prod(X - \mu_j)$. Let

$$D = \prod_i g(\lambda_i) = \prod_{i,j}(\lambda_i - \mu_j).$$

Claim 1: If $|D| < r^{n^2}$, then $g(X)$ is irreducible over K and has a root in $L = K(\lambda)$.

Proof of Claim 1: If $|D| < r^{n^2}$, then at least one of the n^2 factors of D must have absolute value less than r. In other words, there must be a pair (i, j) such that $|\lambda_i - \mu_j| < r$. Since r is the minimum distance between λ_i and its conjugates, we can apply Krasner's Lemma to conclude that $K(\lambda_i) \subset K(\mu_j)$. Hence

$$[K(\mu_j) : K] \geq [K(\lambda_i) : K] = n.$$

But μ_j is a root of a polynomial of degree n, so the only way this can happen is when the polynomial is irreducible and the degree is exactly n. Since both fields are then of degree n, and one is contained in the other, they must in fact be equal.

Thus, we have shown that $g(X)$ is irreducible, and that $K(\lambda_i) = K(\mu_j)$. If $i = 1$, this is what we wanted to prove. If not, there is a little step more: there is certainly an automorphism of $\overline{K}$ that sends λ_i to λ, and this automorphism must send μ_j to some other root $\mu = \mu_{j_1}$ of $g(X)$. Applying the automorphism to the equality $K(\lambda_i) = K(\mu_j)$ gives $L = K(\lambda) = K(\mu)$, so that $g(X)$ has a root μ that belongs to L, which is what we wanted to prove.

Claim 2: There exists a real number $\varepsilon > 0$ such that if $|a_i - b_i| < \varepsilon$, then $|D| < r^{n^2}$.

Proof of Claim 2: We leave this one to the reader. It is a matter of expressing how D depends on the coefficients of the polynomials involved. $\square$

Problem 280 Prove "Claim 2." Suggestion: try to show that the function that maps

$$(a_0, a_1, \ldots, a_{n-1}, b_0, b_1, \ldots, b_{n-1}) \mapsto D$$

is a continuous function. Why does this do the trick?

Problem 281 (For those confident of their abstract algebra) In the last two results we imposed a litany of conditions on our field: "non-archimedean complete valued field of characteristic zero." Which of these conditions are seriously needed? Are there weaker forms of these results which are valid in more generality?

One way of grasping what the corollary says is to think of it as saying that at least some aspect of the "root structure" of a polynomial varies "continuously" as a function of the coefficients of the polynomial. Specifically, it says (or, to be precise, its proof shows) that when two irreducible polynomials are "close enough" (in the sense that the coefficients are close) there will be a root of one "close" to any root of the other. The next problem gives a result in the same spirit.

Problem 282 The point of this problem is to state (and have the reader prove) a version of the statement that "the roots of a polynomial are continuous functions of the coefficients." For this, let $f(X) = \sum a_i X^i$ be a polynomial of degree n whose roots in $\overline{\mathbb{Q}}_p$ are all distinct. Show that given an $\varepsilon > 0$ there exists a $\delta > 0$ such that for any other polynomial $g(X) = \sum b_i X^i$ of degree n such that $|a_i - b_i| < \delta$ for $i = 0, 1, \ldots, n$ and every root λ of $f(X)$ there exists exactly one root μ of $g(X)$ satisfying $|\mu - \lambda| < \varepsilon$. (Ooff! That's quite a mouthful!)

Problem 283 Is the hypothesis (in both the corollary and the previous problem) that the polynomials have the same degree really necessary?

Problem 284 Is the "continuity of the roots as functions of the coefficients" true for polynomials with real and/or complex coefficients?

We now proceed to the big result in this section.

Theorem 6.8.4 *The algebraic closure $\overline{\mathbb{Q}}_p$ is not complete with respect to the (extended) p-adic absolute value.*

PROOF: Proving the theorem is going to require us to come up with a Cauchy sequence in $\overline{\mathbb{Q}}_p$ which does not converge. Since finite extensions of $\mathbb{Q}_p$ are complete, this sequence must involve numbers from bigger and bigger extensions as it proceeds (because otherwise we could find a finite extension containing all the terms of the sequence, and it would have to converge). So this is going to be a complicated sequence!

In fact, it is going to be an infinite series whose general term tends to zero, but which does not converge. That is good enough, of course, since we've already shown that the partial sums of such a series form a Cauchy sequence. We'll construct our example slowly and carefully, in the hope that no one gets lost on the way. Each term in our series will belong to an *unramified* extension of $\mathbb{Q}_p$, so that our example will actually show that the union $\mathbb{Q}_p^{\mathrm{unr}}$ of all the unramified extensions of $\mathbb{Q}_p$ is already not a complete field.

Enough preliminaries: here goes. Remember that one gets unramified extensions of $\mathbb{Q}_p$ by adjoining roots of unity of order prime to p. We begin by putting together a large list of these. Choose integers $f_0, f_1, f_2, \ldots$ such that $f_i < f_{i+1}$ and $f_i | f_{i+1}$. Notice that the f_i will get arbitrarily large as i grows, since $f_{i+1} \geq 2 f_i$. For each i, let $m_i = p^{f_i} - 1$ and let ζ_i be a primitive m_i-th root of unity, so that $\mathbb{Q}_p(\zeta_i)$ is the unique unramified extension of degree f_i.

Now construct the series

$$\sum_{i=0}^{\infty} \zeta_i \, p^i.$$

The partial sums of this series clearly form a Cauchy sequence in $\overline{\mathbb{Q}}_p$ (and in fact even in $\mathbb{Q}_p^{\mathrm{unr}}$). We want to prove that this sequence does not have a limit in $\overline{\mathbb{Q}}_p$.

Well, suppose it did, and call the limit $c \in \overline{\mathbb{Q}}_p$. Whatever it is, c must be a root of some irreducible polynomial over $\mathbb{Q}_p$, since it is an element of the algebraic closure. Say that this polynomial has degree d, so that $[\mathbb{Q}_p(c) : \mathbb{Q}_p] = d$.

Now we want to use Corollary 6.5.6. We have

$$c = \zeta_0 + \zeta_1 p + \zeta_2 p^2 + \ldots,$$

so $c \equiv \zeta_0 \pmod p$. By the Corollary, this implies $\mathbb{Q}_p(\zeta_0) \subset \mathbb{Q}_p(c)$, which implies $d = [\mathbb{Q}_p(c) : \mathbb{Q}_p] \geq [\mathbb{Q}_p(\zeta_0) : \mathbb{Q}_p] = f_0$.

Now let $c_1 = (c - \zeta_0)/p$. Since $\zeta_0 \in \mathbb{Q}_p(c)$, we have $c_1 \in \mathbb{Q}_p(c)$ as well. But clearly $c_1 \equiv \zeta_1 \pmod p$. Using the Corollary again we get $\mathbb{Q}_p(\zeta_1) \subset \mathbb{Q}_p(c)$ and hence $d \geq f_1$.

Continuing this way (or, if you prefer, by induction), we get $d > f_i$ for all i, which is impossible because $f_i \to \infty$ as i grows. $\square$

In fact, since all the ζ_i were roots of unity of order prime to p, we have also proved:

Theorem 6.8.5 *The maximal unramified extension $\mathbb{Q}_p^{\mathrm{unr}}$ of $\mathbb{Q}_p$, obtained by adjoining all the roots of unity of order prime to p, is not complete with respect to the (extended) p-adic absolute value.*

Notice that our proof has constructed an explicit element that is transcendental over $\mathbb{Q}_p$. One can use the basic idea of the proof to obtain a general method for showing that the sums of certain series are transcendental.

Problem 285 Explain the statement: "the expression of c_n as a series is actually its p-adic expansion." Does this observation make the proof/construction easier to understand?

Well, "c'est la vie." Since $\overline{\mathbb{Q}}_p$ is not complete, we need to construct a completion. This is done exactly as in the case of $\mathbb{Q}_p$, by playing with the ring of all Cauchy sequences in $\overline{\mathbb{Q}}_p$. (In fact, our construction in Chapter 3 clearly works in utter generality.) The upshot is the following:

Proposition 6.8.6 *There exists a field $\mathbb{C}_p$ and an absolute value $|\ |$ on $\mathbb{C}_p$ such that:*

- *$\mathbb{C}_p$ contains $\overline{\mathbb{Q}}_p$, and the restriction of $|\ |$ to $\overline{\mathbb{Q}}_p$ coincides with the p-adic absolute value;*

- *$\mathbb{C}_p$ is complete with respect to $|\ |$; and*

- *$\overline{\mathbb{Q}}_p$ is dense in $\mathbb{C}_p$.*

Recall that whenever we have a convergent sequence $x_n \to x \neq 0$ in a non-archimedean field, there exists an N such that $|x_n| = |x|$ for $n \geq N$ (this is Lemma 3.2.10, except that we are saying that it works for any non-archimedean field... which it clearly does). This means that the set of possible absolute values in $\mathbb{C}_p$ is exactly the same as in $\overline{\mathbb{Q}}_p$. In other words,

Proposition 6.8.7 *If $x \in \mathbb{C}_p$, $x \neq 0$, then there exists a rational number $v \in \mathbb{Q}$ such that $|x| = p^{-v}$. In other words, the p-adic valuation v_p extends to $\mathbb{C}_p$, and the image of $\mathbb{C}_p^\times$ under v_p is $\mathbb{Q}$.*

Problem 286 Convince yourself that the last two propositions are true. (In both cases, it is just a matter of seeing that arguments we presented before in the case of $\mathbb{Q}_p$ extend without any difficulty.)

Problem 287 Since we have a valuation, we have a valuation ring, its valuation ideal, etc. Give explicit definitions. Can you describe the residue field? Is the valuation ideal principal? Does the concept of a uniformizer still make sense?

We write $\mathfrak{O}$ for the valuation ring of $\mathbb{C}_p$, i.e.,

$$\mathfrak{O} = \{x \in \mathbb{C}_p : |x| \leq 1\}.$$

This contains the valuation ideal

$$\mathfrak{P} = \{x \in \mathbb{C}_p : |x| < 1\}.$$

As always, $\mathfrak{O}$ is a local ring.

Problem 288 Show that any element of $\mathbb{C}_p$ can be written as a product of (i) a root of unity, (ii) a 1-unit, and (iii) a fractional power of p.

$\mathbb{C}_p$ is an enormous field, gotten by a series of complicated operations: start with $\mathbb{Q}$ and the p-adic absolute value, take a completion, take the algebraic closure of the result, and then complete once again! One might wonder whether the process will ever stop, i.e., whether one might need to take another algebraic closure, and so proceed without ever ending. On the contrary:

Proposition 6.8.8 $\mathbb{C}_p$ *is algebraically closed.*

PROOF: We give a jazzy proof, and ask the reader to come up with a more direct proof in a problem.

Take an irreducible polynomial $f(X)$ with coefficients in $\mathbb{C}_p$. Since $\overline{\mathbb{Q}}_p$ is dense in $\mathbb{C}_p$, we can find polynomials of the same degree and with coefficients in $\overline{\mathbb{Q}}_p$ whose coefficients are as close as we like to the coefficients of $f(X)$. By Corollary 6.8.3, if we choose such an $f_0(X)$ with coefficients close enough to those of $f(X)$, it will be irreducible over $\mathbb{C}_p$, and *a fortiori* also irreducible over $\overline{\mathbb{Q}}_p$. Since $\overline{\mathbb{Q}}_p$ is algebraically closed, this means that $f_0(X)$ will have degree one. Since $f(X)$ and $f_0(X)$ have the same degree, it follows that $f(X)$ has degree one.

This shows that any irreducible polynomial in $\mathbb{C}_p$ must be of degree one, which means that $\mathbb{C}_p$ is algebraically closed. $\qquad\qquad\square$

Problem 289 Here's an idea for a more direct proof (which might be technically more difficult). Take any polynomial $f(X) \in \mathbb{C}_p[X]$; we can assume it has no repeated roots (do you see why?). Build a sequence of polynomials $f_i(X) \in \overline{\mathbb{Q}}_p[X]$, all of the same degree, whose coefficients approach those of $f(X)$. Show, using Problem 282, that one can choose a root of each of the $f_i(X)$ so as to form a Cauchy sequence in $\overline{\mathbb{Q}}_p$ which converges to a root of $f(X)$ in $\mathbb{C}_p$.

Problem 290 Show that $\mathbb{C}_p$ is not locally compact.

In fact, one can show that any locally compact (and therefore complete) valued field of characteristic zero must be isomorphic to either $\mathbb{R}$, $\mathbb{C}$, or a finite extension of $\mathbb{Q}_p$. One can even[16] start the whole thing from here, i.e., start with a locally compact field of characteristic zero, and reconstruct the absolute value from the Haar measure on that field.

Let's conclude with two somewhat surprising remarks. The first is this: if we ignore the absolute value, $\mathbb{C}_p$ is just an algebraically closed field of characteristic 0. It's easy to work out that it has the same cardinality as the field of complex numbers. But there is only one such field up to isomorphism!

Theorem 6.8.9 (Steinitz) *Any two uncountable algebraically closed fields with the same characteristic and the same cardinality are isomorphic.*

The idea of the proof is to show that any such field is isomorphic to the result of starting from $\mathbb{Q}$ or $\mathbb{F}_p$, adjoining a transcendence basis of the right cardinality, and then passing to the algebraic closure. (But since the result is false for *countable* algebraically closed fields, there are subtleties.) The isomorphism comes from choosing a bijection between transcendence bases, so in fact there are infinitely many distinct isomorphisms, but all of them will fix the elements of $\mathbb{Q}$ and all will map algebraic elements to algebraic elements. So we can always, if we want to, choose a field[17] isomorphism $\varphi : \mathbb{C}_p \longrightarrow \mathbb{C}$ and use it to identify the algebraic closure of $\mathbb{Q}$ inside $\mathbb{C}_p$ and the algebraic closure of $\mathbb{Q}$ inside $\mathbb{C}$. This is a very common move in applications to algebraic number theory.

The second surprising remark is that for some applications $\mathbb{C}_p$ is still too small. It turns out that in applications to functional analysis we often want more than just completeness. Instead, we need *spherical completeness*.

---

[16]Mumbo-jumbo alert: this sentence talks about high-powered stuff which we really don't think our readers know about.

[17]*Not* an isomorphism of valued fields, of course.

Suppose are working in a valued field K and we have sequences $a_n \in K$, $r_n \in \mathbb{R}$ for which we get a nested sequence of closed balls $\overline{B}(a_n, r_n)$:

$$\overline{B}(a_1, r_1) \supset \overline{B}(a_2, r_2) \supset \cdots \supset \overline{B}(a_n, r_n) \supset \overline{B}(a_{n+1}, r_{n+1}) \supset \ldots$$

We say that a valued field K is *spherically complete* if any such sequence of closed balls has non-empty intersection. If K is complete and the radii r_n are a decreasing sequence with $\lim_{n \to \infty} r_n = 0$, then we can prove that the intersection is non-empty. (This is Problem 79.) But if we don't require the radii to tend to zero, it is possible to get empty intersection even if K is complete. Indeed, as the reader has probably guessed, $\mathbb{C}$ is spherically complete, but $\mathbb{C}_p$ is not. As a result, for certain applications we need to use an even bigger field. See [53, III.1] for details and for a construction.

7 Analysis in $\mathbb{C}_p$

This chapter tries to give the reader a taste of what analysis in $\mathbb{C}_p$ is like. Rather than attempt to be exhaustive, which would violate the goals of this book, we try to touch on a few remarkable points: the p-adic Weierstrass Preparation Theorem, the description of entire functions, and the theory of Newton polygons. As usual, the first step is to re-appropriate all the results we obtained earlier. We then go on to consider how to extend the p-adic valuation to polynomials and power series. This will yield a norm on the spaces of polynomials and of power series, which will prove to be an important tool. We then go on to proving the main theorems themselves.

Before we start, recall the notation we introduced above: we write

$$\mathfrak{O} = \{x \in \mathbb{C}_p : |x| \leq 1\}$$

for the valuation ring in $\mathbb{C}_p$ (we might want to call it the ring of integers in $\mathbb{C}_p$) and

$$\mathfrak{P} = \{x \in \mathbb{C}_p : |x| < 1\}$$

for the valuation ideal. The ideal $\mathfrak{P}$ is not principal, and the residue field $\mathbb{F} = \mathfrak{O}/\mathfrak{P}$ is an algebraic closure of $\mathbb{F}_p$.

7.1 Almost Everything Extends

As we have already pointed out, most of the results in Chapter 5 did not really depend on the fact that we were working over $\mathbb{Q}_p$; in fact, they hold just as well for more general non-archimedean valued fields. In particular, we can repeat (and enlarge) our list of "things that extend:"

i) A sequence (a_n) in $\mathbb{C}_p$ is Cauchy if and only if

$$\lim_{n \to \infty} |a_{n+1} - a_n| = 0.$$

ii) If a sequence (a_n) converges to a non-zero limit a, then we have $|a_n| = |a|$ for all sufficiently large n.

iii) A series $\sum a_n$ in $\mathbb{C}_p$ converges if and only if its general term tends to zero.

iv) Proposition 5.1.4 holds for double series in $\mathbb{C}_p$.

© Springer Nature Switzerland AG 2020

F. Q. Gouvêa, *p-adic Numbers*, Universitext, https://doi.org/10.1007/978-3-030-47295-5_7

v) A power series $f(X) = \sum a_n X^n$ with coefficients $a_n \in \mathbb{C}_p$ defines a continuous function on an open ball of radius $\rho = 1/\limsup \sqrt[n]{|a_n|}$; the function extends to the closed ball of radius ρ if $|a_n|\rho^n \to 0$ as $n \to \infty$. Note that in contrast to what happens in $\mathbb{Q}_p$ or even in its finite extensions, we can characterize ρ by saying that $\sum a_n x^n$ converges for $|x| < \rho$ and diverges for $|x| > \rho$.

vi) Therefore, given a power series $f(X) = \sum a_n X^n$ with radius of convergence ρ, we can define a function on the open (and perhaps also the closed) ball of radius ρ around $\alpha \in \mathbb{C}_p$ by putting

$$f(x) = \sum a_n (x - \alpha)^n$$

for any $x \in B(\alpha, \rho)$ (or $\overline{B}(\alpha, \rho)$).

vii) Proposition 5.4.2, Theorem 5.4.3, and Problem 159 are true for power series with coefficients in $\mathbb{C}_p$.

viii) Functions defined by power series are differentiable, and their derivatives are defined by the formal derivative of the original series.

ix) If $f(X) = \sum a_n X^n$ and $g(X) = \sum b_n X^n$ are power series with coefficients in $\mathbb{C}_p$, x_m is a convergent sequence contained in the intersection of the disks of convergence of f and g, and $f(x_m) = g(x_m)$ for all m, then $a_n = b_n$ for all n.

x) Strassman's Theorem holds without any change beyond replacing $\mathbb{Q}_p$ with $\mathbb{C}_p$ and $\mathbb{Z}_p$ with $\mathfrak{O}$.

xi) The corollaries to Strassman's Theorem therefore also extend.

xii) The usual power series defines a p-adic logarithm function

$$\log_p : U_1 \longrightarrow \mathbb{C}_p,$$

where

$$U_1 = \{x \in \mathfrak{O} : |x - 1| < 1\} = B(1, 1) = 1 + \mathfrak{P}.$$

This function satisfies the functional equation

$$\log_p(xy) = \log_p(x) + \log_p(y)$$

for any $x, y \in U_1$.

xiii) The usual power series defines an exponential function $\exp_p : D \longrightarrow \mathbb{C}_p$, where

$$D = \{x \in \mathfrak{O} : |x| < p^{-1/(p-1)}\} = B(0, p^{-1/(p-1)}).$$

This function satisfies the functional equation

$$\exp_p(x + y) = \exp_p(x) \exp_p(y)$$

for any $x, y \in D$.

xiv) If $x \in D$, then $|\exp_p(x) - 1| < p^{-1/(p-1)}$, i.e., $\exp_p(x) \in 1 + D$, and we have

$$\log_p(\exp_p(x)) = x.$$

xv) If $|x - 1| < p^{-1/(p-1)}$ (i.e., $x \in 1 + D$), then $\log_p(x) \in D$, and we have

$$\exp_p(\log_p(x)) = x.$$

xvi) The p-adic logarithm gives a homomorphism from the multiplicative group $U_1 = 1 + \mathfrak{P}$ *onto* the additive group of $\mathbb{C}_p$. (Compare with what happens for finite extensions!)

xvii) The p-adic logarithm gives an isometric isomorphism from the multiplicative group

$$1 + D = B(1, p^{-1/(p-1)})$$

to the additive group

$$D = B(0, p^{-1/(p-1)}).$$

xviii) For each $\alpha \in \mathbb{Z}_p$, the binomial series $(1 + x)^\alpha = \mathbf{B}(\alpha, x)$ converges whenever $|x| < 1$ (i.e., for $x \in \mathfrak{P} = B(0, 1)$). In other words, u^α is well defined whenever $u \in B(1, 1) = 1 + \mathfrak{P}$ is a 1-unit in $\mathfrak{O}$ and $\alpha \in \mathbb{Z}_p$ is a p-adic integer.

PROOF: As before, the main change has to do with the image of $\log_p$ on its domain. In $\mathbb{Q}_p$, the image is contained in $p\mathbb{Z}_p$. In a finite extension K, the image is not necessarily contained in $\mathcal{O}_K$ but it is still bounded. In $\mathbb{C}_p$, (*xvi*) says that the image is not only unbounded, but is in fact all of $\mathbb{C}_p$.

The proof is actually fairly easy. Choose $a \in \mathbb{C}_p$. Since $p^n a \to 0$, we can choose $n \in \mathbb{Z}$ large enough so that $|p^n a| < p^{-1/(p-1)}$, i.e., $p^n a \in D$. Let $y = \exp(p^n a)$; we know $y \in 1 + D$. Since $\mathbb{C}_p$ is algebraically closed, there exists an $x \in \mathbb{C}_p$ such that $x^{p^n} = y$.

We want to show $x \in U_1$. Clearly $|x|^{p^n} = |y| = 1$, so $|x| = 1$. Let $\bar{x}$ be the image of x in $\overline{\mathbb{F}}_p = \mathfrak{O}/\mathfrak{P}$. Then $\bar{x}^{p^n} = \bar{y} = 1$. But the only root of $t^{p^n} - 1 = (t - 1)^{p^n}$ in a field of characteristic p is $t = 1$, so we conclude $\bar{x} = 1$, which says that $x \in 1 + \mathfrak{P}$, i.e., $x \in U_1$.

Now we can compute $\log_p(x)$. But

$$p^n a = \log_p(y) = \log_p(x^{p^n}) = p^n \log_p(x),$$

which gives $\log_p(x) = a$. Hence the homomorphism $\log_p : U_1 \longrightarrow \mathbb{C}_p$ is onto. We'll leave the remaining items as an exercise.   □

Problem 291 Make sure you understand how to prove the remaining claims above. Pay particular attention to where the situation in $\mathbb{C}_p$ differs from the ones we considered before. For example, why is the statement characterizing ρ in item (*v*) true?

Problem 292 In assertion (*xvii*), is it true that the additive group of D is isomorphic to the additive group of $\mathfrak{O}$? (The analogous statement *was* true in $\mathbb{Q}_p$ and finite extensions.)

Another important class of results that can be extended to $\mathbb{C}_p$ are the several variants of Hensel's Lemma. The trick here is not to use versions which refer to uniformizers, since there are no uniformizers in $\mathbb{C}_p$. (A uniformizer would be an element with the largest possible absolute value which was still less than one, but in $\mathbb{C}_p$ we have elements with absolute value p^r for any $r \in \mathbb{Q}$, so there is no such thing.) Still, one can either replace "mod p" with "mod $\mathfrak{P}$" throughout, or simply state things in terms of absolute values. So here are two versions of Hensel's Lemma:

Theorem 7.1.1 (Hensel's Lemma in $\mathbb{C}_p$) *Let*

$$F(X) = a_0 + a_1 X + a_2 X^2 + \cdots + a_n X^n$$

be a polynomial whose coefficients are in $\mathfrak{O}$. Suppose that there exists an $\alpha_1 \in \mathfrak{O}$ such that

$$|F(\alpha_1)| < 1 \qquad and \qquad |F'(\alpha_1)| = 1$$

where $F'(X)$ is the (formal) derivative of $F(X)$. Then there exists an $\alpha \in \mathfrak{O}$ such that $|\alpha - \alpha_1| < 1$ and $F(\alpha) = 0$.

PROOF: We need to be just a little bit careful in order to avoid trouble. (The problem is that there is no uniformizer in $\mathbb{C}_p$, and all our proofs up to here depended explicitly on there being one. What we do is to use a convenient element of small valuation to replace the uniformizer.) Let $\delta = |F(\alpha_1)| < 1$, and choose $\pi \in \mathbb{C}_p$ such that $|\pi| = \delta$. Then the argument in the original proof of Hensel's Lemma (Theorem 4.5.2), with p replaced everywhere by π, will allow us to find α_2 such that $|\alpha_1 - \alpha_2| \leq \delta$ and $|F(\alpha_2)| \leq \delta^2$. Proceeding inductively, we get a sequence α_n which converges to the root α. $\qquad\square$

Problem 293 Check the details!

And now the other version, for polynomials: as before, one can state this with absolute values or by talking of reduction modulo the ideal $\mathfrak{P}$. We choose the second path here, simply so that we don't have to talk about "the maximum of the absolute values of the differences of the coefficients" of two polynomials.

Theorem 7.1.2 (Hensel's Lemma for Polynomials over $\mathbb{C}_p$)   *Let* $f(X) \in \mathfrak{O}[X]$ *be a polynomial with coefficients in* $\mathfrak{O}$, *and assume that there exist polynomials* $g_1(X)$ *and* $h_1(X)$ *in* $\mathfrak{O}[X]$ *such that*

  i) $g_1(X)$ *is monic,*

  ii) $g_1(X)$ *and* $h_1(X)$ *are relatively prime modulo* $\mathfrak{P}$, *and*

  iii) $f(X) \equiv g_1(X)h_1(X)$ (mod $\mathfrak{P}$) *(understood coefficient-by-coefficient).*

Then there exist polynomials $g(X)$, $h(X) \in \mathfrak{O}[X]$ *such that*

  i) $g(X)$ *is monic,*

  ii) $g(X) \equiv g_1(X)$ (mod $\mathfrak{P}$) *and* $h(X) \equiv h_1(X)$ (mod $\mathfrak{P}$), *and*

  iii) $f(X) = g(X)h(X)$.

Problem 294 Give a proof of Hensel's Lemma for Polynomials. The same caution we used for the first version will be necessary, and the relevant δ will be the maximum of the absolute values of the differences between coefficients of $f(X)$ and coefficients of $g_1(X)\,h_1(X)$.

Problem 295 In the statement of Hensel's Lemma for Polynomials, we make the assumption that "$g_1(X)$ and $h_1(X)$ are relatively prime modulo $\mathfrak{P}$." Show, using the fact that the residue field $\mathbb{F} = \mathfrak{O}/\mathfrak{P}$ is algebraically closed, that this can be replaced by "$g_1(X)$ and $h_1(X)$ have no common roots in $\mathbb{F}$."

Problem 296 In this version of Hensel's Lemma, can we conclude that $\deg g(X) = \deg g_1(X)$?

7.2   Deeper Results on Polynomials and Power Series

The main goal of this section is to prove a theorem that has become known as the "p-adic Weierstrass Preparation Theorem." This is, of course, the p-adic version of a classical theorem due to Weierstrass which deals with power series in several variables and is an important tool in the theory of functions of several complex variables. One can find the statement in many texts; for example, see [65, p. 22]. There are also versions of the theorem which apply to formal power series in several variables; these versions are useful in algebraic geometry. For this version, see, for example, [67, Vol. 2, VII.1, Thm. 5]. The p-adic version gives fundamental information on p-adic functions defined by power series. Our account of this theorem follows the one in [14].

We will approach power series by way of polynomials. In other words, we will want to think of a power series as a limit of polynomials, and our results will be proved first for polynomials, then for power series.

While we will constantly keep $\mathbb{C}_p$ in mind, the results we will obtain are true, interesting, and useful when we work over other fields, too. In fact, since some of them describe how polynomials factor, they are especially interesting when the field is not algebraically closed. On the other hand, we will often want to interpret what the theorems say in terms of the roots of the polynomials, in which case it's most convenient to place ourselves in $\mathbb{C}_p$. So we'll switch back and forth between these two situations. Just to fix notation, let's let K be some extension of $\mathbb{Q}_p$ which is complete; in practice, we will always be working either with a finite extension of $\mathbb{Q}_p$ or with $\mathbb{C}_p$. Let's write $\mathcal{O}$ for the valuation ring $\mathcal{O} = \{x \in K : |x| \le 1\}$, $\mathfrak{p}$ for its maximal ideal, and $\mathbb{F}$ for the residue field. It might be good to remind ourselves that $\mathbb{F}$ is a finite field if K is a finite extension of $\mathbb{Q}_p$ and that it is the algebraic closure of $\mathbb{F}_p$ when $K = \mathbb{C}_p$.

The first step is to define absolute values (or norms) in the spaces we are interested in studying. Let's look first at polynomials. The most obvious way to define a norm on the space of polynomials is to simply look at the coefficients. This works, but we actually want to do something a little more subtle.

Suppose we are interested in understanding how the values of a polynomial

$$f(X) = a_0 + a_1 X + a_2 X^2 + \cdots + a_n X^n$$

vary when we plug in numbers belonging to the closed ball of radius c around the origin. Then, if $x \in \overline{B}(0,c)$, we have $|x| \le c$, and

$$\begin{aligned}
|f(x)| = |a_0 + a_1 x + a_2 x^2 + \cdots + a_n x^n| \\
\le \max\{|a_0|, |a_1 x|, |a_2 x^2|, \ldots, |a_n x^n|\} \\
\le \max\{|a_0|, |a_1| c, |a_2| c^2, \ldots, |a_n| c^n\} \\
\le \max_i |a_i| c^i.
\end{aligned}$$

If $c = 1$, this last number is just the "obvious" measure of the size of $f(X)$: the absolute value of the largest coefficient. For other values of c, it turns out that we may still use this number as a good measure of the size of $f(X)$.

Theorem 7.2.1 *Let $c > 0$ be an arbitrary positive real number. Define a function $\| \ \|_c : K[X] \longrightarrow \mathbb{R}_+$ as follows: for each polynomial*

$$f(X) = a_0 + a_1 X + a_2 X^2 + \cdots + a_n X^n,$$

set

$$\|f(X)\|_c = \max_i |a_i| c^i.$$

Then we have

i) $\|f(X)\|_c = 0$ *if and only if $f(X)$ is identically zero.*

ii) $\|f(X) + g(X)\|_c \leq \max\{\|f(X)\|_c, \|g(X)\|_c\}.$

iii) $\|f(X)g(X)\|_c = \|f(X)\|_c\|g(X)\|_c.$

iv) If $f(X) = a_0$ is a polynomial of degree zero, then $\|f(X)\|_c = |a_0|.$ In other words, $\| \ \|_c$ induces the p-adic absolute value on the constants.

v) If $|x| \leq c$, then $|f(x)| \leq \|f(X)\|_c.$

vi) The function $\| \ \|_c$ extends to a non-archimedean absolute value on the field of rational functions $K(X)$.

PROOF: A lot of this is easy, and can be safely left to the reader. In fact, the first and second statements follow at once from the properties of the p-adic absolute value, and the last two follow at once from the definition of $\| \ \|_c$. The statement about multiplicativity is the hard one.

Let $f(X)$ and $g(X)$ be two polynomials. Write

$$f(X) = a_0 + a_1 X + a_2 X^2 + \cdots + a_n X^n$$
$$g(X) = b_0 + b_1 X + b_2 X^2 + \cdots + b_n X^n$$

where of course we may very well have some zero coefficients (nobody said $f(X)$ and $g(X)$ had the same degree). Then each coefficient of the product $f(X)g(X)$ looks like a sum

$$\sum_{i+j=k} a_i b_j,$$

and we can estimate the absolute value by

$$\left| \sum_{i+j=k} a_i b_j \right| c^k \leq \max_{i+j=k} |a_i||b_j|c^k = \max_{i+j=k} \left(|a_i|c^i \right)\left(|b_j|c^j \right),$$

which is certainly less than or equal to

$$\left(\max_i |a_i|c^i \right) \left(\max_j |b_j|c^j \right) = \|f(X)\|_c\|g(X)\|_c.$$

This shows that
$$\|f(X)g(X)\|_c \leq \|f(X)\|_c\|g(X)\|_c.$$

Proving the reverse inequality is much trickier. (Part of the reason is that the estimate above could afford to be really "sloppy," since we were looking only for an upper bound. To get the converse, we must be careful about exactly what gets multiplied by what.) Here is a proof.

Let's begin by giving names to things. Choose I so that

$$|a_I|c^I = \|f(X)\|_c \qquad \text{and} \qquad |a_i|c^i < \|f(X)\|_c \quad \text{for } i < I.$$

In other words, I is chosen to be the smallest exponent for which $|a_i|c^i$ achieves its maximum. Similarly, choose J so that $|b_J|c^J$ achieves the maximum:

$$|b_J|c^J = \|g(X)\|_c \quad \text{and} \quad |b_j|c^j < \|g(X)\|_c \quad \text{for } j < J.$$

(These are clearly the coefficients to keep track of!)

Now look at the coefficient of X^{I+J} in the product $f(X)g(X)$. It is given by the horrible formula

$$\sum_{i+j=I+J} a_i b_j.$$

We want to estimate each term of this sum. There are three cases to consider:

Suppose $i < I$. In this case, we know that

$$|a_i|c^i < \|f(X)\|_c \quad \text{and} \quad |b_j|c^j \le \|g(X)\|_c.$$

Putting these two together gives

$$|a_i b_j| < c^{-i-j}\|f(X)\|_c\|g(X)\|_c = c^{-I-J}\|f(X)\|_c\|g(X)\|_c.$$

(Remember that $i + j = I + J$ in our sum!)

Suppose $j < J$. This is similar to the previous case; just switch the roles of i and j to get

$$|a_i b_j| < c^{-i-j}\|f(X)\|_c\|g(X)\|_c = c^{-I-J}\|f(X)\|_c\|g(X)\|_c.$$

Finally, if $i = I$ and $j = J$, we get $|a_I|c^I = \|f(X)\|_c$ and $|b_J|c^J = \|g(X)\|_c$, so we get an equality:

$$|a_I b_J| = c^{-I-J}\|f(X)\|_c\|g(X)\|_c.$$

This means that in the sum

$$\sum_{i+j=I+J} a_i b_j$$

there is one largest term: the one with $i = I$ and $j = J$. Since we are in a non-archimedean field, "the strongest wins": the absolute value of the sum will be equal to the absolute value of the largest term. In other words,

$$\left|\sum_{i+j=I+J} a_i b_j\right| = c^{-I-J}\|f(X)\|_c\|g(X)\|_c$$

which we can rewrite as

$$\left|\sum_{i+j=I+J} a_i b_j\right|c^{I+J} = \|f(X)\|_c\|g(X)\|_c.$$

Now, to compute $\|f(X)g(X)\|_c$, one has to take the maximum over all coefficients of the product; this last inequality says that the $I+J$-th coefficient already gives something equal to $\|f(X)\|_c\|g(X)\|_c$. The maximum can only be bigger. In other words, we have proved

$$\|f(X)g(X)\|_c \geq \|f(X)\|_c\|g(X)\|_c.$$

Putting this together with the opposite inequality (proved just above), we get what we claimed:

$$\|f(X)g(X)\|_c = \|f(X)\|_c\|g(X)\|_c.$$

As promised, the other statements are left to the reader.   □

Problem 297 Prove the remaining statements in the theorem.

Problem 298 Suppose we have two different complete fields $K_1 \subset K_2$. A polynomial $f(X) \in K_1[X]$ also belongs to $K_2[X]$. Show that the value of $\|f(X)\|_c$ does not depend on which ring we put it in. In other words, we have really defined a norm on $\mathbb{C}_p[X]$, and its restriction to $K[X]$ gives the norm for polynomials in $K[X]$.

Problem 299 We can interpret polynomials as $\mathbb{C}_p$-valued functions on K (and also on any extension of K, and even on $\mathbb{C}_p$), and in particular as functions on the closed ball $\overline{B}(0, c) \subset K$. This means we can define a norm on the space of polynomials using the "sup norm" from classical analysis:

$$\|f(X)\| = \sup_{\substack{|x| \leq c \\ x \in K}} |f(x)|.$$

Show that we have $\|f(X)\| \leq \|f(X)\|_c$. Does equality hold? (Hint: the answer is easier to get if $K = \mathbb{C}_p$.)

Problem 300 Now that we have norms on the space of polynomials, it is not difficult to restate the second form of Hensel's Lemma (Theorem 7.1.2) in terms of the $\| \ \|_1$ norm. Do so. Does a version using the $\| \ \|_c$ norm with $c \neq 1$ work?

The existence of the absolute values $\| \ \|_c$ on the ring of polynomials is a useful tool; for example, it can be used to give simpler proofs of some of the results in the previous chapter (such as Lemma 6.3.7 and the Eisenstein Irreducibility Criterion, Theorem 6.3.11). Some examples will also appear below.

Problem 301 Suppose c is a real number of the form p^r with $r \in \mathbb{Q}$, and let α be an element of $\mathbb{C}_p$ such that $|\alpha| = c$. (Why does one exist?) Show that the map $\phi : \mathbb{C}_p[X] \longrightarrow \mathbb{C}_p[X]$ defined by $X \mapsto \alpha X$ satisfies the condition

$$\|f(X)\|_c = \|\phi(f(X))\|_1.$$

Notice that ϕ is clearly a ring isomorphism. What does this tell us about the relation between $\| \ \|_1$ and the various $\| \ \|_c$?

Problem 302 How would one have to restate the previous problem in order to get something that is true over some finite extension of $\mathbb{Q}_p$?

Problem 303 Can one do anything like the previous problems in the case where c is *not* of the form p^r with $r \in \mathbb{Q}$?

Lemma 7.2.2 *Let* $\| \ \| = \| \ \|_c$ *for some* $c > 0$, *and let* $f(X) \in K[X]$ *be any polynomial. Let*

$$g(X) = b_0 + b_1 X + b_2 X^2 + \cdots + b_N X^N$$

be a polynomial of degree N with coefficients in K satisfying the condition

$$\|g(X)\| = |b_N| c^N.$$

(In other words, the maximum of the $|b_n| c^n$ is realized at the very last co-efficient.) Let $q(X)$ and $r(X)$ be the quotient and the remainder which we obtain when we divide $f(X)$ by $g(X)$, so that

$$f(X) = g(X) q(X) + r(X) \qquad and \qquad \deg r(X) < N.$$

Then we have both

$$\|f(X)\| \geq \|q(X)\| \, \|g(X)\| \qquad and \qquad \|f(X)\| \geq \|r(X)\|.$$

PROOF: Rather than plod through a million inequalities (which is elementary but difficult), here is an attempt at a conceptual proof. See [14] for a more direct method. What we will do is handle the case $c = 1$ first, then move on to the general case by means of the ideas in the last three problems.

1. If $c = 1$, then we have $|b_N| = \max |b_i|$. Multiplying $g(X)$ by some element in K if necessary, we may assume that $|b_N| = 1$. Again, we can multiply the whole equation $f(X) = q(X) g(X) + r(X)$ by some element in K in order to get $\max\{\|q(X)\|, \|r(X)\|\} = 1$, which implies $\|f(X)\| \leq 1$. What the theorem says after both reductions is that in fact $\|f(X)\| = 1$.

Well, suppose not. Then every coefficient of $f(X)$ has absolute value less than 1, which means that they belong to the valuation ideal $\mathfrak{p}$. If we use bars to denote reduction modulo $\mathfrak{p}$, we get an equation

$$0 = \bar{f}(X) = \bar{g}(X) \bar{q}(X) + \bar{r}(X).$$

But now, since $|b_N| = 1$ (and here is where we seriously use the assumption about $g(X)$),

$$\deg \bar{g}(X) = N > \deg r(X) \geq \deg \bar{r}(X);$$

this forces $\bar{q}(X) = 0$, which then implies $\bar{r}(X) = 0$, which contradicts the assumption that $\max\{\|q(X)\|, \|r(X)\|\} = 1$. This proves the lemma when $c = 1$.

Before we go on, notice that the polynomials $q(X)$ and $r(X)$ necessarily have coefficients in K. Their norms, however, do not depend on the field, as we pointed out above (Problem 298). Since our theorem is about the norms, we might as well assume $K = \mathbb{C}_p$, and we will. This is essential, because we will want to consider elements with absolute value p^r with $r \in \mathbb{Q}$, and in $\mathbb{C}_p$ we know these exist.

2. **If $c \neq 1$, but c is of the form p^r for $r \in \mathbb{Q}$,** choose an element $\alpha \in \mathbb{C}_p$ with $|\alpha| = c$. Then consider the polynomials $f_1(X) = f(\alpha X)$ and $g_1(X) = g(\alpha X)$. It's easy to see that $\|f_1(X)\|_1 = \|f(X)\|_c$, and similarly for the g's. Applying part 1 to $f_1(X)$ and $g_1(X)$ then gives the inequality we want for the c-norms.

3. **If c is not of the form p^r for $r \in \mathbb{Q}$,** then we have to resort to black magic. Since the numbers of the form p^r are dense in $\mathbb{R}_+$ (can you see why?), there is a sequence c_i of real numbers such that $c_i = p^{r_i}$ with $r_i \in \mathbb{Q}$ and $c_i \to c$ as $i \to \infty$. Then clearly we have $\|f(X)\|_{c_i} \to \|f(X)\|_c$ as $i \to \infty$, so we can get the estimate we want by using part 2 for each of the c_i and taking the limit. $\qquad \square$

Now here's an interesting example of the kind of result we are led to. It is a version for polynomials of the Weierstrass Preparation Theorem.

Proposition 7.2.3 *Let $c > 0$ be some real number, and $\| \ \| = \| \ \|_c$. Let*

$$f(X) = a_0 + a_1 X + a_2 X^2 + \cdots + a_n X^n$$

be a polynomial in $K[X]$, and suppose that there exists an integer N such that $0 < N < n$ for which we have

$$\|f(X)\| = |a_N| c^N \qquad and \qquad \|f(X)\| > |a_j| c^j \quad for \ any \ j > N.$$

Then there exist polynomials $g(X)$, of degree N, and $h(X)$, of degree $n - N$, with coefficients in K, such that $f(X) = g(X)\, h(X)$. Furthermore, we have

$$\|g(X)\| = \|f(X)\| \qquad and \qquad \|h(X) - 1\| < 1.$$

PROOF: For the case $c = 1$, it would not be hard to give a direct proof using Theorem 7.1.2. Instead, we give a general argument (in roughly the same spirit) that works for all choices of c. The point is to start with an approximate factorization and then improve it. If we prove, by induction, that this can always be done, it will produce a convergent sequence of polynomials which, in the limit, give the factorization we want. In order to make the structure of the proof as clear as possible, we separate it into two pieces. The first will describe the general induction step. The next will show how the induction starts. Putting the two pieces together gives the proof.

Step: Let δ be a fixed real number, $\delta < 1$. Suppose that at some stage we have found polynomials $h_i(X)$ and $g_i(X)$ satisfying the following conditions:

i) $\deg g_i(X) = N$ and $\deg h_i(X) \leq n - N$,

ii) $\|f(X) - g_i(X)\|_c \leq \delta \|f(X)\|_c$ and $\|h_i(X) - 1\|_c \leq \delta$,

iii) $\|f(X) - g_i(X)h_i(X)\|_c \leq \delta^i \|f(X)\|_c$, and

iv) $g_i(X) = a_N X^N +$ lower degree terms and $\|g_i(X)\|_c = |a_N| c^N$.

(Never mind, for now, where such things might come from.) Let's describe how to get two polynomials that give a still better approximation.

First of all, because $\| \ \|_c$ satisfies the non-archimedean property ("all triangles are isosceles"), the condition $\|f(X) - g_i(X)\|_c \leq \delta \|f(X)\|_c$ implies (since δ is less than 1) that $\|f(X)\|_c = \|g_i(X)\|_c$. Now we find a way to bring in the estimates in the previous lemma.

If we divide $f(X) - g_i(X)h_i(X)$ by $g_i(X)$, we get

$$f(X) - g_i(X)h_i(X) = q(X)\, g_i(X) + r(X),$$

where $\deg r(X) < N$, and hence $\deg q(X) \leq n - N$. Since we know $g_i(X)$ satisfies condition *(iv)*, the previous lemma gives inequalities for the absolute values of $q(X)$ and of $r(X)$:

$$\|q(X)\|_c \leq \frac{\|f(X) - g_i(X)h_i(X)\|_c}{\|g_i(X)\|_c} = \frac{\|f(X) - g_i(X)h_i(X)\|_c}{\|f(X)\|_c} \leq \delta^i \leq \delta$$

and

$$\|r(X)\|_c \leq \|f(X) - g_i(X)h_i(X)\|_c \leq \delta^i \|f(X)\|_c \leq \delta \|f(X)\|_c.$$

Now let

$$g_{i+1}(X) = g_i(X) + r(X) \qquad \text{and} \qquad h_{i+1}(X) = h_i(X) + q(X).$$

We claim that these will do the job.

To begin with, since $\deg r(X) < N$, we will have $\deg g_{i+1}(X) = N$. Similarly, since $\deg q(X) \leq n - N$, we get $\deg h_{i+1} \leq n - N$. This shows our new polynomials still satisfy our first condition.

Next, we have

$$\begin{aligned}
\|f(X) - g_{i+1}(X)\|_c &= \|f(X) - g_i(X) - r(X)\|_c \\
&\leq \max\{\|f(X) - g_i(X)\|_c, \|r(X)\|_c\} \\
&\leq \delta \|f(X)\|_c,
\end{aligned}$$

since we have shown that $\|r(X)\|_c \leq \delta \|f(X)\|_c$. Similarly,

$$\begin{aligned}
\|h_{i+1}(X) - 1\|_c &= \|h_i(X) - 1 + q(X)\|_c \\
&\leq \max\{\|h_i(X) - 1\|_c, \|q(X)\|_c\} \\
&\leq \delta,
\end{aligned}$$

since we have shown that $\|q(X)\|_c \leq \delta$. This shows that the second condition above still holds.

Next, we check that this gives a better approximate factorization:

$$\begin{aligned}
f(X) - g_{i+1}(X)h_{i+1}(X) &= f(X) - (g_i(X) + r(X))(h_i(X) + q(X)) \\
&= f(X) - g_i(X)h_i(X) - q(X)g_i(X) - r(X)h_i(X) - r(X)q(X) \\
&= r(X) - r(X)h_i(X) - r(X)q(X) \\
&= r(X)\left(1 - h_i(X) - q(X)\right),
\end{aligned}$$

which gives

$$\begin{aligned}
\|f(X) - g_{i+1}(X)h_{i+1}(X)\|_c &= \|r(X)\|_c\|(1 - h_i(X)) - q(X)\|_c \\
&\leq \delta^i\|f(X)\|_c \max\{\|1 - h_i(X)\|_c, \|q(X)\|_c\} \\
&\leq \delta^{i+1}\|f(X)\|_c.
\end{aligned}$$

For the final condition, first notice that since $\deg r(X) < N$, adding it to $g_i(X)$, which has degree N, will not change the leading term, i.e., $g_{i+1}(X) = a_N X^N +$ terms of lower degree. Next, since $\|f(X) - g_{i+1}(X)\|_c \leq \delta\|f(X)\|_c$ and $\delta < 1$, we must have $\|g_{i+1}(X)\|_c = \|f(X)\|_c = |a_N|c^N$.

This means that $g_{i+1}(X)$ and $h_{i+1}(X)$ satisfy all our conditions, with δ^i replaced by δ^{i+1}.

To check that these functions actually form the sort of sequence we want, notice that the inequality $\|q(X)\|_c \leq \delta^i$ translates into

$$\|h_i(X) - h_{i+1}(X)\|_c \leq \delta^i.$$

Similarly, the inequality $\|r(X)\|_c \leq \delta^i\|f(X)\|_c$ translates to

$$\|g_i(X) - g_{i+1}(X)\|_c \leq \delta^i\|f(X)\|.$$

Since $\delta < 1$, these inequalities show that both the sequence of the $g_i(X)$ and the sequence of the $h_i(X)$ are Cauchy sequences with respect to the $\|\ \|_c$ norm.

Start: To start the process, we need to find a $\delta < 1$ and an initial pair $g_1(X)$ and $h_1(X)$. But those are relatively easy to find. The assumption is that $\|f(X)\|_c = |a_N|c^N$ and that the terms of higher degree are smaller. This means that if we subtract off the part of $f(X)$ up to degree N the remaining polynomial will have smaller norm. In other words, we will have

$$\left\|f(X) - \sum_{i=0}^N a_i X^i\right\|_c < \|f(X)\|_c.$$

Let δ be a measure of how much smaller:

$$\left\|f(X) - \sum_{i=0}^N a_i X^i\right\|_c = \delta\|f(X)\|_c.$$

Then, of course, $0 < \delta < 1$.

Then let

$$g_1(X) = \sum_{i=0}^{N} a_i X^i = a_0 + a_1 X + a_2 X^2 + \cdots + a_N X^N$$

and let $h_1(X) = 1$. It is very easy to check that all of our conditions are satisfied.

Convergence: We're almost there. We've got a Cauchy sequence of polynomials *of bounded degree*. And that's enough, by the next problem, to guarantee convergence. Taking the limit, we get $g(X)$ and $h(X)$ as specified by the proposition. □

Problem 304 Show that a Cauchy sequence of polynomials of bounded degree is convergent with respect to the $\| \ \|_c$. Show that the hypothesis that the degree is bounded is essential.

Problem 305 The limit of a sequence of polynomials of degree N is a polynomial of degree at most N. Why can we assert that $g(X)$ is actually of degree N?

Problem 306 Show that the polynomial $g(X)$ obtained in our proof has the property that $\|f(X) - g(X)\|_c < \|f(X)\|_c$. This can sometimes be a useful extra bit of information.

Problem 307 Can we re-apply the Proposition to factor $g(X)$ itself?

Problem 308 What can be said about the speed of the convergence of the sequences of polynomials $g_i(X)$ and $h_i(X)$?

Problem 309 Suppose $c = 1$. Give a proof of the proposition for this case that is just a direct application of Theorem 7.1.2.

From the ring $K[X]$ of polynomials we now move on to the ring $K[[X]]$ of power series with coefficients in K. Of course, convergence questions now become important. It will be useful to remember that the power series

$$\sum a_n X^n$$

converges for $|x| \leq c$ if and only if $\lim |a_n| c^n = 0$. This suggests that the same idea used for polynomials will make sense here. We first define appropriate subrings of the ring of power series with coefficients in K.

Definition 7.2.4 *Let $c > 0$ be an arbitrary positive real number. We define A_c to be the ring of power series $\sum a_n X^n \in K[[X]]$ which satisfy the condition $\lim |a_n| c^n = 0$.*

Notice that, if $f(X) \in A_c$, then $f(X)$ converges for x in the closed unit ball of radius c. For that reason, the rings A_c are often called *rings of convergent power series*. The next couple of problems check that this works as advertised.

Problem 310 Show that A_c is indeed a ring, and that it is also a vector space over K.

Problem 311 We have avoided indicating the base field in the notation for A_c so that the notation does not become too heavy. For this problem, however, write $A_c(K)$ for the ring we get when the field of coefficients is K. It's clear that if $K_1 \subset K_2$, then $A_c(K_1) \subset A_c(K_2)$. Show that in fact we have

$$A_c(K_1) = A_c(K_2) \cap K_1[[X]].$$

(In other words: the fact that the series is in A_c is independent of the field to which we think its coefficients belong.) We will use this fact, as before, to move up and down between smaller and bigger fields.

Problem 312 Suppose c can be written as a rational power of p, i.e., that there exists $r \in \mathbb{Q}$ such that $c = p^r$. Show that a power series $f(X)$ belongs to A_c if and only if it converges in the closed ball in $\mathbb{C}_p$ with center 0 and radius c. Is the same true without the assumption on c?

Problem 313 If $c_1 > c_2$, show that $A_{c_1} \subset A_{c_2}$.

Now we put norms on our spaces:

Theorem 7.2.5 *Let $c > 0$ be an arbitrary positive real number. Define a function $\| \ \|_c : A_c \longrightarrow \mathbb{R}_+$ as follows: for each power series*

$$f(X) = a_0 + a_1 X + a_2 X^2 + \cdots + a_n X^n + \ldots$$

belonging to A_c, set

$$\|f(X)\|_c = \max_n |a_n| c^n.$$

Then we have

i) $\|f(X)\|_c = 0$ *if and only if $f(X)$ is identically zero.*

ii) $\|f(X) + g(X)\|_c \leq \max\{\|f(X)\|_c, \|g(X)\|_c\}.$

iii) $\|f(X)g(X)\|_c \leq \|f(X)\|_c \|g(X)\|_c.$

iv) $\| \ \|_c$ *induces the p-adic absolute value on the constant power series.*

v) *If $|x| \leq c$, then $|f(x)| \leq \|f(X)\|_c$.*

PROOF: This is all easy. Notice that the definition makes sense because we know that $|a_n| c^n$ tends to zero as n grows, and hence there must be a maximum. $\qquad\square$

Problem 314 Prove the theorem. (It is all very straightforward.)

Problem 315 The game we played above relating the sup-norm for functions on $\overline{B}(0, c)$ with the c-norm on the polynomials also makes sense for power series in A_c. Does anything change?

Problem 316 Once again, the norm does not depend on the base field: show that if $K_1 \subset K_2$ then $\| \ \|_c$ on $A_c(K_2)$ restricts to $\| \ \|_c$ on $A_c(K_1)$. In particular, when we are interested only in computing the norm of some power series, we might as well think of it as having coefficients in $\mathbb{C}_p$.

Problem 317 Suppose $K = \mathbb{C}_p$. Use the idea we used above for polynomials to show that if c is of the form p^r for some rational number r then there is an isomorphism $\phi : A_c \longrightarrow A_1$ which is an isometry, i.e., which satisfies $\|\phi(f(X))\|_1 = \|f(X)\|_c$. What happens when c is not of this form?

Problem 318 Suppose $c_1 > c_2$. Consider the map

$$\iota : A_{c_1} \longrightarrow A_{c_2}$$

that maps each $f(X)$ to itself (this makes sense because, as we showed in a previous problem, $A_{c_1} \subset A_{c_2}$ in this case). Give A_{c_1} the topology defined by $\| \ \|_{c_1}$ and give A_{c_2} the topology defined by $\| \ \|_{c_2}$. Is the map ι continuous?

We are now ready to begin to work toward the proof of the main result in this section, the Weierstrass Preparation Theorem. This can be viewed as a direct extension of Strassman's Theorem from Chapter 5. The goal is to get a very close relation between functions defined by power series and functions defined by polynomials.

We will work with the norms described above, but at first will stick to $c = 1$. Hence, we will be working with power series which converge in the closed unit ball around the origin, i.e., for any x such that $|x| \le 1$. A series $\sum a_n X^n$ will have this property if and only if $a_n \to 0$ as $n \to \infty$, and this is what we assume. (Notice that this was also what we needed for Strassman's Theorem.)

Theorem 7.2.6 (p-adic Weierstrass Preparation Theorem) *Let*

$$f(X) = \sum a_n X^n$$

be a power series with coefficients in K such that $a_n \to 0$ as $n \to \infty$, so that $f(x)$ converges for $x \in \mathcal{O}$. Let N be the number defined by the conditions

$$|a_N| = \max |a_n| \qquad and \qquad |a_n| < |a_N| \quad for\ all\ n > N.$$

Then there exists a polynomial

$$g(X) = b_0 + b_1 X + \cdots + b_N x^N$$

of degree N and with coefficients in K, and a power series

$$h(X) = 1 + c_1 X + c_2 X^2 + \dots$$

with coefficients in K, satisfying:

i) $f(X) = g(X) h(X)$,

ii) $|b_N| = \max |b_n|$, *i.e.*, $\|g(X)\|_1 = |b_N|$,

iii) $\lim\limits_{n \to \infty} c_n = 0$, *so that $h(x)$ converges for $x \in \mathcal{O}$ (i.e., $h(X) \in A_1$),*

iv) $|c_n| < 1$ *for all $n \geq 1$, i.e., $\|h(X) - 1\|_1 < 1$, and*

v) $\|f(X) - g(X)\|_1 < 1$.

In particular, $h(X)$ has no zeros in $\mathcal{O}$.

This clearly is closely related to Strassman's Theorem. Since $h(X)$ has no zeros in $\mathcal{O}$, it is clear that the zeros of $f(X)$ in $\mathcal{O}$ are exactly the same as the zeros of $g(X)$. Since $g(X)$ is a polynomial of degree N, there are at most N of these, and we get Strassman's Theorem. If we move to $\mathbb{C}_p$ we can say more: since $\mathbb{C}_p$ is algebraically closed, we get that, counting multiplicities, $g(X)$ has exactly N zeros in $\mathbb{C}_p$, and the condition on its coefficients means that all of them are in $\mathcal{O}$ (see the problem below). So we know that, counting multiplicities, $f(X)$ has exactly N zeros in $\mathcal{O}$, which gives a stronger version of Strassman's Theorem.

Problem 319 Suppose the polynomial $g(X) = b_0 + b_1 X + \dots + b_N X^N$ satisfies the condition in the theorem: $|b_N| = \max |b_n|$. Show that if $g(\alpha) = 0$, then $|\alpha| \leq 1$.

Proving the Weierstrass Preparation Theorem will take a while and will require some effort. We will do it by means of a series of lemmas of various kinds. To begin to set everything up, recall that A_1 is the ring of power series that converge in $\mathcal{O}$; in other words, a power series

$$f(X) = \sum a_n X^n = a_0 + a_1 X + a_2 X^2 + \dots$$

belongs to A_1 if and only if $a_n \to 0$ as $n \to \infty$. We already know that defining

$$\|f(X)\| = \|f(X)\|_1 = \max_n |a_n|$$

gives a norm on A_1 (we will drop the subscript 1, since this is the only norm we'll be working with in this proof). The first step of the proof is to prove that A_1, with this norm, has very nice properties.

Lemma 7.2.7 *A_1 is complete with respect to the norm $\| \ \|$.*

PROOF: We have to show that a Cauchy sequence in A_1 (with respect to the norm $\| \ \|$) converges. So consider a sequence of power series

$$f_i(X) = a_{i0} + a_{i1}X + a_{i2}X^2 + a_{i3}X^3 + \cdots$$

Saying this sequence is Cauchy amounts to saying that for each $\varepsilon > 0$ there exists an M such that we have $\|f_i(X) - f_j(X)\| < \varepsilon$ whenever $i, j > M$. Translating that inequality, we get that

$$\max_n |a_{in} - a_{jn}| < \varepsilon \qquad \text{whenever } i, j > M,$$

which certainly implies that

$$|a_{in} - a_{jn}| < \varepsilon \qquad \text{for each } n, \text{ whenever } i, j > M.$$

In other words, each of the sequences $(a_{in})_i$ is Cauchy. Since K is complete, that means they are all convergent.

So, for each n, let

$$a_n = \lim_{i \to \infty} a_{in},$$

and consider the series

$$g(X) = a_0 + a_1 X + a_2 X^2 + a_3 X^3 + \cdots$$

We obviously want to say that it is the limit of the sequence of series. To see why, we need two things: first, we need to estimate

$$\|f_i(X) - g(X)\| = \max_n |a_{in} - a_n|$$

and show that it goes to zero; next, we need to show that $g(X)$ is actually in A_1.

The first part is easy: we know that if $i, j > M$ we have $|a_{in} - a_{jn}| < \varepsilon$ for every n. Letting $j \to \infty$, it follows that if $i > M$ we have $|a_{in} - a_n| \leq \varepsilon$ for all n, which means that if $i > M$ we have

$$\|f_i(X) - g(X)\| = \max_n |a_{in} - a_n| \leq \varepsilon,$$

so that $f_i(X) \to g(X)$ with respect to $\| \ \|$.

For the second part, we use what we have just proved. For $i > M$, we know that $|a_{in} - a_n| < \varepsilon$ for every n. Now, since $f_i(X) \in A_1$, we know that $a_{in} \to 0$ as $n \to \infty$, i.e., that for each i there exists an M_i such that $|a_{in}| < \varepsilon$ for all $n > M_i$. Choose any $i > M$. Then if n is greater than the corresponding M_i we have

$$|a_n| \leq |a_{in} - a_n| + |a_{in}| < \varepsilon + \varepsilon.$$

It follows that $a_n \to 0$ as $n \to \infty$.

Thus, $f_i(X) \to g(X)$ and $g(X) \in A_1$; since this works for any Cauchy sequence of power series, it shows that A_1 is complete. $\qquad \square$

Problem 320 Go through that proof and make sure it works as advertised. Make sure you understand the different roles of i and n.

Problem 321 Consider the sequence $f_n(X) = 1 + X + X^2 + \cdots + X^n$. Clearly all the $f_n(X)$ belong to A_1. Does this sequence converge?

Problem 322 Is A_c complete with respect to the norm $\| \ \|_c$?

The complete space A_1 contains the polynomials as a subspace. It is natural to guess that they are in fact a dense subspace. This does turn out to be the case:

Lemma 7.2.8 *The space of polynomials $K[X]$ is dense in A_1.*

PROOF: We need to show that any power series is the limit of a sequence of polynomials. Let

$$f(X) = a_0 + a_1 X + a_2 X^2 + a_3 X^3 + \ldots$$

be a power series in A_1, so that $a_n \to 0$ as $n \to \infty$. We need to get a sequence of polynomials which approximate $f(X)$ (with respect to the $\| \ \|_1$ norm). The obvious choice is to take the truncations of $f(X)$. So let

$$f_0(X) = a_0$$
$$f_1(X) = a_0 + a_1 X$$
$$f_2(X) = a_0 + a_1 X + a^2 X^2$$
$$\cdots$$
$$f_k(X) = a_0 + a_1 X + a_2 X^2 + \cdots + a_k X^k.$$

Then we have

$$\|f(X) - f_k(X)\| = \max_{n>k} |a_n|,$$

which tends to zero as $k \to \infty$ because $a_n \to 0$ as $n \to \infty$. Then $f_k(X) \to f(X)$, so that $f(X)$ is a limit of polynomials.   □

Problem 323 Check that if we have $a_n \to 0$, then also

$$\lim_{k\to\infty} \max_{n>k} |a_n| = 0.$$

Problem 324 Will this proof work if we replace $\| \ \|_1$ by $\| \ \|_c$?

We will often use the fact that the polynomials are dense in the space of convergent power series to prove things about power series by using facts about polynomials. The main issue in such a proof will be to check that the properties we are interested in are preserved when taking limits. The next lemma is an example of this: it is a version for series of a lemma we proved for polynomials:

Lemma 7.2.9 *Let $f(X) \in A_1$ be a power series converging in the closed unit disk, and let*

$$g(X) = b_0 + b_1 X + \cdots + b_N X^N$$

be a polynomial with coefficients in K satisfying

$$|b_N| = \max_i |b_i|.$$

Then there exist a power series $q(X) \in A_1$ and a polynomial $r(X) \in K[X]$, of degree less than N, such that

$$f(X) = g(X)\, q(X) + r(X)$$

where $q(X)$ and $r(X)$ satisfy

$$\|f(X)\| \geq \|g(X)\|\, \|q(X)\| \qquad and \qquad \|f(X)\| \geq \|r(X)\|.$$

PROOF: The idea of the proof is to use the statement for polynomials to obtain the statement for series, using the fact that any power series in A_1 is the $\|\ \|_1$-limit of polynomials. So let $f_k(X)$ be a sequence of polynomials converging to $f(X)$. By Lemma 7.2.2, one can find polynomials $q_k(X)$ and $r_k(X)$ such that

$$f_k(X) = g(X)\, q_k(X) + r_k(X) \qquad and \qquad \deg r(X) < \deg g(X)$$

and which satisfy the conditions

$$\|f_k(X)\| \geq \|q_k(X)\|\, \|g(X)\| \qquad and \qquad \|f_k(X)\| \geq \|r_k(X)\|.$$

We need to show that as $k \to \infty$ the sequences $q_k(X)$ and $r_k(X)$ converge. Since we have already shown that the space A_1 is complete, what we need to do is show that these sequences are Cauchy.

To see that, consider the equations

$$f_k(X) = g(X)\, q_k(X) + r_k(X) \quad and \quad f_{k+1}(X) = g(X)\, q_{k+1}(X) + r_{k+1}(X).$$

Subtracting one from the other gives

$$f_{k+1}(X) - f_k(X) = g(X)\, (q_{k+1}(X) - q_k(X)) + (r_{k+1}(X) - r_k(X)).$$

Now, since both $r_k(X)$ and $r_{k+1}(X)$ have degree less than N, so does their difference. What that means is that $q_{k+1}(X) - q_k(X)$ is the quotient and $r_{k+1}(X) - r_k(X)$ is the remainder when we divide $f_{k+1}(X) - f_k(X)$ by $g(X)$. Using Lemma 7.2.2 yields the estimates

$$\|q_{k+1}(X) - q_k(X)\| \leq \|g(X)\|^{-1} \|f_{k+1}(X) - f_k(X)\|$$

and

$$\|r_{k+1}(X) - r_k(X)\| \leq \|f_{k+1}(X) - f_k(X)\|.$$

Finally, remember that the sequence $f_k(X)$ is convergent, hence Cauchy, so that

$$\lim_{k \to \infty} \|f_{k+1}(X) - f_k(X)\| = 0.$$

It follows that

$$\lim_{k \to \infty} \|r_{k+1}(X) - r_k(X)\| = \lim_{k \to \infty} \|q_{k+1}(X) - q_k(X)\| = 0,$$

which, since the norm is non-archimedean, means that both sequences are Cauchy, hence convergent. Letting $r(X) = \lim r_k(X)$ and $q(X) = \lim q_k(X)$ gives the equation we want; furthermore, since each $r_k(X)$ is a polynomial of degree less than N, so is $r(X)$. Finally, the estimates on the norms are clearly preserved by passing to the limit, so that we are done.   □

We now see how to prove the Weierstrass Preparation Theorem. The point is to notice that it is just the power series version of Proposition 7.2.3. The proof of that proposition was a direct application of the lemma preceding it, whose power series version is the lemma we have just proved. Hence...

PROOF OF THE WEIERSTRASS PREPARATION THEOREM: Mimic the proof of Proposition 7.2.3 replacing calls to Lemma 7.2.2 with calls to Lemma 7.2.9.□

Problem 325 Make sure you understand how to prove the theorem. How does one prove the various statements about $g(X)$ and $h(X)$?

Problem 326 Why is it, in the statement of the Weierstrass Preparation Theorem, that the conditions on the power series $h(X)$ imply that it has no zeros in $\mathcal{O}$?

The power of the Weierstrass Preparation Theorem will only become clear from its applications. To get some idea of how it is used, let

$$f(X) = a_0 + a_1 X + a_2 X^2 + a_3 X^3 + \dots$$

be a power series converging in the closed unit disk, so that $a_n \to 0$ (in other words, $f(X) \in A_1$). Let N be chosen as in the theorem:

$$|a_N| = \max_n |a_n| \quad \text{and} \quad |a_N| > |a_j| \quad \text{if } j > N.$$

Then, according to the theorem, $f(X)$ can be factored as $g(X) h(X)$, where $g(X)$ is of degree N and $h(X)$ is a power series with no zeros of absolute value ≤ 1. We want to consider the roots of $g(X)$, so we move, for a while, to $\mathbb{C}_p$. Since $\mathbb{C}_p$ is algebraically closed, we can factor $g(X)$ as a product

$$g(X) = b_0 + b_1 X + b_2 X^2 + \dots + b_N X^N$$
$$= b_N(X - \alpha_1)(X - \alpha_2)\dots(X - \alpha_N),$$

where $\alpha_1, \alpha_2, \dots, \alpha_N$ are the roots of $g(X)$ (counted with multiplicities). This shows that $f(X)$ will have exactly N zeros in $\mathfrak{D}$, counted with multiplicities, and also gives a precise sense to the "multiplicity" of a zero of a

p-adic power series converging on $\mathfrak{O}$: it is just the multiplicity of that zero in the polynomial appearing in the Weierstrass factorization.

Problem 327 This problem (taken from [14]) gives an alternative definition for the multiplicity of a zero. Let $f(X) \in A_1$ be a power series converging on $\mathfrak{O}$. Consider the successive derivatives $f(X)$, $f'(X)$, $f''(X)$, $\ldots$, $f^{(n)}(X)$, $\ldots$ Show that for any $x \in \mathfrak{O}$ there must exist an n such that

$$f(x) = f'(x) = \cdots = f^{(n-1)}(x) = 0 \qquad \text{but} \qquad f^{(n)}(x) \neq 0.$$

Show that n is equal to the multiplicity, as defined above, of x as a zero of $f(X)$. Conclude that the sum of the multiplicities of all the zeros is exactly N. Why is Cassels' definition nicer than the one given above?

Just as we did for Strassman's Theorem, we can easily apply the Weierstrass Preparation Theorem to functions defined on bigger or smaller balls around zero by scaling the variable appropriately.

Problem 328 Let $c = p^r$ for some $r \in \mathbb{Q}$. Let $f(X)$ be a power series converging in the closed ball of radius c around zero. Explain how to use the Weierstrass Preparation Theorem to count the number of zeros of $f(X)$. Does anything change if we take more general values for c?

While the previous problem shows that the Weierstrass Preparation Theorem, as given above, can be applied in a large number of situations, it is tidier to find a version that applies not only to the $\| \ \|_1$ norm but also to the other norms $\| \ \|_c$. The statement is not hard to find:

Theorem 7.2.10 (p-adic Weierstrass Preparation Theorem) *Let c be a positive real number, and let*

$$f(X) = \sum a_n X^n$$

be a power series with coefficients in K such that $|a_n| c^n \to 0$ as $n \to \infty$, so that $f(x)$ converges for $x \in \overline{B}(0, c)$. Let N be the number defined by the conditions

$$|a_N| c^N = \max_n |a_n| c^n = \|f(X)\|_c \quad \text{and} \quad |a_n| c^n < |a_N| c^N \text{ for all } n > N.$$

Then there exists a polynomial

$$g(X) = b_0 + b_1 X + \cdots + b_N x^N$$

of degree N and with coefficients in K, and a power series

$$h(X) = 1 + d_1 X + d_2 X^2 + \ldots$$

with coefficients in K, satisfying:

i) $f(X) = g(X) h(X)$,

ii) $|b_N|c^N = \max |b_n|c^n$, *so that* $\|g(X)\|_c = |b_N|c^N$,

iii) $h(X) \in A_c$,

iv) $|d_n|c^n < 1$ *for all* $n \geq 1$, *so that* $\|h(X) - 1\|_c < 1$, *and*

v) $\|f(X) - g(X)\|_c < 1$.

In particular, $h(X)$ has no zeros in $\overline{B}(0, c)$.

Problem 329 Prove the generalized p-adic Weierstrass Preparation Theorem. (Hint: you will need to generalize Lemma 7.2.9 to arbitrary c; for that, you might imitate the trick used in the proof of Lemma 7.2.2.)

Problem 330 Let

$$f(X) = \log(1 + X) = X - \frac{X^2}{2} + \frac{X^3}{3} + \dots$$

Count the number of zeros in various balls. (We can clearly divide this by X in order to get a series whose zeroth term is 1. The resulting series converges in the open ball of radius 1, so we need to consider various closed balls of smaller radius.) How many zeros does the series have (in $\mathbb{C}_p$, of course) in the open unit ball around zero?

7.3 Entire Functions

One of the applications of the Weierstrass Preparation Theorem is the description of "entire" p-adic power series, i.e., power series which converge in all of $\mathbb{C}_p$. This is actually quite easy, but offers a nice enough example that we decided it deserved its own section.

For this section, then, let $f(X)$ be a power series which converges in all of $\mathbb{C}_p$. If we write it out,

$$f(X) = a_0 + a_1 X + a_2 X^2 + a_3 X^3 + \dots$$

we must have $|a_n|c^n \to 0$ for *any* $c \in \mathbb{R}$. We can also write this in terms of v_p: if $v_p(c) = k$ it says that $v_p(a_n) - kn \to +\infty$ for any $k \in \mathbb{Q}$. It's nicer, however, to find a condition that doesn't depend on k.

Lemma 7.3.1 *The series*

$$f(X) = a_0 + a_1 X + a_2 X^2 + a_3 X^3 + \dots$$

defines an entire function if and only if

$$\lim_{n \to \infty} \frac{v_p(a_n)}{n} = +\infty.$$

PROOF: If $v_p(a_n) - kn \to \infty$, then in particular $v_p(a_n) > kn$ for all sufficiently large n and so $v_p(a_n)/n > k$ for all sufficiently large n. Since $k \in \mathbb{Q}$ can be arbitrarily large, this says that $v_p(a_n)/n \to \infty$.

Conversely, suppose $\lim\limits_{n\to\infty} \dfrac{v_p(a_n)}{n} = \infty$. For any $k \in \mathbb{Q}$, choose $M > k$. Then we know $v_p(a_n)/n > M$ for all sufficiently large n, so

$$\frac{v_p(a_n) - kn}{n} = \frac{v_p(a_n)}{n} - k > M - k$$

for all sufficiently large n. Multiplying through, we see that $v_p(a_n) - kn > (M-k)n \to \infty$. Since k was arbitrary, $f(X)$ is entire. □

In other words, a power series will be entire if $v_p(a_n)$ tends to infinity faster than linearly.

Very well, suppose we have such a power series. If $a_0 = 0$, we can factor out a power of X so that the remaining series has zeroth coefficient not equal to zero. So, for our purposes, we might as well assume that $a_0 \neq 0$; in that case, we can divide by a_0 and assume that the zeroth coefficient is equal to one. So let

$$f(X) = 1 + a_1 X + a_2 X^2 + a_3 X^3 + \dots$$

and assume that $v_p(a_n)/n \to \infty$, so that $f(X)$ is entire. Our plan to understand the zeros of $f(X)$ is to apply the Weierstrass Preparation Theorem in larger and larger balls around zero.

So begin with the closed unit ball: a straight application of the theorem says that we can factor $f(X)$ as $g_0(X) h_0(X)$ where $g_0(X)$ is a polynomial (whose degree N is given precisely in the theorem, but that won't matter all that much here) and where $h_0(X)$ is a power series of the form $1 + b_1 X + b_2 X^2 + \dots$ with $|b_i| < 1$. Since $\mathbb{C}_p$ is algebraically closed, $g_0(X)$ factors into a bunch of linear terms, which we can write as follows:

$$g_0(X) = \prod_{i=1}^{N}(1 - \lambda_i X),$$

since we are assuming that $a_0 = 1$. Notice that the λ_i are not the roots of $g_0(X)$, but rather the *reciprocals* of the roots of $g_0(X)$; the reason for this particular bit of perverseness will become clear soon. In any case, the upshot is that

$$f(X) = h_0(X) \cdot \prod_{i=1}^{N}(1 - \lambda_i X),$$

where $\|h_0(X) - 1\|_1 < 1$.

Problem 331 Show that any polynomial $g(X) \in \mathbb{C}_p[X]$ satisfying $g(0) = 1$ can be written in the form

$$g(X) = \prod(1 - \lambda X),$$

where λ runs through the reciprocals of the roots of $g(X)$.

Now what happens when we look at a bigger ball? Consider, say, the closed ball of radius p around the origin. To apply the ($\|\ \|_1$-form[1] of the) Weierstrass Preparation Theorem, we need to change variables: let $f_1(Y) = f(Y/p)$. Then plugging $x \in \overline{B}(0, p)$ into $f(X)$ amounts to plugging $y = px$ into $f_1(Y)$, so the roots of $f(X)$ in the ball we are looking at correspond to roots of $f_1(Y)$ in the closed unit ball. Finally, it's clear that $f_1(Y)$ is still entire (if $f(\alpha)$ converges for every α then so does $f(\alpha/p)$), and that the first coefficient is still equal to 1. Applying the theorem gives

$$f_1(Y) = g_1(Y)(1 + c_1 Y + c_2 Y^2 + \cdots),$$

with, as before, $g_1(Y)$ a polynomial and $|c_i| < 1$. To get back to $f(X)$, we just replace Y by pX to get

$$f(X) = f_1(pX) = g_1(pX)(1 + d_1 Y + d_2 Y^2 + \cdots)$$

with $|d_i| = |p^i c_i| < 1/p^i$. Now $g_1(pX)$, whatever it is, is just another polynomial, whose roots give the roots of $f(X)$ in the closed ball of radius p. (It is therefore divisible by $g_0(X)$; do you see why?) So we can repeat the trick:

$$g_1(pX) = \prod_{i=1}^{N_1} (1 - \lambda_i X),$$

where now the λ_i are the reciprocals of the roots of $f(X)$ in the closed ball of radius p. In other words, we've got $f(X)$ written as

$$f(X) = h_1(X) \cdot \prod_{i=1}^{N_1} (1 - \lambda_i X).$$

The inequalities on the d_i show that we have $\|h_1(X) - 1\|_p < 1$ and also $\|h_1(X) - 1\|_1 < 1/p$. Notice that this inequality implies that if $x \in \overline{B}(0, p)$ then we must have $|h_1(x) - 1| < 1$, which implies that $h_1(x) \neq 0$; in other words, the inequality shows that $h_1(X)$ has no zeros in the closed ball of radius p.

Problem 332 Explain why $g_0(X)$ is a factor of $g_1(pX)$.

We can now understand why it's nice to use the reciprocals of the roots: they do two things for us. First, they give a clean way to write a product expression for a polynomial whose independent coefficient is 1 (and whose top coefficient might be anything). Second, and more interesting, notice that as our ball grows, the λ_i get smaller: if the root has absolute value p, say, then $|\lambda_i| = 1/p$, which is cheering if we're looking for convergence.

And we are. The reader can probably see what's coming by now: we work in bigger and bigger balls. The polynomial part gives a longer and longer

---

[1] Using Theorem 7.2.10 would also work well.

product expression (since the balls are nested in each other, the roots that appear for a disk reappear in any bigger disk, so that the product is indeed growing longer rather than just changing). The λ_i that appear in the product expression are reciprocals of roots with larger and larger absolute value, so they get smaller and smaller. The other factor gets closer and closer to 1. In the limit, we get just the product! So we've proved:

Proposition 7.3.2 *Let $f(X) = 1 + a_1 X + a_2 X^2 + \ldots$ be a power series defining an entire function on $\mathbb{C}_p$. Then $f(X)$ has a finite number of zeros in any closed ball around the origin, and a countable number of zeros in $\mathbb{C}_p$. The reciprocals of these zeros form a sequence λ_i tending to zero, and $f(X)$ can be written as an infinite product*

$$f(X) = \prod_{i=1}^{\infty}(1 - \lambda_i X)$$

(with convergence in the $\| \ \|_c$ metric for any c).

PROOF: We've done it all except for the remark on convergence. What we showed was that the infinite product converged to $f(X)$ in the $\| \ \|_1$ metric. But $f(X)$ is entire, and as usual we can change variables to handle the $\| \ \|_c$ metrics. $\square$

Problem 333 Have we really "done it all"? Make sure you see that the proof is indeed to be found in the text above.

Problem 334 Explain the cryptic remark at the end of the proof. How would one prove that the product converges to $f(X)$ in the $\| \ \|_c$ topology?

Problem 335 One might also want to understand in what sense the functions given by the partial products converge to the function defined by $f(X)$. Show that the convergence is uniform in any closed ball around zero (this is easy if you know about uniform convergence). One almost wants to say that the convergence is "uniform on compact sets"... if it weren't for the slight detail that closed balls in $\mathbb{C}_p$ are not compact!

Passing from power series whose initial term is 1 to general power series is easy:

Corollary 7.3.3 *Let $f(X)$ be a power series defining an entire function on $\mathbb{C}_p$. Then $f(X)$ can be written as an infinite product*

$$f(X) = aX^r \prod_{i=1}^{\infty}(1 - \lambda_i X),$$

where $a \in \mathbb{C}_p$, r is an integer, $r \geq 0$, and λ_i ranges through the reciprocals of the nonzero roots of $f(X)$, which form a sequence tending to zero.

This is very similar to, but also simpler than, a classical result about complex entire functions; see, for example, [2, Chapter 5, Section 2.3]. It is the starting point for any serious study of p-adic entire functions.

Problem 336 We haven't really met any (non-polynomial) entire functions. Can you give an example?

Problem 337 Show that the product expansion can be used to construct entire functions: take a sequence λ_i tending to zero; does it make sense to define a function by

$$f(x) = \prod_{i=1}^{\infty} (1 - \lambda_i x)?$$

For which x does this converge? Can the resulting function be expressed as a power series? (What we are aiming for, of course, is a converse of Corollary 7.3.3.)

7.4   Newton Polygons

One of the best ways to understand the theory of polynomials and power series with coefficients in a complete p-adic field K is to introduce the concept of the Newton polygon of a polynomial (and later of a power series). This gives us a clear geometric picture that encodes much of the information we have collected about the zeros of polynomials and power series.

We begin, once again, by considering polynomials. We will define the Newton polygon and then explore its meaning in a leisurely way. As before, we will work in a field K, which will either be a finite extension of $\mathbb{Q}_p$ or equal to $\mathbb{C}_p$ (in particular, K is complete with respect to the p-adic valuation). So let $f(X) \in K[X]$ be a polynomial. Since we are mostly interested in understanding the zeros of $f(X)$ we may as well factor out any powers of X which divide $f(X)$. In other words, we may assume that $f(0) \neq 0$. Then, dividing through by $f(0)$, we may also assume that $f(0) = 1$.

Thus, we take a polynomial

$$f(X) = 1 + a_1 X + a_2 X^2 + \cdots + a_n X^n$$

with $a_i \in K$. On a set of axes, we plot the points $(0,0)$ and, for each i between 1 and n, $(i, v_p(a_i))$. (There is one caveat: if $a_i = 0$ for some i, it is not clear what $v_p(a_i)$ is to be; we just take it to be $+\infty$, and think of the point as "infinitely high." In practice this just means that we ignore that value of i.) The polygon we want to consider is, in fancy terms, the lower boundary of the convex hull of this set of points. In less fancy terms, we can think of it this way:

   i) Start with the vertical half-line which is the negative part of the y-axis (i.e., the points $(0, y)$ with $y \leq 0$).

ii) Rotate that line counter-clockwise until it hits one or more of the points we have plotted.

iii) "Break" the line at the rightmost point that was hit, and continue rotating the remaining part until another point(s) is(are) hit.

iv) Continue until all the points have either been hit or lie strictly above a portion of the polygon.

(One may or may not want to think of the polygon as ending with an infinitely long vertical line going upwards; we will prefer to simply cut off the polygon at its last vertex.)

The resulting polygon is called the *Newton polygon* of the polynomial $f(X)$. Notice that, in the same spirit as before, the polygon depends only on the $v_p(a_i)$, which do not depend on which field we think the a_i belong to. In other words, the polygon belongs to the polynomial, rather than to the polynomial as an element of $K[X]$.

It may be that an example helps more at this point than any number of words. Let's take $p = 5$ and consider the polynomial

$$f(X) = 1 + 5X + \frac{1}{5}X^2 + 35X^3 + 25X^5 + 625X^6.$$

The points we want to work with are

$$(0,0) \quad (1,1) \quad (2,-1) \quad (3,1) \quad (5,2) \quad (6,4)$$

(as agreed, we simply ignore the missing term of degree 4, or think of its point as "very, very high up" and hence irrelevant for the construction). Plotting these points gives Figure 7.1. The process with the rotating line gives the polygon in Figure 7.2.

It's a nice picture, but what does it mean? It will turn out that the Newton polygon encodes a lot of information about the roots of the polynomial. The crucial features of the polygon will be:

i) the slopes of the line segments appearing in the polygon—we will call these the "Newton slopes" of $f(X)$;

ii) the "length" of each slope, by which we mean the length of the projection of the corresponding segment on the x-axis;

iii) the "breaks," i.e., the values of i such that the point $(i, v_p(a_i))$ is a vertex of the polygon.

In our example, the slopes are $-1/2$, 1, and 2, of lengths 2, 3, and 1, respectively, and the breaks happen when $i = 0, 2, 5, 6$. Notice that the sum of all the lengths will always be equal to the degree, and that $(0,0)$ and $(n, v_p(a_n))$ will always be vertices. It is also clear from the "rotating line" construction that the slopes will form an increasing sequence.

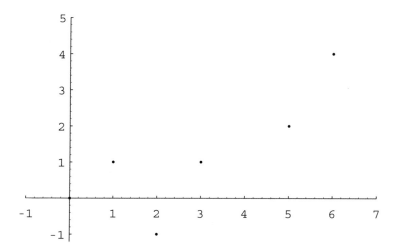

Figure 7.1: Points for $f(X) = 1 + 5X + \frac{1}{5}X^2 + 35X^3 + 25X^5 + 625X^6$

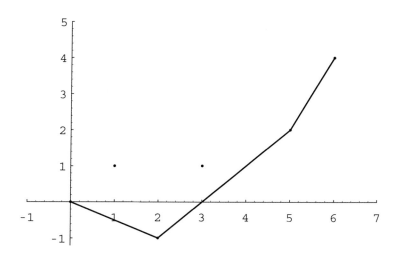

Figure 7.2: Newton polygon for $f(X) = 1+5X+\frac{1}{5}X^2+35X^3+25X^5+625X^6$

Problem 338 Let p=5. Work out the Newton polygon of the following polynomials:

i) $1 + X + X^2 + X^3 + X^4 + 2X^5 + 100X^6$

ii) $1 + X + X^2 + X^3 + X^4 + 2X^5 + \frac{1}{100}X^6$

iii) $3 + 5X + 4X^2 + 35X^3 + 40X^4 + 1250X^5 + 100X^6$ (Remember that in our discussion above we normalized things so that $f(0) = 1$. That means you must divide through by 3 before making the polygon... but must you really?)

iv) $3 + 5X + 4X^2 + 35X^3 + 40X^4 + 1250X^5 + 100X^6 + 5X^{10}$ (How does this relate to the previous one?)

Problem 339 Suppose a polynomial $F(X)$ satisfies the condition in the Eisenstein irreducibility criterion over $\mathbb{Q}_p$ (i.e., it is an "Eisenstein polynomial"). Let $f(X)$ be the polynomial obtained by dividing $F(X)$ by whatever number is necessary so that $f(0) = 1$. Describe the Newton polygon of $f(X)$.

Problem 340 It might be useful to generalize the definition in order to remove the condition $f(0) = 1$, and just assume $f(0) \neq 0$. How would the definition change? What would be the relation between the polygons of $f(X)$ and of $af(X)$ (for $a \in K^\times$)?

In order to begin to see what information is hidden in the Newton polygon of a polynomial, let's begin by seeing the significance of the breaks. What we want to do is to consider a polynomial $f(X) = 1 + a_1 X + a_2 X^2 + \cdots + a_n X^n$ and look at its norms $\|f(X)\|_c$ for many different c. Since the norm corresponding to c is essentially the sup-norm on the closed ball of radius c centered at 0 (in $\mathbb{C}_p$), it is easy to see that they satisfy

$$\text{If } c_1 > c_2, \text{ then } \|f(X)\|_{c_1} \geq \|f(X)\|_{c_2}.$$

(It's also very easy to give a direct proof of this.) So, if we start with a very small c and gradually increase it, the norms $\|f(X)\|_c$ will also increase. This observation will help interpret the breaks in the Newton polygon.

Let's look at the first segment of the Newton polygon. If this segment has slope m, it connects the point $(0, 0)$ to some other point (i, mi). (So that the first Newton slope is m, and it has length i.) Let's think about what that means. First, it means that there are no points below the line $y = mx$; in other words, $v_p(a_j) \geq mj$ for every j. Second, the point (i, mi) itself tells us that $v_p(a_i) = mi$. Third, the fact that there is a break tells us that the subsequent points are really *above* the line; in other words, $v_p(a_j) > mj$ if $j > i$.

Translating from valuations to absolute values, we get

- $|a_j| \leq p^{-mj} = (p^{-m})^j$ for all j, which we can rewrite as $|a_j|(p^m)^j \leq 1$ for all j,

- $|a_i| = p^{-mi}$, which we can rewrite as $|a_i|(p^m)^i = 1$, and

- $|a_j| < p^{-mj}$ if $j > i$, which we can rewrite as $|a_j|(p^m)^j < 1$ if $j > i$.

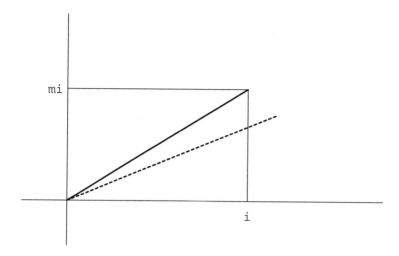

Figure 7.3: The first segment

If we now let $c = p^m$, we can read these conditions in terms of the c-norm. They say:

- $\|f(X)\|_c = 1$, and

- i is the largest integer such that $\|f(X)\|_c = |a_i|c^i$.

In other words, the fact that the first break is at (i, mi) means that if we take $c = p^m$ then $\|f(X)\|_c = 1$ and i is the distinguished number that appears in Proposition 7.2.3. In particular, we see that if i is less than the degree of $f(X)$, then $f(X)$ is divisible by a polynomial of degree i.

Just this is already quite nice. Let's follow Cassels in making the following definition:

Definition 7.4.1 *A polynomial $f(X) \in K[X]$ is called* pure *if its Newton polygon has only one slope. If this slope is m, we will say $f(X)$ is* pure of slope m.

Then we can state what we have just observed as:

Proposition 7.4.2 *Irreducible polynomials are pure.*

PROOF: A break at (i, mi) yields, by the discussion above, a factor of degree i; hence, if there is a break at $i \neq 0$, n, the polynomial is reducible.   □

In fact, we can go further, by noticing that a polynomial $h(X) = 1 + b_1 X + b_2 X^2 + \cdots + b_n X^n$ will be pure of slope m, according to the discussion above, exactly when it has the property that, for $c = p^m$, $\|f(X)\|_c = |b_n|c^n = 1$, i.e., the maximum occurs at the end and is equal to 1.

Problem 341 Prove that a polynomial $h(X) = 1 + b_1 X + b_2 X^2 + \cdots + b_n X^n$ is pure of slope m if and only if we have $\|f(X)\|_{p^m} = |b_n| p^{mn} = 1$.

Using this, we can push the analysis further:

Proposition 7.4.3 *Let*

$$f(X) = 1 + a_1 X + \cdots + a_n X^n \in K[X],$$

and assume the Newton polygon of $f(X)$ has its first break at (i, mi). Then there exist polynomials $g(X)$, $h(X) \in K[X]$ satisfying:

i) $f(X) = g(X)h(X)$,

ii) $g(X)$ *has degree i and is pure of slope m,*

iii) $h(X)$ *has no zeros in the closed ball of radius p^m around 0.*

PROOF: This has all been proved already; it's just a matter of putting all the pieces together.   □

Problem 342 Put all the pieces together.

The connection between "pureness" and polynomial factorization is important. Here is another data-point:

Problem 343 Let $f(X)$ and $g(X)$ both be pure polynomials of slope m. Show that their product is also pure of slope m.

We are still not done thinking of the meaning of the first break... What we still need to do is understand the significance of the *slope* of the first segment. That isn't hard to do.

Lemma 7.4.4 *Let $f(X) = 1 + a_1 X + a_2 X^2 + \cdots + a_n X^n \in K[X]$, and assume that the first break of the Newton polygon of $f(X)$ occurs at the point (i, mi). Let c be any positive real number less than p^m. Then we have $\|f(X)\|_c = 1$ and $\|f(X) - 1\|_c < 1$.*

PROOF: If $\|f(X) - 1\|_c < 1$, then we must have $\|f(X)\|_c = 1$ by the usual "all triangles are isosceles" yoga, so we only need to prove the inequality.

A line through the origin with slope $m_1 < m$ (e.g., the dotted line in Figure 7.3) passes below all the points on the polygon, touching it only at $(0, 0)$. This means that $v_p(a_j) > m_1 j$ for every $j > 0$. Translating to absolute values, this means that $|a_j| < p^{-m_1 j}$, or $|a_j|(p^{m_1})^j < 1$ for any $j > 0$. Since $a_0 = 1$, the zeroth coefficient of $f(X) - 1$ is just 0, and therefore also $|a_0 - 1| < 1$. It follows that $\|f(X) - 1\|_c < 1$ for $c < p^{m_1}$.   □

This already gives one way to characterize the first slope, as the next problem shows:

Problem 344 Show that if the first break happens at (i, mi), and $c_1 > p^m$, then $\|f(X)\|_{c_1} > 1$.

We can read this as saying that $c = p^m$ is the largest value of c such that $\|f(X)\|_c = 1$, which gives an interpretation of the slope of the first segment. The more interesting interpretation, however, has to do with obtaining information about zeros:

Lemma 7.4.5 *Let* $f(X) = 1 + a_1 X + a_2 X^2 + \cdots + a_n X^n \in K[X]$, *and assume that the first break of the Newton polygon of* $f(X)$ *occurs at the point* (i, mi). *Let* c *be any positive real number less than* p^m. *Then* $f(X)$ *has no zeros in the closed ball in* $\mathbb{C}_p$ *of radius* c *around* 0.

PROOF: The previous lemma says that $\|f(X) - 1\|_c < 1$. From that, it follows that for any x such that $|x| \le c$ we have $|f(x) - 1| < 1$, which certainly implies that $f(x) \ne 0$. Thus, $f(X)$ has no zeros in $\overline{B}(0, c)$. $\qquad\square$

So let's put it together: if the first break is at (i, mi), then

- If $c < p^m$, $f(X)$ has no roots in the closed ball of radius c.

- $f(X)$ factors as the product of a pure polynomial $g(X)$ of slope m and a polynomial which has no roots in the closed ball of radius p^m.

What about the roots of $g(X)$? Well, any root of $g(X)$ is a root of $f(X)$, so we know that $g(X)$ has no roots of absolute value less than p^m. On the other hand, if α_1, α_2, ..., α_i are the roots of $g(X)$ (in $\mathbb{C}_p$, with multiple roots listed repeatedly), then we must have

$$g(X) = (1 - \alpha_1^{-1} X)(1 - \alpha_2^{-1} X) \ldots (1 - \alpha_i^{-1} X),$$

and hence the top coefficient of $g(X)$ is equal to $(\alpha_1 \alpha_2 \ldots \alpha_i)^{-1}$. Since $g(X)$ is pure, this must have valuation mi; since we already know $v_p(\alpha_j) \le -m$, it follows that all of the α_j have valuation exactly equal to $-m$. Translating back, *all the roots of* $g(X)$ *have absolute value* p^m.

Problem 345 Generalize the argument above to show that all the roots of any pure polynomial of slope m have absolute value p^m (or, equivalently, valuation $-m$).

Putting all the pieces together, we get:

Proposition 7.4.6 *Let* $f(X) = 1 + a_1 X + a_2 X^2 + \cdots + a_n X^n \in K[X]$, *and assume that the first break of the Newton polygon of* $f(X)$ *occurs at the point* (i, mi). *Then* $f(X)$ *has no roots with absolute value less than* p^m *and has exactly* i *roots (counting multiplicities, in* $\mathbb{C}_p$) *with absolute value* p^m.

Very well, let's move on to the second segment. In other words, let's assume that there are breaks at (i, mi) and at $(k, mi + m'(k - i))$, so that the first slope is m and has length i and the second slope is m' and has length

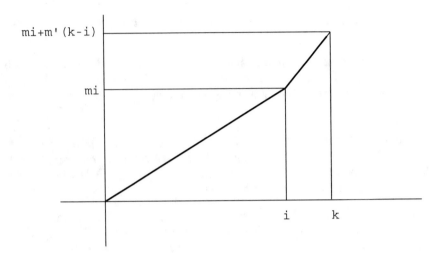

Figure 7.4: The second segment

$k - i$. All that we have obtained about the first segment still works, of course, so what we want to understand is the second segment.

For that, we first translate the fact that the second statement has slope m'. The line through (i, mi) with slope m' has equation

$$y = mi + m'(x - i),$$

and we know the following things.

- (i, mi) and $(k, mi + m'(k - i))$ are on the line; in other words, $v_p(a_i) = mi$ and $v_p(a_k) = mi + m'(k - i)$.

- All the points between i and k are on or above the line. In other words, $v_p(a_j) \geq m'(j - i) + mi$ if $i < j < k$.

- All the points beyond k are strictly above the line. In other words, $v_p(a_j) > m'(j - i) + mi$ if $j > k$. The same inequality holds for $j < i$, as is very easy to see (draw a picture!).

Now we translate all this to absolute values. Our inequalities say that

- $|a_k| = p^{-m'(k-i)-mi} = p^{-m'k}p^{(m'-m)i}$,

- for $i \leq j \leq k$, we have $|a_j| \leq p^{-m'(j-i)-mi} = p^{-m'j}p^{(m'-m)i}$,

- for $j < i$ and for $j > k$, we have $|a_j| < p^{-m'(j-i)-mi} = p^{-m'j}p^{(m'-m)i}$,

which we rewrite once again by taking $c = p^{m'}$:

- $|a_j|c^j \leq p^{(m'-m)i}$ for all j,

- the equality holds for $j = k$, and

- the inequality is strict for $j > k$.

In other words, we get, for $c = p^{m'}$, that $\|f(X)\|_c = p^{(m'-m)i}$ (and notice that since $m' > m$ this is bigger than 1) and that k is the distinguished number in Proposition 7.2.3 (i.e., the maximum is realized at the degree k term). Using the proposition, we again find a factor $g(X)$, which now need not be pure (why?). In any case, we can go through a process completely analogous to what we did before to conclude that $f(X)$ has exactly k roots in the closed ball of radius $p^{m'}$, i of which have absolute value p^m (we knew that already), and $k - i$ of which have absolute value $p^{m'}$. Of course, we can go through a similar argument at the other breaks, and get roots with bigger absolute values. In the end, we'll get all the roots, and we'll know exactly what their absolute values should be:

Theorem 7.4.7 *Let $f(X) = 1 + a_1 X + a_2 X^2 + \cdots + a_n X^n \in K[X]$ be a polynomial, and let $m_1, m_2, \ldots, m_r$ be the slopes of its Newton polygon (in increasing order). Let $i_1, i_2, \ldots, i_r$ be the corresponding lengths. Then, for each k, $1 \leq k \leq r$, $f(X)$ has exactly i_k roots (in $\mathbb{C}_p$, counting multiplicities) of absolute value p^{m_k}.*

PROOF: Just repeat the arguments we went through above at each break in the polygon.   □

Notice that since the sum of all the lengths is equal to the degree, the theorem accounts for all the roots of the polynomial.

Problem 346 Fill in the complete details of the analysis of the second break, and convince yourself that the argument will indeed work at the other breaks.

Notice that one of the things that follows from the theorem is the fact that the factor $g(X)$ of $f(X)$ whose existence follows from the existence of a break (together with Proposition 7.2.3) has the same Newton polygon as $f(X)$ up to that break. This shows that the polygons and the factors they tell us about are really tightly connected.

Problem 347 Suppose the Newton polygon of $f(X)$ has breaks, as above, at i and k, with slopes m and m' of length i and $k - i$, respectively. Our discussion shows that there exists a polynomial $g_1(X)$ of degree i which is pure of slope m and divides $f(X)$, and a polynomial $g_2(X)$ of degree k whose Newton polygon coincides with that of $f(X)$ up to k. Show that $g_1(X)$ is a divisor of $g_2(X)$. Let $h(X)$ denote the quotient, so that $g_2(X) = g_1(X)h(X)$. Is $h(X)$ pure?

Problem 348 Suppose the Newton polygon of $f(X)$ starts with a segment of slope m. Let λ be a root of $f(X)$ with absolute value p^m (one exists, by the discussion above). Let $h(X)$ be the polynomial such that $f(X) = (1 - \lambda^{-1}X)h(X)$ (it exists, since λ is a root). Can you relate the Newton polygons of $f(X)$ and of $h(X)$?

Problem 349 Go back to the polygons you drew above, and explain what they tell you about the roots of their polynomials.

Problem 350 Consider the polynomials $f(X) = 1 + X + p^{300}X^{100}$ and $g(X) = 1 + X + p^{100}X^{100}$. These polynomials are "very close," since we have $\|f(X) - g(X)\|_1 = p^{-100}$. Are their Newton polygons close? What is similar in the two polygons? What is different?

Problem 351 The previous problem showed that two polynomials can be very close with respect to the $\|\ \|_1$-norm, and still have different numbers of roots in balls of radius larger than one. What condition would you need in order to be able to conclude that $f(X)$ and $g(X)$ have the same number of roots in the closed ball of radius c?

The moral of the story so far is that Newton polygons codify quite a lot of information about the zeros of polynomials. That should encourage us in the next step, which is to consider the Newton polygon of a power series.

The definition is formally identical: given a power series of the form

$$f(X) = 1 + a_1 X + a_2 X^2 + \cdots + a_n X^n + \cdots$$

we plot the points

$$(i, v_p(a_i)) \qquad \text{for } i = 0, 1, 2, \ldots,$$

ignoring, as before, any points where $a_i = 0$. The Newton polygon of $f(X)$ is again obtained by the "rotating line" procedure. In this case, however, things are more complicated than in the case of polynomials. An example will do more to explain what can happen than any number of generic descriptions.

Consider the power series

$$f(X) = 1 + pX + pX^2 + pX^3 + \cdots + pX^n + \cdots$$

The points we get are

$$(0,0), \quad (1,1), \quad (2,1), \quad (3,1), \quad \cdots$$

Now, clearly the line can sweep unbroken until it is horizontal, but then we have the following curious situation:

- *none* of the points $(i, 1)$ are on our line (so there is nowhere to "break" it), but

- if we rotate the line ever so slightly, some points will be left behind. More precisely, for any positive slope ε there exists an i such that $\varepsilon i > 1$, so that the point $(i, 1)$ is below the line $y = \varepsilon x$.

This means that we must amend our rules for obtaining the Newton polygon to account for this possibility. So here are revised rules:

Start with the vertical half-line which is the negative part of the y-axis (i.e., the points $(0, y)$ with $y \leq 0$). Rotate that line counter-clockwise until one of the following happens:

$i)$ The line simultaneously "hits" infinitely many of the points we have plotted. In this case, stop, and the polygon is complete.

$ii)$ The line reaches a position where it contains only one of our points (the one currently serving as the center of rotation) but can be rotated no further without leaving behind some points. In this case, stop, and the polygon is complete.

$iii)$ The line hits a finite number of the points. In this case, "break" the line at the last point that was hit, and begin the whole procedure again. Notice that the segment beginning at the last point hit may find itself immediately in the situation of case (ii), so that there may be no further change.

This procedure pretty much assumes that the power series is really a series, rather than a polynomial. To handle the case of a polynomial in a unified way, we would have to add one further stopping procedure: if the line reaches the vertical position (after rotating 180 degrees), we stop. The Newton polygon of a polynomial will then end with an infinite vertical segment.

Notice that there are only three ways for the procedure to end:

$i)$ the last segment contains an infinite number of points,

$ii)$ the last segment contains a finite number of points, but can be rotated no further,

$iii)$ there is an infinite sequence of segments of finite length.

Problem 352 Can it happen that "the line can be rotated no further" from the very beginning, so that the Newton polygon gets reduced simply to the negative half of the y-axis?

Let's look at a simple example. Take the power series

$$f(x) = \frac{1}{1 - pX} = 1 + pX + p^2 X^2 + p^3 X^3 + \cdots + p^n X^n + \cdots$$

The points $(i, v_p(a_i))$ are just

$$(0,0), \quad (1,1), \quad (2,2), \quad (3,3), \quad \ldots, \quad (i,i), \quad \ldots$$

We are in the first case above, and the polygon comes out to be a line of slope one which contains infinitely many points (Figure 7.5).

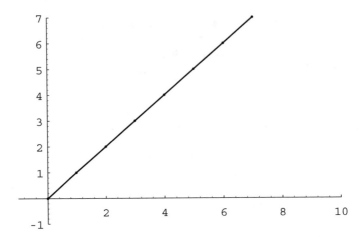

Figure 7.5: Newton polygon for $1 + pX + p^2X^2 + p^3X^3 + \cdots + p^nX^n + \cdots$

To work out the radius of convergence of this series, we need to compute

$$\limsup_{n \to \infty} \sqrt[n]{|a_n|} = \limsup_{n \to \infty} \sqrt[n]{p^{-n}} = 1/p.$$

It follows that the series converges for $|x| < p$ and diverges for $|x| > p$. To handle the remaining case, notice that if $|x| = p$, then clearly $|p^n x^n| = p^{-n}p^n = 1$, and the series does not converge. As we will soon prove, the fact that the Newton polygon ends in (in fact, is) a line of slope 1 is connected with the fact that the region of convergence is the open ball of radius p^1.

We have already seen an example of a series whose Newton polygon falls into case (ii) above:

$$f(X) = 1 + pX + pX^2 + pX^3 + \cdots + pX^n + \cdots$$

In this case, the Newton polygon is a horizontal line (see Figure 7.6). Notice that in this case

$$\limsup_{n \to \infty} \sqrt[n]{|p|} = \lim_{n \to \infty} p^{-1/n} = 1,$$

so that the radius of convergence is $1 = p^0$. Checking the special case shows that if $|x| = 1$, then the series does not converge, so that once again the region of convergence is an open ball, this time of radius 1.

We clearly need an example where the region of convergence is a closed ball. To get one, let's define a function $\ell(n) = \lfloor \log n \rfloor$, where $\lfloor \cdot \rfloor$ is the "greatest integer" or "floor" function (in words, $\ell(n)$ is the greatest integer which is less than or equal to $\log n$). Then consider the power series

$$1 + \sum_{n=1}^{\infty} p^{\ell(n)}X^n = 1 + X + X^2 + pX^3 + \cdots$$

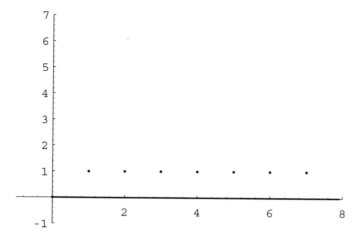

Figure 7.6: Newton polygon for $1 + pX + pX^2 + pX^3 + \cdots + pX^n + \cdots$

The points we want to plot are

$$(0,0), \quad (1,0), \quad (2,0), \quad (3,1), \quad \ldots, \quad (n, \ell(n)), \quad \ldots$$

If we use the rotating line procedure, we can certainly rotate the line unbroken until it becomes horizontal (at which point it hits the first three points in our list).

We claim the line can rotate no further. To see this, consider a line through the point $(2,0)$ of some small positive slope ε; this will have equation $y = \varepsilon(x - 2)$. We want to see that there is a point $(n, \ell(n))$ below this line; this translates to the assertion that we have $\ell(n) < \varepsilon(n - 2)$ for some n. To see that this is indeed the case, notice that for $n > 2$ we have

$$0 \le \frac{\ell(n)}{n - 2} \le \frac{\log n}{n - 2},$$

and remember that

$$\lim_{n \to \infty} \frac{\log n}{n - 2} = 0.$$

It follows that

$$\lim_{n \to \infty} \frac{\ell(n)}{n - 2} = 0,$$

which means that given any ε we can find an n_0 such that $\ell(n_0)/(n_0 - 2) < \varepsilon$. Rearranging, this says that $\ell(n_0) < \varepsilon(n_0 - 2)$, which is what we wanted to prove: any line of positive slope has some (most, in fact) of our points below it. The conclusion is that the Newton polygon of our series is once again a horizontal line. It is easy to see that the radius of convergence of this series is 1, and that it *does* converge when $|x| = 1$, so that in this case the region of convergence is the closed unit ball.

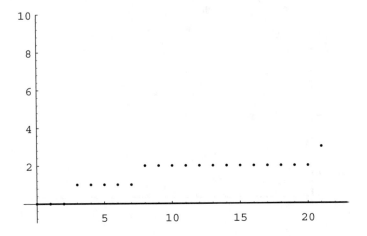

Figure 7.7: Newton polygon for $1 + \sum p^{\ell(n)} X^n$

Problem 353 Modify this last series slightly by changing the first few coefficients:

$$g(X) = 1 + \frac{1}{p}X + \frac{1}{p}X^2 + \sum_{n=3}^{\infty} p^{\ell(n)} X^n.$$

What does the Newton polygon now look like?

The next lemma gives the connection (which we have been hinting at) between the slope of the final segment and the radius of convergence.

Lemma 7.4.8 *Let m be the sup of the slopes appearing in the Newton polygon of a series $f(X) = 1 + a_1 X + a_2 X^2 + \cdots$ (so that m is either a number or is $+\infty$). Then the radius of convergence of the series is p^m (which we understand as $+\infty$ if $m = +\infty$).*

PROOF: Let $|x| = p^b$ with $b < m$. Let's show directly that

$$f(x) = 1 + a_1 x + a_2 x^2 + \cdots + a_n x^n + \cdots$$

converges. For that, we need to prove that $|a_i x^i|$ goes to zero as $i \to \infty$. Since $|x| = p^b$, we have $|a_i x^i| = |a_i| p^{bi}$; translating to valuations, this says $v_p(a_i x^i) = v_p(a_i) - bi$. To show $|a_i x^i|$ goes to zero is the same as to show its valuation goes to infinity, so we want to show that $v_p(a_i)$ gets arbitrarily larger than bi as i grows.

Now superpose the line $y = bx$ on the Newton polygon of the series (see Figure 7.8). Since the slope of the polygon eventually becomes larger than b, the polygon eventually passes and then gets farther and farther above the

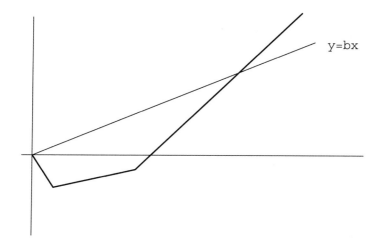

Figure 7.8: A Newton polygon and a line of slope b

line $y = bx$. The points $(i, v_p(a_i))$ are on or above the polygon, so it follows that

$$v_p(a_i) - bi \to \infty \qquad \text{as } i \to \infty,$$

and the series converges.

If $m = +\infty$, we are done. If not, to show that p^m is actually the radius of convergence, we need also to check that if $|x| = p^b$ with $b > m$ the series does not converge. We leave that to the reader (just use the same idea). $\square$

Problem 354 Complete the proof of the lemma.

Of course, we would like to know the exact region of convergence: is it the open or the closed disk of radius p^m? That turns out to be a bit harder to decide, so we'll content ourselves with a partial answer:

Lemma 7.4.9 *Let m be the sup of the slopes appearing in the Newton polygon of a series $f(X) = 1 + a_1X + a_2X^2 + \cdots$. Then:*

i) If the polygon ends in an infinite segment of slope m which contains infinitely many of the points $(i, v_p(a_i))$, then the region of convergence is the open ball of radius p^m.

ii) If the polygon contains an infinite number of segments of finite length, then the region of convergence is the open ball of radius p^m.

PROOF: Suppose, first, that the polygon ends in an infinite segment of slope m which contains an infinite number of the points $(i, v_p(a_i))$. This means that there is a subsequence $i_1, i_2, \ldots, i_j, \ldots$ such that $v_p(a_{i_j}) = k + mi_j$, where k is some fixed constant. In absolute value notation, this says

$$|a_{i_j}| = p^{-k}p^{-mi_j}.$$

To see that the region of convergence is the open disk of radius p^m, what we need to show is that the series fails to converge if $|x| = p^m$. So suppose $|x| = p^m$. Then, along our subsequence, we would have

$$|a_{i_j}x^{i_j}| = p^{-k}p^{-mi_j}p^{mi_j} = p^{-k}.$$

Since this does not converge to zero, the series $f(x)$ does not converge, and we are done.

Now suppose the polygon has infinitely many line segments. Since the sup of all the slopes is m, all of the segments will have slopes less than m, and the slopes will form an increasing sequence converging to m. To handle this case, we can use an argument similar to the one in the previous lemma: the series will converge at a point x with $|x| = p^m$ if we have

$$\lim_{i \to \infty} |a_i|p^{mi} = 0,$$

or, in valuation notation, if $v_p(a_i) - mi$ goes to infinity as i goes to infinity. This would mean that the points in our polygon get arbitrarily far above the line $y = mx$. But that clearly cannot happen. $\square$

Problem 355 Convince yourself that it "clearly cannot happen."

Problem 356 Do the proofs we gave apply to the case where the polygon is just the negative y-axis (i.e., where the "rotating line" can't even leave its starting point)? What conclusion should we get in that case?

Problem 357 The reader will have noticed that we avoided saying what the exact region of convergence would be if the final segment does not contain infinitely many of the points $(i, v_p(a_i))$. This case is complicated, as the two examples above show. Try to come up with a criterion to decide what happens in this case.

Problem 358 Work out the Newton polygon and the region of convergence for each of the series

i) $1 + pX + p^4X^2 + p^9X^3 + \cdots + p^{n^2}X^n + \cdots$

ii) $1 + X^p + pX^{p^2} + p^2X^{p^3} + \cdots + p^{n-1}X^{p^n} + \cdots$

iii) $1 + X + 2X^2 + 3X^3 + \cdots + nX^n + \cdots$

iv) $1 + X + \frac{1}{4}X^2 + \frac{1}{9}X^3 \cdots + \frac{1}{n^2}X^n + \cdots$

Problem 359 (More examples.) Find the Newton polygons for the power series defining the p-adic logarithm (you'll need to divide by X first in order to get the zeroth coefficient to be 1), for the power series defining the exponential, and for the series for $(1 + X)^{1/2}$ (you'll want to assume $p \neq 2$ for this one).

We now want to go on to obtain power series versions of the results describing how the Newton polygon carries information about the zeros of a power series. The crucial insight, here, will be to notice that the arguments we obtained for polynomials all work without change for power series: all we need to do is to replace references to Proposition 7.2.3 with references to the Weierstrass Preparation Theorem (more precisely, to Theorem 7.2.10).

Rather than simply send the reader back to check that our arguments do work, let's re-examine the discussion of the first segment of the Newton polygon. So let

$$f(X) = 1 + a_1 X + a_2 X^2 + a_3 X^3 + \cdots$$

be a power series, and suppose that its Newton polygon has a first segment of length i and slope m. Since we are dealing with series, we need to be careful about what we want to assume about what goes on *after* this initial segment, so let's make the necessary assumptions specific. We assume that:

i) The points $(0,0)$ and (i, mi) are on the polygon, and the segment connecting them is part of the polygon, and

ii) *either* the polygon has a "break" at (i, mi) (i.e., it continues with a different slope) *or* it continues with an infinite segment of slope m which does not contain any more of the points $(j, v_p(a_j))$. In the former case, we know that the series will converge on the closed ball of radius p^m; in the latter case, we will *assume* that it does.

The reason for these assumptions is really clear: we want to relate the segment of slope m to the zeros on the closed ball of radius p^m. The assumptions simply describe the two situations in which the series converges on that closed ball.

One way to think about our special assumptions for the case when there is an infinite line of slope m is that they give a definition for the length of that segment. In other words, if the Newton polygon of a series ends in an infinite portion of slope m we will say the length of that portion is ℓ if ℓ is the distance between the x-coordinates of the first and last of the points $(n, v_p(a_n))$ which are on the line, provided that the series converges on the closed ball of radius p^m. (Recall that the convergence assumption implies that there *is* a last such point.) Otherwise, we may want to say that the length corresponding to slope m is zero.

Once we have made these assumptions we have the following. First, $f(X)$ converges on the closed ball of radius p^m. Next all the points $(j, v_p(a_j))$ are on or above the line $y = mx$, and the ones where $j > i$ are strictly above it. This translates to

• $|a_j|(p^j)^m \le 1$ for all j,

• $|a_i|(p^i)^m = 1$, and

- $|a_j|(p^m)^j < 1$ if $j > i$.

This says that $\|f(X)\|_{p^m} = 1$ and that the maximum is last realized at degree i. In other words, it puts us exactly in the same position as in the case of polynomials: we can use the Weierstrass Preparation Theorem to conclude that there is a polynomial $g(X)$ of degree i and a power series $h(X)$, satisfying the inequality $\|h(X) - 1\|_{p^m} < 1$, such that $f(X) = g(X)h(X)$. Furthermore, we know that $\|f(X) - g(X)\|_{p^m} < 1$, which implies that $\|g(X)\|_{p^m} = \|f(X)\|_{p^m} = 1$, so that $g(X)$ is pure of slope m. Then, using what we know about Newton polygons of polynomials, it follows that the zeros of $f(X)$ in the closed ball of radius p^m coincide with those of $g(X)$, which we already know are all of absolute value p^m. So we've got the same result as for polynomials: if the first segment is of length i and slope m, then $f(X)$ has exactly i zeros of absolute value p^m, and no zeros of smaller absolute value.

Thinking about what we just did suggests the following:

Proposition 7.4.10 *Let*

$$f(X) = 1 + a_1 X + a_2 X^2 + a_3 X^3 + \cdots$$

be a power series. Let $m_1, m_2, \ldots, m_k$ *be the first k slopes of the Newton polygon of $f(X)$, and assume that $f(X)$ converges on the closed ball of radius $c = p^{m_k}$. Let N be the x-coordinate of the right endpoint of the k-th segment of the Newton polygon. Then there exist a polynomial $g(X)$ of degree N and a power series $h(X)$ such that*

i) $f(X) = g(X)h(X)$,

ii) $\|f(X) - g(X)\|_c < 1$,

iii) $h(X)$ *converges on the closed ball of radius c,*

iv) $\|h(X) - 1\|_c < 1$, *and*

v) *the Newton polygon of $g(X)$ is the same as to the portion of the Newton polygon of $f(X)$ contained in the region $0 \leq x \leq N$.*

PROOF: By induction on k:

If $k = 1$, then we have the situation above, and we have already proved the existence of $g(X)$ and $h(X)$.

Now assume the proposition is true for $k - 1$. Then we know there is a polynomial $g_1(X)$ which is a factor of $f(X)$ and whose Newton polygon coincides with the first $k - 1$ segments of the polygon for $f(X)$. We have $f(X) = g_1(X)h_1(X)$, and we know $h_1(X)$ has no zeros on the closed ball of radius $p^{m_{k-1}}$. Let's go on, then, to consider the k-th segment.

First of all, the fact that the k-th segment ends at $x = N$ says that for any $i > N$ the point $(i, v_p(a_i))$ lies above the line of slope m_k through the point $(N, v_p(a_N))$. As in our analysis of "the second segment" of the Newton polygon of a polynomial, it is easy to see that this means that $\|f(X)\|_c = |a_N|c^N$ and that $|a_N|c^N > |a_i|c^i$ for any $i > N$. Therefore, we can apply the Weierstrass Preparation Theorem to get a polynomial $g(X)$. It is then easy to see that $g(X)$ is divisible by $g_1(X)$, and that its Newton polygon coincides with the relevant portion of the Newton polygon of $f(X)$.   □

Problem 360 Flesh out the details of the proof. The crucial point is that any zero of $g(X)$ must be either a zero of $g_1(X)$, and we know about those, or a zero of $h_1(X)$ (and hence outside the ball of radius p^{m_k-1}). One needs to show that there are no zeros of absolute value *less* than p^{m_k}, and the rest falls into place.

Corollary 7.4.11 *Let*

$$f(X) = 1 + a_1 X + a_2 X^2 + a_3 X^3 + \cdots$$

be a power series which converges on the closed ball of radius $c = p^m$. Let $m_1, m_2, \ldots, m_k$ be the slopes of the Newton polygon of $f(X)$ which are less than or equal to m, and let $i_1, i_2, \ldots, i_k$ be their lengths. Then, for each j, $f(X)$ has i_j zeros with absolute value p^{m_j}, and there are no other zeros in the closed ball of radius p^m.

PROOF: Clear, because we know this for polynomials, and the proposition says that the relevant part of the Newton polygon of $f(X)$ *is* the Newton polygon of the polynomial $g(X)$. Since $g(X)$ is a factor of $f(X)$ and the quotient $h(X)$ clearly has no zeros in the closed ball of radius p^m, the conclusion follows.   □

Problem 361 Is the following version of the last proposition true?
Possible Proposition Let

$$f(X) = 1 + a_1 X + a_2 X^2 + a_3 X^3 + \cdots$$

be a power series which converges on the closed ball of radius $c = p^m$. Let N be the integer defined by the conditions

$$\|f(X)\|_c = |a_N|c^N \qquad \text{and} \qquad |a_n|c^n < |a_N|c^N \quad \text{if } n > N.$$

Then there exist a polynomial $g(X)$ of degree N and a power series $h(X)$ such that

i) $f(X) = g(X)h(X)$,

ii) $\|f(X) - g(X)\|_c < 1$,

iii) $h(X)$ converges on the closed ball of radius c,

iv) $\|h(X) - 1\|_c < 1$, and

v) the Newton polygon of $g(X)$ is the same as the portion of the Newton polygon of $f(X)$ contained in the region $0 \leq x \leq N$.

Furthermore, all the slopes in this portion of the Newton polygon of $f(X)$ will be less than or equal to m.

8 Fun With Your New Head

We've gone about as far as we want to go, but the reader may enjoy exploring further. In this final chapter we collect various problems that might be worth exploring. Several of these require more background than we have assumed throughout the book. Just ignore the ones you don't understand.

No hints will be supplied for these (it would spoil the fun!) beyond remarking that some of them are very much harder than others...

1 We first described p-adic numbers as base p representations that are allowed to be infinitely long "to the left." What happens if we do that for base ten expansions? What can you say about the ring of 10-adic integers?

2 Suppose K is a field with a nontrivial non-archimedean valuation. Let $\mathcal{O}_K$ be the valuation ring and $\mathfrak{p}$ be the maximal ideal in $\mathcal{O}_K$. Show that either $\mathfrak{p}$ is principal or $\mathfrak{p}^2 = \mathfrak{p}$ and $\mathcal{O}_K$ is not Noetherian.

3 When we discussed ways of visualizing $\mathbb{Q}_p$, we focused on topological equivalence. Explore the following alternative point of view. In Chapter 1, we drew pictures representing a p-adic integer as the limit of a coherent sequence of integers: see Figures 1.1 and 1.3. To see all of $\mathbb{Z}_p$ this way, we draw a tree: from the root there are p branches (one for each possible choice of $\alpha_1 \in \mathbb{F}_p$), then each vertex has p branches (one for each lift of α_1 to $\mathbb{Z}/p^2\mathbb{Z}$), etc. Then a p-adic integer is an infinite path down that tree. Explore this idea. Can we see the topology? The metric? Can we extend the picture to all of $\mathbb{Q}_p$? (This is the approach discussed in [37].)

4 Let $x \in \mathbb{Q}_p$, $x \neq 0$. Show that $x \in \mathbb{Z}_p^\times$ if and only if x^{p-1} is an n-th power for infinitely many $n \geq 1$. Use this to show that there are no nontrivial field automorphisms $\mathbb{Q}_p \longrightarrow \mathbb{Q}_p$. (The point is that the first result gives a characterization of the p-adic units in $\mathbb{Q}_p$ in terms of field operations only.) If you need help, see [53, I.6.8].

5 We have done very little linear algebra over a p-adic field. Of course, the elementary theory doesn't change. It might be interesting to study invertible matrices over $\mathbb{Q}_p$ or $\mathbb{Z}_p$. These are groups but they inherit topological structures as well. Find out more about them. (One advanced source is [58].)

© Springer Nature Switzerland AG 2020
F. Q. Gouvêa, *p-adic Numbers*, Universitext, https://doi.org/10.1007/978-3-030-47295-5_8

6 (This problem was proposed in the *American Mathematical Monthly* by Nicholas Strauss and Jeffrey Shallit. A solution by Don Zagier, using 3-adic methods, appeared in the January, 1992 issue.)

If k is a positive integer, let $v(k) = v_3(k)$ be the 3-adic valuation. For each positive integer n, let

$$r(n) = \sum_{i=0}^{n-1} \binom{2i}{i}.$$

Prove that

- $v(r(n)) \geq 2v(n)$, and

- $v(r(n)) = v\left(\binom{2n}{n}\right) + 2v(n)$.

Zagier's solution generalizes and extends this statement, and even formulates a conjecture at the end, so make sure to check it out after you've solved the problem.

7 On page 99 we raised the possibility of a version of Hensel's Lemma for Polynomials that still worked when the approximate factors were not relatively prime modulo p. What would that theorem say? Can you prove it? (There is a sketch in [52].)

8 Prove that for every positive integer k, we have

$$\sum_{n=0}^{\infty} n^k p^n \in \mathbb{Q}.$$

(The assertion is that the series converges, and that the sum is a rational number.)

9 Ostrowski's Theorem 3.1.4 tells us all the absolute values over $\mathbb{Q}$, and the product formula 3.1.5 tells us that there are some relations between different absolute values. The *Strong Approximation Theorem* is a result in the opposite direction: it says that we can always find a rational number that satisfies finitely many arbitrary local conditions and is a p-adic integer at all other primes.

Let S be a finite set of primes, and for each $p \in S$ let $a_p \in \mathbb{Q}_p$. For any $\varepsilon > 0$ there exists an $a \in \mathbb{Q}$ such that

i) For any prime $p \in S$ we have $|a - a_p|_p < \varepsilon$.

ii) For any prime $p \notin S$ we have $|a|_p \leq 1$.

Find a proof. It might also be interesting to research the *weak* approximation theorem to see what the differences are.

10 In Chapter 5 we saw examples of rational squares $x = y^2$ such that the binomial series $\mathbf{B}(1/2, x)$ converges to y in $\mathbb{R}$ and to $-y$ in $\mathbb{Q}_p$. How far can one push this?

i) Can you find an example of a series that converges in $\mathbb{Q}_p$ for all $p \leq \infty$, but whose sum is different each time?

ii) Can you find an example of a series that converges in $\mathbb{Q}_p$ for all $p \leq \infty$, but whose sum is the same rational number in every case? (Of course, a polynomial works, but that's not a fair example.)

11 Let $f(X)$ be a formal power series

$$f(X) = 1 + a_1 X + a_2 X^2 + a_3 X^3 + \ldots$$

with $a_n \in \mathbb{Q}_p$. Consider the power series

$$\frac{(f(X))^p}{f(X^p)} = 1 + b_1 X + b_2 X^2 + b_3 X^3 + \ldots.$$

Show that the following are equivalent:

i) $a_n \in \mathbb{Z}_p$ for all $n \geq 1$;

ii) $b_n \in p\mathbb{Z}_p$ for all $n \geq 1$.

This was first proved by Dieudonné and later generalized by Dwork.

12 Let $\mathbf{exp}(X)$ be the formal power series for the exponential and let

$$\mathbf{h}(X) = \sum_{n=0}^{\infty} \frac{X^{p^n}}{p^n}.$$

Define a new series $\mathrm{AH}(X)$ as the formal composition

$$\mathrm{AH}(X) = \mathbf{exp}(\mathbf{h}(X)).$$

Show that $\mathrm{AH}(X)$ has coefficients in $\mathbb{Z}_p$, so that $\mathrm{AH}(x)$ converges when $|x| < 1$.

One way to show this is to use the lemma in the previous problem. The function $\mathrm{AH}(x)$ is known as the Artin–Hasse exponential, and it is discussed in more detail in many of the more advanced texts. For example, there is an extended discussion in [53, VII.2].

13 We might consider a slightly more complicated function related to the Artin–Hasse exponential $AH(x)$ considered in the previous problem. Define a formal power series

$$e_p(X) = \exp(X + X^p/p).$$

If x is small enough, $e_p(x) = \exp_p(x)\exp_p(x^p/p)$, but the region of convergence of $e_p(X)$ is larger than that of the exponential. Figure out what it is.

The functions $e_p(x)$ were first studied by Dwork. The function in Problem 161 is $e_2(-2x)$.

14 Suppose X is a metric space. We say X has the Baire property if any countable union K of closed subsets K_n without interior points cannot have an interior point. (Remember that a point $x \in K$ is interior if there exists an open ball such that $x \in B(x, r) \subset K$.)

  i) Show that any complete metric space has the Baire property.

  ii) Show that any locally compact metric space has the Baire property.

  iii) Show that $\overline{\mathbb{Q}}_p$ does not have the Baire property.

This alternative way of showing $\overline{\mathbb{Q}}_p$ is neither complete nor locally compact is used in [53, III.1.4].

15 We mentioned that for a given degree n there are only finitely many field extensions $K/\mathbb{Q}_p$ of degree n. (Specifically, in a fixed algebraic closure $\overline{\mathbb{Q}}_p$ there are finitely many subfields of degree n over $\mathbb{Q}_p$.) Use Theorem 6.8.3 to prove this. (This is the proof given in [53, III.1.6]. See also [49].)

16 Consider the function $g : \mathbb{Z}_+ \longrightarrow \mathbb{Z}$ defined by $g(n) = n!$.

  i) Show that there is no continuous function $\tilde{g} : \mathbb{Z}_p \longrightarrow \mathbb{Z}_p$ that extends g, i.e., such that $\tilde{g}(n) = n!$ when n is a positive integer.

  ii) Modify the function g by "removing the p-part", i.e., define a new function that sends n to the product of the integers less than n that are not divisible by p:
$$g_p(n) = \prod_{\substack{1 \leq k \leq n \\ p \nmid k}} k.$$

  Can $g_p(n)$ be extended to $\mathbb{Z}_p$? If not, would a "twist" help?

17 In $\mathbb{C}$, it is trivial to see that any analytic function (even any continuous function) is bounded on any closed ball, because closed balls in $\mathbb{C}$ are compact. In $\mathbb{C}_p$, closed balls are no longer compact. Nevertheless, the boundedness result is still true: show that if $f(X)$ is a power series converging on a closed ball of radius r, then $f(X)$ is bounded on $\overline{B}(0, r)$. In fact, show that $f(X)$ has a maximum (rather than just a sup) on $\overline{B}(0, r)$.

18 Let $f(X)$ be a power series converging on the closed ball of radius r. By the previous problem, $f(X)$ is bounded. Show that

$$\max_{x \in \overline{B}(0,r)} |f(x)| = \max_{|x|=r} |f(x)|.$$

We might want to read this as "the maximum occurs at the boundary," even though we know that the sphere is not the boundary of the closed ball. (This is the p-adic analogue of the "maximum modulus principle.")

19 Corollary 5.6.5 showed that there are no non-constant periodic p-adic analytic functions. How about a multiplicative version of periodicity: can a p-adic analytic function satisfy $f(qx) = f(x)$ for all x and some fixed "period" q?

20 Suppose $f_n(X)$ is a family of power series satisfying:

i) All of the $f_n(X)$ converge in the closed ball of radius $\rho > 1$ around the origin.

ii) There exists a bound B such that $\|f_n(X)\|_\rho \le B$ for all n.

iii) There exists a power series $f(X)$ such that the series $f_n(X)$ converge to $f(X)$ with respect to the norm $\|\ \|_1$ (or, what is the same, coefficient-by-coefficient).

Show that $f(X)$ converges in the open ball of radius ρ, and that the $f_n(X)$ converge to $f(X)$ with respect to the norm $\|\ \|_c$ for any $c < \rho$.

21 How close do two power series need to be in order to allow us to conclude that they have the same number of zeros in the closed ball of radius r around 0? (This question is deliberately open-ended.)

22 We focused on functions defined by power series, but we could have also considered functions defined by Laurent series

$$f(X) = \sum_{n \ge n_0} a_n X^n.$$

Of course, if $n_0 < 0$ we can no longer evaluate this at $X = 0$, but we can consider convergence in an annulus $\{x \mid r < |x| < R\}$. What would such a theory look like? What would correspond to an entire function would be a series that converges on $\mathbb{C}_p^\times$; are there any?

23 Prove that $2^{p-1} \equiv 1 \pmod{p^2}$ if and only if p divides the numerator of

$$\sum_{j=1}^{p-1} \frac{(-1)^j}{j}.$$

24 (From [48].)

i) Let $f(X) = 1 - X^{p-1}$, and define

$$m(f, k) = \sup\{|f(x)| : x \in \mathbb{Q}_p, |x| = p^k\}.$$

Compute $m(f, k)$ for each $k \in \mathbb{Z}$. Does the answer change if we let $x \in \mathbb{C}_p$ instead?

ii) Find a sequence of integers $h_1, h_2, \ldots, h_k, \ldots$ such that if we set

$$f_k(X) = f(X) \cdot (f(pX))^{h_1} \cdot (f(p^2 X))^{h_2} \cdots (f(p^k X))^{h_k},$$

then we have

$$\sup\{|f_k(X)| : x \in \mathbb{Q}_p, |x| \leq p^k\} = 1.$$

iii) Use this to construct an example of an entire function which is bounded on $\mathbb{Q}_p$. What happens if we go to $\mathbb{C}_p$?

The point, of course, is that in classical complex analysis Liouville's Theorem says that there are no non-constant bounded entire functions.

25 Consider the function (stolen from [45]) $f : \mathbb{Z}_p \longrightarrow \mathbb{Z}_p$ which maps

$$x = a_0 + a_1 p + a_2 p^2 + a_3 p^3 + \cdots + a_n p^n + \cdots$$

to

$$f(x) = a_0 + a_1 p + a_2 p^4 + a_3 p^9 + \cdots + a_n p^{n^2} + \cdots$$

Is this function continuous? Is it differentiable? Can you extend it to $\mathbb{Q}_p$? To $\mathbb{C}_p$?

26 (Also from [45], but originally due to Dieudonné; see [24].) In the same spirit as the previous problem, consider $g : \mathbb{Z}_p \longrightarrow \mathbb{Z}_p$ which maps

$$x = a_0 + a_1 p + a_2 p^2 + a_3 p^3 + \cdots + a_n p^n + \cdots$$

to

$$g(x) = a_0^2 + a_1^2 p + a_2^2 p^2 + a_3^2 p^3 + \cdots + a_n^2 p^n + \cdots$$

(Notice that this will not be a "p-adic expansion," because the coefficients are not necessarily between 0 and $p-1$, but it clearly does converge, so that the definition makes sense.) Show that if $p \neq 2$ the function g is continuous but not differentiable on $\mathbb{Z}_p$. What happens if instead of squaring we use some other function to change the digits?

27 Suppose f is a continuous function on $\mathbb{Z}_p$. Consider the values

$$a_n = \sum_{k=0}^{n} (-1)^k \binom{n}{k} f(n - k).$$

- Explain the significance of the a_n. (Notice that they depend only on $f(n)$ for n a positive integer.)

- Define a formal series

$$f^*(X) = \sum_{n=0}^{\infty} a_n \binom{X}{n}.$$

Show that if m is a positive integer, then $f^*(m) = f(m)$.

- Suppose that $a_n \to 0$ as $n \to \infty$. Show that $f^*(x)$ converges uniformly for $x \in \mathbb{Z}_p$, and is a continuous function of x. Conclude that in this case $f^* = f$.

- Show that if f is continuous on $\mathbb{Z}_p$, then $a_n \to 0$ as $n \to \infty$. (This is quite hard; see Mahler's book for a detailed proof.)

- You know several functions that are continuous on $\mathbb{Z}_p$. Can you determine the numbers a_n for those functions?

This problem gives an approach to the interpolation problem developed by Mahler in [45]. If we know $f(n)$ for n a positive integer and we can show that the a_n tend to zero, then it gives a way of constructing an interpolating function.

28 In the classical situation, Rolle's Theorem tells us that there is some relation between the number of zeros of a function and the number of zeros of its derivative. Of course, Rolle's Theorem is a special case of the Mean Value Theorem, so it is not true in the p-adic context. But maybe we have a chance with functions defined by power series. Suppose $f(x)$ is a function defined by a power series convergent on some closed ball centered at 0 and let $f'(x)$ be its derivative. Compare the number of zeros of $f(x)$ and $f'(x)$ on that ball. (This is made-to-order for the Weierstrass Preparation Theorem and/or Newton polygons.)

29 The classical Weierstrass Approximation Theorem says that any continuous function on a closed interval can be uniformly approximated by polynomials. Is there a (true) p-adic analogue?

30 We didn't explore Galois theory at all. Suppose K is a finite extension of $\mathbb{Q}_p$. Are there any restrictions on the Galois group $\mathrm{Gal}(K/\mathbb{Q}_p)$? Or is any finite group possible?

A Sage and GP: A (Very) Quick Introduction

Mathematical software is part of the standard toolkit of mathematicians. Throughout this book we use the programs Sage and GP to show how to do computations with p-adic numbers. The goal of this appendix is to provide some basic orientation on the two programs, including where to get them. Both programs are free, both can be used online or installed on your own machine, both are very powerful, and both take some time to learn.

The two programs are different in several ways. GP is mostly intended for use by number theorists, while Sage wants to tackle all kinds of mathematics. Indeed, Sage has incorporated the functionality of many other free mathematics programs, including Maxima, Octave, R, GAP, and even GP itself. Sage usually runs in a browser window, while GP is more old-fashioned and usually runs in a terminal window. I learned GP first, and it's still my go-to program for quick computations, but lately I have been trying to teach my students to use Sage.

Of course, there are other mathematical software tools. Both *Maple* and *Mathematica* are well-known and quite powerful. The people who make *Mathematica* are also behind Wolfram Alpha, which can do lots of mathematics as well. These can all do some of the things we do in this book, but I have decided to focus on GP and Sage because they are free and I know how to use them.

A.1 Pari and GP

The interactive calculator GP was designed to serve the needs of people working in number theory. It is actually a front end for a software library called Pari, which can be used to create mathematical programs in C and C++. (Such programs are faster than using GP, but for most things GP is fast enough.)

Pari and GP were created by Henri Cohen and his team in 1985 and it has continued to grow since them. The current chief developer is Karim Belabas, who has many collaborators. The Pari-GP system is still actively developed, with new features being added and bugs being fixed all the time. The home web site for Pari-GP is `pari.math.u-bordeaux.fr`. That's where you go to download it, but you can also find a lot of information there.

© Springer Nature Switzerland AG 2020

F. Q. Gouvêa, *p-adic Numbers*, Universitext, https://doi.org/10.1007/978-3-030-47295-5

The normal way to use GP is to download and install it on your computer. The installer will usually create an icon on your desktop; clicking that icon opens a terminal window where you will see something like this, with the obvious variations as to version and operating system:

```
GP/PARI CALCULATOR Version 2.11.2 (released)
amd64 running mingw (x86-64/GMP-6.1.2 kernel) 64-bit version
compiled: Apr 28 2019, gcc version 6.3.0 20170516 (GCC)
threading engine: single
(readline v6.2 enabled, extended help enabled)

Copyright (C) 2000-2018 The PARI Group

PARI/GP is free software, covered by the GNU General Public
License, and comes WITHOUT ANY WARRANTY WHATSOEVER.

Type ? for help, \q to quit.
Type ?17 for how to get moral (and possibly technical)
support.

parisize = 8000000, primelimit = 500000
gp >
```

Some things to note:

- The last line is a *prompt*: GP expects you to type in some instruction and then hit the enter/return key. Prompts are configurable; on my machine it is gp >

- If you type \q and hit enter/return you will exit the program.

- If you type a question mark or a question mark followed by a number you will get (some) help.

Let's try something simple.

```
(15:15) gp > 25!
%1 = 15511210043330985984000000
```

Notice that GP understands the factorial notation and is not afraid of computing large numbers. Also notice that it uses %1 to label the result. You can later refer to this number by that name if you want to. Or you can give it a name.

```
gp > a=44/5
%2 = 44/5
```

The basic philosophy of GP is that it assumes the numbers you enter are in the simplest setting that makes sense. In this case, GP assumes that you mean the *rational number* 44/5. If you want the decimal version (i.e., the *real number*), you can either make it a decimal from the beginning or multiply it by 1.0.

```
gp > b=44.0/5
%3 = 8.8000000000000000000000000000000000000
gp > a*1.0
%4 = 8.8000000000000000000000000000000000000
```

Real numbers are presented with a default precision, in this case 38 significant digits. You can enter polynomials in the usual way.

```
gp > p=x^2-2*x+1
%5 = x^2 - 2*x + 1
gp > q = x^2 - 3*x + 2
%6 = x^2 - 3*x + 2
gp > p*q
%7 = x^4 - 5*x^3 + 9*x^2 - 7*x + 2
gp > p/q
%8 = (x - 1)/(x - 2)
```

Notice that the quotient of two polynomials is a rational function, given in lowest terms. What if we try something strange?

```
gp > log(p)
%9 = -2*x - x^2 - 2/3*x^3 - 1/2*x^4 - 2/5*x^5 - 1/3*x^6
     - 2/7*x^7 - 1/4*x^8 - 2/9*x^9 - 1/5*x^10 - 2/11*x^11
     - 1/6*x^12 - 2/13*x^13 - 1/7*x^14 - 2/15*x^15 + O(x^16)
```

As usual, GP makes the most reasonable interpretation of what you want, and returns the Taylor series. (The output is actually one long line, which I have broken up to make it easier to read.)

We can enter numbers modulo m and do operations with them.

```
gp > Mod(234,37)
%10 = Mod(12, 37)
gp > %10^16
%11 = Mod(9, 37)
gp > 1/%10
%12 = Mod(34, 37)
gp > Mod(5,12)/Mod(3,12)
  ***    at top-level: Mod(5,12)/Mod(3,12)
  ***                            ^----------
  *** _/_: impossible inverse in Fl_inv: Mod(3, 12).
```

Notice that GP understands "dividing" by a mod m, but of course, the number you are dividing by needs to be invertible in $\mathbb{Z}/m\mathbb{Z}$.

```
gp > Mod(12,37)^36
%13 = Mod(1, 37)
```

Ah, it's good to see that Fermat's Little Theorem is still true.

All GP constructions are iterative, so you can construct polynomials with coefficients in $\mathbb{Z}/37\mathbb{Z}$ and so on. Things can break if you overdo it (for example, power series in x whose coefficients are power series in y can be problematic), but generally it all works well. Polynomials in two variables are handled in this spirit: elements of $R[x, y]$ are handled by default as elements of $R[y][x]$:

```
gp > (1+x)*(1+y)^2
%14 = (y^2 + 2*y + 1)*x + (y^2 + 2*y + 1)
```

This means that x has "higher priority" than y by default. The priority order can be adjusted; see the manual.

Factoring is straightforward, even for reasonably large numbers:

```
gp > factor(42)
%15 =
[2 1]
[3 1]
[7 1]
gp > factor(2^67-1)
%16 =
[   193707721 1]
[761838257287 1]
```

If the input of factor is a polynomial, that's fine too.

```
gp > factor(x^3+x)
%17 =
[      x 1]
[x^2 + 1 1]
```

Here's an example of a polynomial with coefficients in $\mathbb{Z}/37\mathbb{Z}$. It's usually easier to write the polynomial with integer coefficients and then multiply by 1 mod 37 to make the coefficients be in $\mathbb{Z}/37\mathbb{Z}$

```
gp> factor(Mod(1,37)*(x^3+x))
%18 =
[                Mod(1, 37)*x 1]
[ Mod(1, 37)*x + Mod(6, 37) 1]
[Mod(1, 37)*x + Mod(31, 37) 1]
```

Notice that the output of factor is a matrix whose first column gives the prime divisors and whose second column gives the multiplicities.

What if we ask for help?

```
gp > ?
```
Help topics: for a list of relevant subtopics, type ?n for n in
 0: user-defined functions (aliases, installed and user
  functions)
 1: PROGRAMMING under GP
 2: Standard monadic or dyadic OPERATORS
 3: CONVERSIONS and similar elementary functions
 4: functions related to COMBINATORICS
 5: NUMBER THEORETICAL functions
 6: POLYNOMIALS and power series
 7: Vectors, matrices, LINEAR ALGEBRA and sets
 8: TRANSCENDENTAL functions
 9: SUMS, products, integrals and similar functions
10: General NUMBER FIELDS
11: Associative and central simple ALGEBRAS
12: ELLIPTIC CURVES
13: L-FUNCTIONS
14: MODULAR FORMS
15: MODULAR SYMBOLS
16: GRAPHIC functions
17: The PARI community
Also:
 ? functionname (short on-line help)
 ?\   (keyboard shortcuts)
 ?.   (member functions)
Extended help (if available):
 ??   (opens the full user's manual in a dvi previewer)
 ?? tutorial / refcard / libpari
     (tutorial/reference card/libpari manual)
 ?? refcard-ell
     (or -lfun/-mf/-nf: specialized reference card)
 ?? keyword (long help text about "keyword" from the user's
     manual)
 ??? keyword (a propos: list of related functions).

The next step is to make a more specific request for help, say with ?5 or ?gcd. As that list shows, GP can do a lot.

While the default way to use GP is in a terminal window, a browser-based version also exists. You can access it at pari.math.u-bordeaux.fr/gp.html It is also possible to use the Sage Cell Server (see below) in GP-mode to do quick computations.

There is a lot more to say, but that should get you started. At the Pari-GP home page you can find a tutorial and a user's manual. There are also email lists where you can ask for help if necessary.

A.2   Sage

Sage is an ambitious attempt to create powerful mathematical software that is free and open-source. The ultimate ambition, says the Sage home page at www.sagemath.org, is to create "a viable free open source alternative to *Magma, Maple, Mathematica* and MATLAB." (I'd say that they are very close to that, and in some aspects well beyond.) The development approach emphasizes openness: while William Stein is the leader of the team, contributions have come from across the mathematical community.

Sage can be used through a web interface, without needing to download and install the program. There are two ways to do that: either the Sage Cell Server at sagecell.sage.org or the more elaborate interface based on projects and notebooks offered by *CoCalc*, which is located at cocalc.com. It is also possible to download and install the program on your own computer. When you do, you can run Sage in a terminal window or you can run it in your browser. The latter is much like using *CoCalc*, but the program is running on your local machine.

Of the two web interfaces, the Sage Cell Server is particularly easy to use for small computations. It presents the user with a big blank rectangle where one can type in Sage commands. Below is a button labeled "Evaluate," which does exactly that. The output appears below. The downside is that Sage will not remember what you did, so if you define a symbol and then press the "Evaluate" button, you cannot use it again without repeating the definition.

There are two very nice features of the Cell Server that deserve note. First, it works on a tablet or phone. Second, because Sage incorporates GP and other open-source mathematical software, the Cell Server can be put into GP mode to do computations in GP. I have also used it in R mode. The mode switcher is at the bottom right of the input window.

CoCalc requires creating an account. Once you log in to your account, you can create projects, and each project can contain many notebooks. Notebooks allow you to enter lines of Sage code, which are evaluated when you hit Shift-Enter. Definitions and results are remembered within each session.

If you are going to use the program a lot, then of course the right thing to do is to download and install it. It takes quite a bit of space, but having it on your own machine avoids connectivity issues.

In mathematical terms, there is an important philosophical difference between GP and Sage. In GP, as we noted, the program assumes (or guesses) the mathematical context for the objects you create, i.e., whether they are integers, rational numbers, polynomials, etc. Sage, on the other hand, prefers to be told. You do that by creating a ring or field (or something else) in which you then do computations. Several standard rings and fields have preset names: the integers, rationals, reals, and complexes are, respectively ZZ, QQ, RR, CC, for example. Finite fields are GF(p^n) and the p-adic numbers are Qp(p), where in each case you put in actual numbers for p and n (and p must be prime).

Time to show some examples. You can find many more in *A Tour of Sage*, which is available online.

The basic "calculator" commands work as expected. Entering

```
3 + 5
57.1^100
```

into the Sage Cell Server and hitting "Evaluate" will produce

```
8
4.60904368661394403311007477775359103369 E175
```

You can get some space between the two lines of output by adding `print(" ")` between the two commands. Here we find the inverse of a matrix:

```
matrix([[1,2],[3,4]])^(-1)
```

results in

```
[  -2    1]
[ 3/2 -1/2]
```

Notice that you can enter a matrix by providing a list in brackets containing the rows as lists in brackets. (There are other ways; most things in Sage can be done in several different ways.) I didn't specify the ring in which the coefficients live; Sage assumed they were integers. To tell it otherwise, you would do something like this

```
A=matrix(GF(11),[[1,2],[3,4]])
print(A.inverse())
[9 1]
[7 5]
```

Let's try some calculus:

```
x = var('x')
f=integrate(sqrt(x)*sqrt(1+x), x)
```

That may surprise you: there would be no output at all. Sage has been told x is a variable and to assign the symbol f to the answer. It does, but doesn't print anything out. To see the answer you need to say

```
x = var('x')
f=integrate(sqrt(x)*sqrt(1+x), x)
print(f)
```

The result now is

```
1/4*((x + 1)^(3/2)/x^(3/2)
+ sqrt(x + 1)/sqrt(x))/((x + 1)^2/x^2
- 2*(x + 1)/x + 1) - 1/8*log(sqrt(x + 1)/sqrt(x) + 1)
+ 1/8*log(sqrt(x + 1)/sqrt(x) - 1)
```

That answer is very hard to read. It's good to know that you can also do it like this:

```
x = var('x')
f=integrate(sqrt(x)*sqrt(1+x), x)
show(f)
```

That gives something like this:

$$\frac{\frac{(x+1)^{\frac{3}{2}}}{x^{\frac{3}{2}}} + \frac{\sqrt{x+1}}{\sqrt{x}}}{4\left(\frac{(x+1)^2}{x^2} - \frac{2(x+1)}{x} + 1\right)} - \frac{1}{8}\log\left(\frac{\sqrt{x+1}}{\sqrt{x}} + 1\right) + \frac{1}{8}\log\left(\frac{\sqrt{x+1}}{\sqrt{x}} - 1\right)$$

There is also `latex(f)`, which is what I did to get the LaTeX code to typeset the result. While `show` produces output that is nicer to look at, the output of `print` is easier to cut-and-paste. In general, it's best to ask Sage to either `print` or `show` the outputs you want to see.

I should remark that there are subtle differences between writing `f=` as above and writing `f(x)=`. The Sage Tutorial is helpful on this issue.

As mentioned above, Sage likes to know in what ring it is working. Indeed, it will assume one if you don't give it one.

```
M=matrix([[1,2], [3,4]])
print(M.base_ring())
```

results in

```
Integer Ring
```

That is, Sage has assumed your matrix lives in $M_2(\mathbb{Z})$. This affects some of the results you can get, though, as you saw above, Sage is happy to move to $M_2(\mathbb{Q})$ when you ask for the inverse matrix.

You can tell Sage to work in a specific ring by adding its name to the command that creates the matrix:

```
M=matrix(CC,[[1,2], [3,4]])
print(M.base_ring())
print(M)
```

results in

```
Complex Field with 53 bits of precision
[1.00000000000000 2.00000000000000]
[3.00000000000000 4.00000000000000]
```

Notice that a complex number is $a + bi$ with $a, b \in \mathbb{R}$, so that both a and b will be printed as real numbers, i.e., as decimal expansions.

If you are going to work for a while in a given ring, or you don't want to have to write out something like `IntegerModRing` more than once, you can give your ring a name and use it:

```
R=IntegerModRing(24)
M=matrix(CC,[[1,2], [3,4]])
print(M.base_ring())
print(M)
```

gives

```
Ring of integers modulo 24
[1 2]
[3 4]
```

Many commands in Sage use the object-oriented `A.something()` format as above; sometimes you need to put something into the parentheses, but often they just need to be there. One advantage of *CoCalc* and the terminal interface is that if you type `A.` and then hit the Tab key, you will get a list of possible continuations. Here's a selection of matrix commands with the corresponding output after a line break:

```
M=matrix(QQ,[[1,2], [3,4]])
print(M)

[1 2]
[3 4]

M.characteristic_polynomial()

x^2 - 5*x - 2

M.column_space()

Vector space of degree 2 and dimension 2 over Rational Field
Basis matrix:
[1 0]
[0 1]
```

Without the `QQ`, the `column_space` function would return a free $\mathbb{Z}$-module of rank two rather than a vector space.

```
M.determinant()

-2
```

If you ask for eigenvalues of a matrix defined over $\mathbb{Q}$, Sage may well give them as real or complex numbers, however.

```
M.eigenvalues()

[-0.3722813232690144?, 5.372281323269015?]
```

Or you can ask it to find eigenvectors too, but with a catch:

```
M.eigenspaces_right()
```

```
[
(-0.3722813232690144?, Vector space of degree 2 and dimension 1
over Algebraic Field
User basis matrix:
[                     1 -0.6861406616345072?]),
(5.372281323269015?, Vector space of degree 2 and dimension 1
over Algebraic Field
User basis matrix:
[                     1 2.186140661634508?])
]
```

The problem is that Sage can think of matrices acting on vectors on the right
or on the left, and the eigenspaces (but not the eigenvalues) are different. If
you just call M.eigenspaces(), you get an error. Similarly, there are both
M.left_kernel() and M.right_kernel().

That's probably enough to start. To learn more, start with the Sage
Tutorial, which is at doc.sagemath.org/html/en/tutorial. See also [68],
which is available online as well as in print. There's a lot more helpful
documentation online.

B Hints, Solutions, and Comments on the Problems

This appendix contains hints, solutions, and comments of several kinds for the various problems set in the main text. There are only a few complete solutions here; rather, the intention is to provide a jump-off point for a solution, and perhaps to discuss the implications of some of the problems. Full solutions are given only when they are particularly tricky. Some of the comments even suggest further problems!

The hints and partial solutions become sketchier as we move toward the latter part of the book, in the expectation (or hope) that the experience and ability of the reader will increase. In many cases we have given references to where more details can be found.

1 The formula for the sum of the geometric series says that

$$1 + a + a^2 + a^3 + a^4 + \cdots = \frac{1}{1-a}$$

provided that $|a| < 1$. This can be used directly for the first expansion. For the third, write $X - 1 = 1 + (X - 2)$ and use $a = -(X - 2)$ in the geometric series.

These are, of course, easy to do with Sage. To do the third one, for example, the command would be `taylor(x/(x-1),x,2,12)`. (Note, however, that Sage can't resist writing $2 - (x - 2)$ as $4 - x$; the other powers of $(x - 2)$ are not expanded.) And yes, this works for the series centered at 1 as well.

2 If the expansion is finite, it will certainly become a polynomial after we multiply by $(X - \alpha)^m$, where m is the biggest exponent appearing in a denominator.

3 The sum is easy:

$$\sum_{i \geq n_0} a_i(X - \alpha)^i + \sum_{i \geq n_0} b_i(X - \alpha)^i = \sum_{i \geq n_0} (a_i + b_i)(X - \alpha)^i$$

(why can we assume the two series start at n_0?). The product takes only a little more work:

$$\left(\sum_{i \geq n_0} a_i(X - \alpha)^i\right)\left(\sum_{i \geq n_0} b_i(X - \alpha)^i\right) = \sum_{i \geq 2n_0} c_i(X - \alpha)^i,$$

© Springer Nature Switzerland AG 2020
F. Q. Gouvêa, *p-adic Numbers*, Universitext, https://doi.org/10.1007/978-3-030-47295-5

where the new coefficients are given by

$$c_i = \sum_{i_1+i_2=i} a_{i_1} b_{i_2},$$

which is a finite sum because the negative exponents only go back so far. Most of the field properties are easy to check; for the existence of inverses, one has to show that the equations for the coefficients of the product can be solved to find the coefficients of the inverse.

4 Well, y and $-y$ have to add up to zero, so we just make an expansion that does that. Let's do an example first.

Let $p = 5$ and $y = 120 = 440 = 0 + 4 \times 5 + 4 \times 5^2$. The zeroth digit of $-y$ will be 0 again, so that the sum has zero last digit. The first digit will have to produce a zero when we add 4, so it will be a 1, since $4 + 1 = 10$. So far we have $-y = 0 + 1 \times 5 + \ldots$. When we add y and $-y$, the first digit will be 0, and we will have to carry a 1 to the second (i.e., the 5^2) place. So now we want to set the digit so the sum is 4, so that adding the carry makes $5 = 10$. So the second digit is 0, and $-y = 0 + 1 \times 5 + 0 \times 5^2 + \cdots = \ldots 010$. Now we have run out of digits of y, but we have a 1 to carry to the third place. So we need to put a 4 for that digit of $-y$, so that adding the carry we get 5 again. And we'll have to do that over and over. So

$$-y = 0 + 1 \times 5 + 0 \times 5^2 + 4 \times 5^3 + 4 \times 5^4 + \cdots = \ldots 44444010.$$

The rule should now be clear. Leave any leading zeros alone. If a_k is the first (i.e., lowest power of p) nonzero digit, replace it by $p - a_k$. Then for all $\ell > k$ replace a_ℓ by $p - 1 - a_\ell$. So the recipe is: take the p-complement of the first nonzero digit and the $(p-1)$-complement of all higher digits. If the expansion is finite, we think of it having infinitely many zero digits, so the negative will start with an infinite string of $(p-1)$s.

5 Imitate the definitions for Laurent series, but watch out for carrying. That's doable, but proving that the result is a field gets tangled up in the trickiness of the carries.

So here are two suggestions of ways to make it easier. One idea, which is in fact the way Hensel did it originally, is to define an "irregular p-adic number" to be *any* expansion of the form

$$a_0 + a_1 p + a_2 p^2 + \cdots + a_n p^n + \ldots,$$

with no restriction on the a_i except that they be non-negative integers. It's then very easy to define the sum and product of irregular p-adic numbers by using the same formulas as with Laurent series. Then we need a theorem that says that any irregular p-adic number can be reduced to a regular p-adic number. The process would be something like this: beginning with a_0, write out each coefficient in base p, and rearrange the series accordingly. What needs to be proved is that this is all well-defined. (A harder question is whether the process can actually be done in a finite amount of time!)

Another option is to work like this: start with a p-adic number x, and factor out a power of p so that we have $x = a_0 + a_1 p + \ldots$ (a_0 may be equal to 0, of course). It's clear that it is enough to define the sum and product of such numbers (because we then factor the powers of p back in). Now, given such an x, let x_n be its truncation at p^n, so that

$$x_n = a_0 + \cdots + a_n p^n.$$

Now note that x_n is an integer, and we know how to add and multiply integers! Then define the sum and the product of two p-adic numbers by the rule

$$(x + y)_n = (x_n + y_n)_n \qquad \text{and} \qquad (x \cdot y)_n = (x_n \cdot y_n)_n$$

(i.e., multiply the truncations, and truncate the result). Once you've checked that all these truncations of $x+y$ and $x{\cdot}y$ "match," you can put them together to get a p-adic number. You need to check that this number is uniquely defined, i.e., that two p-adic numbers which have the same n-truncations for every n must be equal. This gives the operations. (You've still got to check the field properties!)

Note: what is really going on here is that we want to deduce the operations in $\mathbb{Q}_p$ from operations in $\mathbb{Q}$ (and even in $\mathbb{Z}$); truncation is the lazy way to do it, since it frees us from having to work out the carrying business explicitly, but it forces a bit of mumbo-jumbo. The difficulty of proving the field properties when we use a formal definition of $\mathbb{Q}_p$ is one of the reasons for the more conceptual theory we'll develop in the next chapter.

Note²: Of course, the computer programs that know how to compute with p-adic numbers "know" a way to do it. But those algorithms are allowed to assume that $\mathbb{Q}_p$ is a field, while here we are trying to use the algorithms to prove it is. Possible, but a pain.

6 Suppose first that the expansion is purely periodic, so that there is a repeating block $a_0 + a_1 p + \cdots + a_{k-1} p^{k-1}$ with k digits. Let $A = a_0 + a_1 p + \cdots + a_{k-1} p^{k-1}$. Then, taking the expansion of y in blocks of k digits, we get

$$y = A + A p^k + A p^{2k} + \ldots$$

From that, $y - p^k y = A$, which gives $y = A/(1 - p^k)$, which is rational. Notice that A is positive and $0 \leq A < p^k$ and $-p^k < 1 - p^k < 0$, so y is a negative rational number between -1 and 0.

In the same spirit, suppose the expansion is eventually periodic. Then $y = B + p^\ell y_1$, where B is the non-repeating block and y_1 is purely periodic. Since $y_1 = A/(1 - p^k)$ as above, we see that $y \in \mathbb{Q}$ as well.

7 If $y \geq 0$, the p-adic expansion is just the expression for y in base p and therefore is "finite," i.e., it ends in an infinite string of 0s. If $y < 0$, using problem 4 we see that the expansion of y ends in an infinite string of $(p-1)$s. So both are eventually periodic and the repeating block has length one.

8 We already know -1 and 0 have periodic expansions, so we can focus on the case when $b > 0$ and $-b < a < 0$. First we work modulo b; since $p \nmid b$ and p is prime, we have $\gcd(p, b) = 1$. By Euler's generalization of Fermat's Little Theorem, we know that there exists a positive integer k such that $p^k \equiv 1 \pmod{b}$. Then $1 - p^k = mb$ for some (negative) integer m. Multiply both numerator and denominator by m to get

$$y = \frac{a}{b} = \frac{ma}{mb} = \frac{A}{1 - p^k} = A \sum_{n=0}^{\infty} p^{nk}.$$

Multiplying out we get

$$y = A + Ap^k + Ap^{2k} + \dots$$

Now notice that $0 < A < p^k$ (because $-1 < y < 0$), so that $A = a_0 + a_1 p + \dots + a_{k-1} p^{k-1}$, plugging that into the formula we get a periodic p-adic expansion with a repeating block of length k.

9 This is the hard one! Tellingly, most of the textbooks leave it as an exercise. This solution is inspired by the method used by Keith Conrad in one of the many expository notes he has posted online. See [21].

First of all, we can focus only on $y = a/b$ where neither a nor b are divisible by p, since multiplying by a power of p will only shift the digits left or right. Second, we already know that the result is true for integers, by problem 7. So we assume $b \neq 1$. Since taking the negative is done by finding $(p-1)$-complements of the digits, the negative of an eventually periodic expansion is eventually periodic. So we can fix the sign of y. *We assume $y < 0$.*

So now we know $y = a/b$ is a negative non-integer with denominator not divisible by p. Then there exists a positive integer B such that $-1 < y+B < 0$; the denominator won't change when we add an integer, so it is still b and still not divisible by p. By the previous problem, the expansion of $y + B$ is purely periodic, so

$$y + B = \sum_{n=0}^{\infty} a_n p^n$$

with the sequence a_n periodic (and not all zeros, since $y \notin \mathbb{Z}$).

Thus far our argument is exactly the one Hensel gave in 1904, but he thought it was obvious that adding two eventually periodic expansions would give an eventually periodic expansion. I don't think it's *that* obvious, and here is where Conrad's idea helps.

We need a way to show that subtracting B from the periodic expansion on the right of our equation won't mess up more than a finite number of digits. For that, look at the truncations

$$C_k = \sum_{n=0}^{k} a_n p^n,$$

which are positive integers, $0 < C_k < p^{k+1}$. Since the a_n can't all be zero and are periodic, as k grows the positive integer C_k gets bigger (there must always be one more nonzero a_n). Therefore there is a k such that $C_k > B$. Now we can write

$$y + B = C_k + \sum_{n=k+1}^{\infty} a_n p^n,$$

so

$$y = (C_k - B) + \sum_{n=k+1}^{\infty} a_n p^n.$$

Now notice that $C_k - B$ is a positive integer (because $C_k > B$) which is smaller than p^{k+1} (because $C_k - B < C_k < p^{k-1}$), so its base-p expansion looks like

$$C_k - B = b_0 + b_1 p + b_2 p^2 + \cdots + b_k p^k.$$

Plugging that in gives

$$y = b_0 + b_1 p + b_2 p^2 + \cdots + b_k p^k + \sum_{n=k+1}^{\infty} a_n p^n,$$

which is eventually periodic.

An alternative approach is to focus on the process of creating the expansion and notice that at some stage we will be down to a finite number of choices, which will force the expansion to eventually start repeating; see, for example, [39, p. 31] and [53, pp. 39–40].

10 First of all, any ideal J in $\mathbb{C}[X]$ is principal, and will be maximal when its generator is irreducible. Since $\mathbb{C}$ is algebraically closed, the only irreducible polynomials are those of degree one, and we can always divide by the degree one coefficient (an invertible element of $\mathbb{C}[X]$, so the ideal doesn't change) to get a polynomial of the form $X - \alpha$. So, to a maximal ideal J we can attach the number α, and conversely. Part (ii) is even easier: just remember that $f(\alpha) = 0$ if and only if $f(X)$ is divisible by $X - \alpha$. Part (iii) is also standard.

For the rational numbers, just follow the hints as given. The order of the "pole at p" will be the largest power of p dividing the denominator of x. As to whether this is a reasonable thing to do, it turns out to be a very useful point of view in modern algebraic geometry, so I guess it must be "reasonable"... It certainly does make the analogy a little bit tighter.

11 This should be straightforward: $6 \times 2 = 11$. Write down 1, carry a 1. $5 \times 2 + 1 = x + 1 = 10$, so write down 0 and carry a one, rinse, repeat.

12 Same deal. Make sure you can do it.

13 The addition is as before. For the multiplication, remember that you need to multiply $\ldots$xxxxxx by each of the digits of **102**. Luckily, 0 and 1 are easy. Multiplying by 2 gives $\ldots$xxxxx9. Now add:

$$\ldots\text{xxxxxxx9}$$
$$\ldots\text{0000000}$$
$$\underline{\ldots\text{xxxxxx}}$$
$$\ldots\text{xxxx9x9}$$

14 These can be done either by repeated division by 11 or by following the solution of problem 8.

i) Let's try repeated division:

$$-\frac{2}{3} = 11 \times \frac{-1}{3} + 3$$
$$-\frac{1}{3} = 11 \times \frac{-2}{3} + 7$$

and now it will repeat

Therefore

$$-\frac{2}{3} = 3 + 7 \times 11 + 3 \times 11^2 + 7 \times 11^3 + \cdots = \overline{73}.$$

ii) This one is easy since $11 \equiv 1 \pmod 5$, so the k in problem 8 is $k = 1$. We get

$$-\frac{3}{5} = \ldots 6666666 = \overline{6}.$$

iii) This one is much harder because the smallest exponent is $k = 6$, so the repeating block will have six digits:

$$-\frac{4}{9} = \overline{498612}.$$

iv) This one shows that the repeating block can have a leading 0, i.e., the expansion is purely periodic also when the numerator is divisible by 11.

$$-\frac{11}{12} = 0 + 10 \times 11 + 0 \times 11^2 + 10 \times 11^3 + \cdots = \overline{x0}.$$

15 We need to divide by 3 over and over until things start to repeat. After the first line, every quotient will be will be $\frac{-n}{17}$ with n between 1 and 16, so eventually things will have to repeat.

$$\frac{24}{17} = 3 \times \frac{8}{17} + 0$$

$$\frac{8}{17} = 3 \times \frac{-3}{17} + 1$$

$$\frac{-3}{17} = 3 \times \frac{-1}{17} + 0$$

$$\frac{-1}{17} = 3 \times \frac{-6}{17} + 1$$

$$\frac{-6}{17} = 3 \times \frac{-2}{17} + 0$$

$$\frac{-2}{17} = 3 \times \frac{-12}{17} + 2$$

$$\frac{-12}{17} = 3 \times \frac{-4}{17} + 0$$

$$\frac{-4}{17} = 3 \times \frac{-7}{17} + 1$$

$$\frac{-7}{17} = 3 \times \frac{-8}{17} + 1$$

$$\frac{-8}{17} = 3 \times \frac{-14}{17} + 2$$

$$\frac{-14}{17} = 3 \times \frac{-16}{17} + 2$$

$$\frac{-16}{17} = 3 \times \frac{-11}{17} + 1$$

$$\frac{-11}{17} = 3 \times \frac{-15}{17} + 2$$

$$\frac{-15}{17} = 3 \times \frac{-5}{17} + 0$$

$$\frac{-5}{17} = 3 \times \frac{-13}{17} + 2$$

$$\frac{-13}{17} = 3 \times \frac{-10}{17} + 1$$

$$\frac{-10}{17} = 3 \times \frac{-9}{17} + 1$$

$$\frac{-9}{17} = 3 \times \frac{-3}{17} + 0$$

$$\ldots$$

And now we know that the digits will repeat, since we have seen $-3/17$ before. So the last two digits are **10** and the repeating part is **0112021221102010** (sixteen digits, because the smallest k such that $3^k \equiv 1 \pmod{17}$ is $k = 16$, i.e., 3 is a primitive root mod 17). In other words, we have $24/17 = \overline{011202122110201010}$. Part (b) is just a long exercise in elementary multiplication, which should convince you that computing by hand quickly loses its appeal.

16 This is standard business; see any basic text on number theory. To do it yourself, note that saying that $x^2 \equiv 25 \pmod{p^n}$ is the same as saying that p^n divides $x^2 - 25 = (x-5)(x+5)$. Then it's a matter of showing that it's not possible for both factors to be divisible by p. For $p = 2, 5$ there can be more than two roots. For example, $X^2 \equiv 25 \pmod{25}$ has roots 0, 5, 10, 15, 20 (mod 25). Describing the general behavior in the "bad" cases is harder than in the case $p \neq 2, 5$.

17 When we are working in base p, reducing a number modulo p^k just means erasing all but the first k digits. So a coherent sequence is a sequence (α_n) of positive integers so that

  i) In base p, α_n has at most $n + 1$ digits.

  ii) For every k, deleting all but the last $k + 1$ digits of α_n gives α_k.

So putting the α_n together we get an infinite sequence of digits from which all the α_k are obtained by truncation.

18 $X^2 \equiv 49 \pmod{5^n}$ goes just like the example in the text. Similarly, $X^3 \equiv 27 \pmod{2^n}$ goes smoothly (there is only one root).

19 Just work it out. For every large enough n you should find *four* roots, only two of which "continue" on to the next n.

20 Standard number theory business. For $n = 1$, you are looking at an equation of degree 2 in a field. Then show that solutions modulo p^n always lift uniquely to solutions modulo p^{n+1}. Alternatively, imitate problem 16.

21 See the previous problem. This can be found in most books, too, but it's easy anyway. If we have a solution $a \pmod{7^n}$ then a "lift" $\tilde{a} \pmod{7^{n+1}}$ has to be of the form $\tilde{a} = a + x7^n$, with $x = 0$ or 1 or ... or 6. Now plug $\tilde{a}$ into the equation, and show that one can always solve (uniquely) for x.

22 One idea is to use truncations, as in Problem 5 above: show that $(x_1^2)_n = (2)_n$ for every n, just by tracing through where we got x_1. Of course, you can also check with GP.

23 If $x = a_0 + a_1 5 + \ldots$ and $x^2 = 2$, then $a_0^2 \equiv 2 \pmod 5$.

24 For the negative statement, it's enough to check that $X^2 + 1 = 0$ has no solutions modulo 7. For the positive statement, start from the fact that it *does* have solutions modulo 5, and then use the methods we've been playing with.

25 This is just a generic version of Problem 22; the "truncation" method will work.

26 Read carefully over the last several problems, and write up your methods as a general result.

27 As long as $p \neq 2$, what we have already done solves the problem: one can always find an m which is not a square in $\mathbb{Q}$ but is a quadratic residue modulo p, hence is a square in $\mathbb{Q}_p$. For $p = 2$, consider either cubes, or roots of polynomials of the form $X^2 + X + m = 0$.

28 Repeat the last problem in reverse. For any $p \neq 2$, there is an m which is *not* a quadratic residue modulo p, hence is not a square in $\mathbb{Q}_p$, which is therefore not algebraically closed. The same workaround as before handles $p = 2$.

29 The usual proof works: multiply the sum by p and subtract.

30 I certainly can't, but see Problem 10.10 in [48] and its solution. For a not-so-elementary solution, see problem 175.

31 Let $|\ |$ be an absolute value on $\Bbbk$. We have $|0| = 0$ by the definition. The equation $1 = 1 \cdot 1$ forces $|1| = |1| \cdot |1|$. Since $|1|$ is a strictly positive real number, it follows that $|1| = 1$. Now take any element $x \in \Bbbk$, $x \neq 0$. Since $\Bbbk$ is a finite field, there exists an integer q such that $x^q = x$ (we can take q to be equal to the number of elements in $\Bbbk$). Taking absolute values, we get $|x|^q = |x|$; since $|x|$ is real and positive, this forces $|x| = 1$. Thus, $|\ |$ must be trivial.

32 If $a/b = c/d$, then $ad = bc$. By unique prime factorization, the highest power of p that divides ad is just the sum of the highest powers dividing a and d; thus, $v_p(ad) = v_p(a) + v_p(d)$. Similarly, $v_p(bc) = v_p(b) + v_p(c)$. Then, if $ad = bc$, we have

$$v_p(a) + v_p(d) = v_p(ad) = v_p(bc) = v_p(b) + v_p(c).$$

Now rearrange.

33 It's just a matter of dividing by p: $v_5(400) = 2$, $v_7(902) = 0$, $v_3(123/48) = 0$, $v_5(180/3) = 1$. Try a large number: what is $v_{11}(452, 298)$?

34 First consider the case when x and y are both integers. Write $x = p^a x'$ and $y = p^b y'$ where both x' and y' are integers not divisible by p. Since we may interchange x and y if necessary, we can assume that $a \leq b$. Then $xy = p^{a+b} x' y'$, which shows (i), and

$$x + y = p^a x' + p^b y' = p^a (x' + p^{b-a} y'),$$

which shows that $v_p(x+y) \geq a$, and so proves (ii). This proves both statements when x and y are integers. To get it for fractions, let $x = t/q$, $y = r/s$. Then

$$v_p(xy) = v_p\left(\frac{tr}{qs}\right) = v_p(tr) - v_p(qs)$$

$$= v_p(t) + v_p(r) - v_p(q) - v_p(s)$$

$$= v_p\left(\frac{t}{q}\right) + v_p\left(\frac{r}{s}\right).$$

This proves part (i) for the general case. Part (ii) is similar (in other words, do it!).

35 $1/7$, $1/7$, 1, 343, respectively. In this case, there are no Sage or GP commands.

Notice that with respect to this absolute value, $3/686$ is big, while 35 is small. So 2 is closer to 37 in $\mathbb{Q}_7$ than it is to $2 + \frac{3}{686}$.

36 There's nothing much to do here: just straight translation. Remember that the elements of K look like a/b where a and b are in A (and $b \neq 0$), and that $a/b = c/d$ if and only if $ad = bc$. Then follow your nose.

37 First, $v(1) = 0$, so 0 is in the image. If α and β are in the image, then we must have $\alpha = v(x)$ and $\beta = v(y)$ for some non-zero x, $y \in \Bbbk$. But then $\alpha + \beta = v(xy)$ and $-\alpha = v(1/x)$, so that we have a subgroup. In the case of the p-adic valuation, the value is always an integer by definition, so the value group is $\mathbb{Z}$.

38 This is easy to see, since we defined $|p^n| = p^{-n}$, so that $|p^n| \to 0$ as $n \to \infty$. The more a number is divisible by p, the smaller it is in the p-adic world.

39 This is straightforward, since passing from the valuation properties of v_p to the properties of the absolute values is just a matter of taking powers. The obvious conjecture is that it does not matter what value of c is used, in the sense that the resulting absolute values for varying c are "similar" enough that we might as well treat them as being the same. That's exactly what happens. Why $c = p$ is a good choice is more subtle—see ahead for the Product Formula.

40 Yes, it is enough to check for polynomials (see Problem 36). For polynomials, both equations are well-known, if we restate them in terms of the degree. (Notice that the sum of two polynomials of the same degree can have smaller degree, so that after changing signs the $\geq$ is indeed necessary.)

41 A rational function f/g will be small with respect to $\vert\ \vert_\infty$ when $v_\infty(f/g)$ is *big*, hence when the degree of g is much bigger than the degree of f. Hence, polynomials are never small: if f is a polynomial, then $\deg(f) \geq 0$ gives $v_\infty(f) \leq 0$ and so $\vert f\vert_\infty \geq 1$. The higher the degree, the bigger the absolute value.

42 Boring but easy. Just run through the definition of the p-adic absolute value and check that everything works. For concreteness, play with the case where $\mathsf{F} = \mathbb{R}$ and $p(t) = 1 + t^2$, or the case $\mathsf{F} = \mathbb{C}$, $p(t) = t - 4$.

43 In every case, it turns out to be the trivial absolute value. For $\vert\ \vert_\infty$, just notice that any non-zero constant has degree zero, and hence absolute value 1. For the $p(t)$-adic absolute values, notice that the constants in $\mathsf{F}[t]$ are not divisible by any irreducible polynomial.

44 Every polynomial of degree n with coefficients in $\mathbb{C}$ has n roots, so that we can always write it as a product of linear terms. Hence, the only irreducible polynomials are the ones of degree one, $p(t) = t - \lambda$. The $p(t)$-adic valuation of a polynomial $f(t)$ just measures the multiplicity of λ as a root of $f(t)$. We are very close indeed to Hensel's original idea.

45 This is very hard, and, as advertised, depends on the choice of the field F. If an archimedean absolute value on $\mathsf{F}(t)$ can be found, its restriction to the subfield of constants F will have to be an absolute value on F, and it will have to be archimedean (if you can't see why, wait till the next section). So it can't be done if (a) there are no archimedean absolute values on F, nor if (b) we require that the restriction to F be the (non-archimedean) trivial absolute value.

Here's the sneaky bit: take $\mathsf{F} = \mathbb{Q}$, and choose a transcendental number, say π. Since π is not a root of any polynomial over $\mathbb{Q}$, the rings $\mathbb{Q}[\pi]$ and $\mathbb{Q}[t]$ are isomorphic. (Map $\mathbb{Q}[t] \to \mathbb{Q}[\pi]$ by $t \mapsto \pi$; this is obviously onto, and what can possibly be in the kernel?) It follows that the fields $\mathbb{Q}(\pi)$ and $\mathbb{Q}(t)$ are isomorphic. Now, $\mathbb{Q}(\pi)$ is contained in $\mathbb{R}$, so we can restrict the archimedean absolute value on $\mathbb{R}$ to $\mathbb{Q}(\pi)$, and then pull it back to $\mathbb{Q}(t)$ via the isomorphism to get an archimedean absolute value! (We know it'll be archimedean by computing $\vert 2\vert$.)

I said it was sneaky.

46 Consider a polynomial $f(t) = a_n t^n + \cdots + a_1 t + a_0$, with $a_n \neq 0$, so that its degree is really n. We have $v_\infty(f) = -n$, by the definition above. Now, as a polynomial in $1/t$, we have

$$f(t) = t^n \left(a_n + \cdots + a_1 \left(\frac{1}{t}\right)^{n-1} + a_0 \left(\frac{1}{t}\right)^n \right)$$

$$= \frac{a_n + \cdots + a_1 (\frac{1}{t})^{n-1} + a_0 (\frac{1}{t})^n}{(\frac{1}{t})^n}$$

so that $v_1(f) = -n$ (the numerator above is clearly not divisible by $1/t$).

47 The construction of the π-adic valuation v_π should be routine by now. Some hints: the "good" value for c will depend on π; if π divides a rational prime p, then we want to choose either p or p^2 (can you come up with a reason?). For the last question, reading the three cases above, we'll get that $v_\pi(p)$ will equal zero for all primes except the (unique) one that is divisible by π, in which case it will always be equal to one, except if $p = 2$. (Now that is a convoluted sentence!) What is $v_{1+i}(2)$? If $\pi = x + iy$ and $\overline{\pi} = x - iy$ are two primes as in case (iii), what is the relation between v_π and $v_{\overline{\pi}}$?

48 This is all rather easy: the point is that $|x|$ is always a positive real number. For (ii), just note that if $\lambda^n = 1$ and λ is a positive real number, then $\lambda = 1$. Statement (iii) is just (ii) with $n = 2$; statement (iv) then follows from $|-x| = |-1| \cdot |x|$. Finally, in a finite field with q elements, we have $x^{q-1} = 1$ whenever $x \neq 0$, and applying (ii) shows that any absolute value must then be trivial.

49 Suppose $\sup\{|n| : n \in \mathbb{Z}\} = C$, and $C > 1$. Then there must exist an integer m whose absolute value is bigger than 1. But then $|m^k| = |m|^k$ gets arbitrarily large as k grows, so that C cannot be finite. It follows that $C \leq 1$, and since $|1| = 1$, this means $C = 1$, so that $|\ |$ is non-archimedean.

50 This is straight translation from the properties of absolute values.

51 All of these are proved for the standard absolute value in most texts on real analysis.

For (i), notice that

$$d(x + y, x_0 + y_0) = |(x + y) - (x_0 + y_0)| = |(x - x_0) + (y - y_0)|$$

and use the triangle inequality. For (ii), the trick is to write

$$xy - x_0 y_0 = x(y - y_0) + y_0(x - x_0)$$

and remember that $|x|$ is bounded. For (iii), use

$$\frac{1}{x} - \frac{1}{x_0} = \frac{x_0 - x}{x x_0}$$

and the fact that you can assume x is bounded away from 0. For (iv), the key inequality is

$$\left| |x| - |y| \right| \leq |x - y|$$

(where the outside absolute value on the left is the one in $\mathbb{R}$), which follows from applying the triangle inequality to both $(x-y)+y = x$ and $x+(y-x) = y$. This gives uniform continuity because nothing depends on what $x, y \in \mathbb{k}$ are.

You should work these out carefully or look them up in an analysis textbook if you find them troublesome.

52 The hints in the text should be enough to suggest the proof. For the second statement, notice that for any two positive real numbers α and β we have $\alpha + \beta \geq \max\{\alpha, \beta\}$.

53 We have

$$x - y = -\frac{1}{15} \qquad y - z = -\frac{4}{15} \qquad x - z = -\frac{5}{15} = -\frac{1}{3};$$

the first two sides have length 5, the third has length 1.

54 Open balls first. If $x \in B(a, r)$, then let $\delta = |x - a| < r$. We need to show that a small enough ball around x is completely contained in $B(a, r)$. Consider the ball around x with radius $\varepsilon = r - \delta$. If a point y belongs to this ball, then $|y - x| < \varepsilon$. But if that is the case, then

$$|y - a| \leq |y - x| + |x - a| < \varepsilon + \delta = r - \delta + \delta = r,$$

so that $y \in B(a, r)$. Make a picture if this is not clear!

Do something similar for closed balls (if you made a picture, this should be easy).

55 The missing parts are easily done by imitating the parts given in the text. We need the $r \neq 0$ condition because a closed ball of radius 0 is a point, which is *not* an open set unless the absolute value is trivial. (Why not?) By contrast, open balls of radius zero are just empty sets, which are always both closed and open anyway. (Why?)

56 $\overline{B}(0, 1)$ is the set of fractions a/b such that $|a/b|_p \leq 1$, which means that $v_p(a/b) \geq 0$. This will happen when, after putting the fractions into lowest terms, the denominator is not divisible by p. So the closed unit ball around 0 consists of the fractions a/b where p does not divide b.

$B(3,1)$ is the set of fractions a/b such that $a/b - 3$ has absolute value less than one. Reasoning as before, this means that the denominator is not divisible by p, but the numerator *is*. If we assume that a/b is in lowest terms and (so that a and b have no common factors), then

$$\frac{a}{b} - 3 = \frac{a - 3b}{b}$$

will also be in lowest terms (check!), so the conditions are that p does not divide b but does divide $a - 3b$. To find out what integers satisfy this condition we set $b = 1$; the first condition is automatically true, and the second condition says that p divides $a - 3$, i.e., that $a \equiv 3 \pmod{p}$.

57 Well, as we found out in Problem 56, the closed unit ball is the set of all fractions a/b where p does not divide b. Look at the numbers

$$a, \quad a - b, \quad a - 2b, \quad a - 3b, \quad \ldots \quad a - (p-1)b.$$

It is easy to see that exactly one of these numbers will be divisible by p (the easiest way to see it is to note that these are p integers, and no two of them can be congruent modulo p, because if p divides the difference of any two, then it divides b, which it doesn't). But if $a - ib$ is divisible by p, then

$$\left| \frac{a}{b} - i \right| = \left| \frac{a - ib}{b} \right| < 1,$$

so that a/b is in the open ball $B(i,1)$ of center i and radius 1. This proves the equality. The disjointness amounts to the statement that *only one* of the numbers listed can be divisible by p.

As noted, this gives another proof that the closed ball is open, but this one is specific to the p-adic absolute value, while Proposition 2.3.7 is true for any ultrametric space.

58 The point is that the 5-adic absolute value can only take values of the form 5^n, where n is an integer. So saying that a 5-adic absolute value is less than one, less than $1/2$, or less than or equal to $1/5$ all amount to the same thing. This is not something we should expect for all non-archimedean absolute values.

59 To see that the sphere is closed, notice that it is the intersection of the closed unit ball with the complement of the open unit ball. These sets are both closed, and so their intersection is closed. (Notice that this part doesn't depend on the absolute value being non-archimedean.)

To see that the sphere is also open, take x in the sphere, so that $|x-a| = r$, and choose $\varepsilon < r$. Then if $|x - y| < \varepsilon$ we must have $|y - a| = r$ because all triangles are isosceles. Hence, the open ball around x of radius ε is completely contained in the sphere. (Notice that this part does.)

For the sphere to be the boundary of the open ball B, any small open ball centered on a point on the sphere should intersect B. But that can't happen if the sphere is an open set.

60 To go one way, we need to take $A = S \cap U_1$ and $B = S \cap U_2$. To go the other, we need to find two open sets; we might try to let U_1 be the complement of the closed set $\overline{B}$, and U_2 be the complement of the closed set $\overline{A}$. Check that this works.

61 The intervals.

62 The ball is both a closed and open set. Take a smaller ball inside it. It is also closed and open, so that the complement of the smaller ball in the bigger ball is a closed and open set. This gives us the decomposition we want. It doesn't matter whether we use open or closed balls, since they're all clopen.

63 Suppose a set S contains both x and another point y; we will show S cannot be connected. To simplify the notation, let $r = |x - y|$. To show S is disconnected, we need to find the sets U_1 and U_2 in the definition. Remember that balls are clopen. For U_1 we take the open ball of radius $r/2$ around x; this contains x and not y. For U_2 we take the complement of U_1, which is open because U_1 is closed; this contains y but not x. The union of U_1 and U_2 is the whole space, so this does what we want.

64 First of all, the trivial case: the empty set and all of $\mathbb{Q}$ are both clopen sets. But there are other clopen sets in $\mathbb{Q}$. For example, consider the set of rational numbers a/b whose square is less than two:

$$S = \left\{ \frac{a}{b} \mid \left(\frac{a}{b}\right)^2 < 2 \right\} = \left\{ \frac{a}{b} \mid -\sqrt{2} < \frac{a}{b} < \sqrt{2} \right\}.$$

This is clearly open, but it is also closed (can you check that?). It's not hard to see that this means that $\mathbb{Q}$ is totally disconnected also with respect to the usual absolute value. The problem, of course, is that the only possible boundary points are not in $\mathbb{Q}$.

On the other hand, there are no nontrivial clopen sets in $\mathbb{R}$, and $\mathbb{R}$ is connected; prove it. Notice that this happens because we have "filled in" all the missing points *and* the metric is archimedean. As we saw in the text, ultrametric spaces are *always* totally disconnected.

65 We showed in a previous problem that the closed unit ball around 0 was the disjoint union of $p - 1$ open balls of radius 1. Scaling and translating, we see that the same will be true for any closed ball. For open balls, we can use a dirty trick. Again, look at the open ball of radius 1 around 0; for the reasons explained in Problem 58, this is *equal* to the closed ball of radius $1/p$

around 0; by the argument above, this is the disjoint union of open balls! This proves our claim for the open ball of radius 1 around 0, but again we can scale and translate to get the general result.

The fact that we had to use the special property that the values of the p-adic absolute value are all of the form p^n with n an integer should be a hint that this will *not* work in general. Nevertheless, it's pretty hard (at this point) to come up with a counter-example (the algebraic closure of $\mathbb{Q}$ will work, but defining an absolute value on that field is a non-trivial task).

66 (Very much a set of hints rather than a solution.) To show that $\mathcal{O}$ is a subring, we need to show that it contains 0 and 1 and is closed under addition, multiplication, and change of sign. This is all easy, and you only need to use the non-archimedean property to show closure under addition. (Do it.)

To show that $\mathfrak{P}$ is an ideal, we need to check that it is closed under addition (easy), contains 0 (clear), and that if $x \in \mathcal{O}$ and $y \in \mathfrak{P}$, then $xy \in \mathfrak{P}$. The two assumptions say that $|x| \le 1$ and $|y| < 1$; since $|xy| = |x||y|$, it follows that $|xy| < 1$, i.e., $xy \in \mathfrak{P}$.

If $x \in \mathcal{O}$ but $x \notin \mathfrak{P}$, then we must have $|x| = 1$. It then follows that $|1/x| = 1$, so that $1/x \in \mathcal{O}$, which means that x is invertible in $\mathcal{O}$. Any ideal in $\mathcal{O}$ strictly containing $\mathfrak{P}$ would have to contain such an x. By what we have just shown, such an ideal will contain an invertible element, and would therefore be all of $\mathcal{O}$; this shows $\mathfrak{P}$ is maximal.

67 The jazzy proof: we have an injective homomorphism $\mathbb{Z} \hookrightarrow \mathbb{Z}_{(p)}$, and it maps $p\mathbb{Z}$ into $p\mathbb{Z}_{(p)}$ (do you see that?). By the usual hocus-pocus (e.g., [33, 5.3]), this gives an injective map

$$\mathbb{Z}/p\mathbb{Z} \longrightarrow \mathbb{Z}_{(p)}/p\mathbb{Z}_{(p)}$$

(it's injective because an integer a maps to zero only if $a/1 \in p\mathbb{Z}_{(p)}$, which happens only if $p \mid a$). To see that it is also onto, we use an argument involving congruences: if $p \nmid b$, then there exists an integer b_1 such that $bb_1 \equiv 1 \pmod{p}$. Then for any a/b in $\mathbb{Z}_{(p)}$, the integer ab_1 maps to the class of a/b in $\mathbb{Z}_{(p)}/p\mathbb{Z}_{(p)}$.

Notice that we have in fact showed that

$$\mathbb{Z}/p\mathbb{Z} \cong \mathbb{Z}_{(p)}/p\mathbb{Z}_{(p)},$$

i.e., moving from the integers $\mathbb{Z}$ to the p-integers $\mathbb{Z}_{(p)}$ does not change the quotient.

68 This is very similar to the calculations for $\mathbb{Q}$, and we leave it for the reader to puzzle over at leisure. Hints: the valuation rings all look like sets of rational functions with restrictions on the numerators and denominators, and the residue fields are often (but not always) equal to F itself. If $\mathsf{F} = \mathbb{C}$, then all the residue fields are isomorphic to $\mathbb{C}$.

69 The open ball of radius 1 around a is the coset $a + \mathfrak{P}$ of the ideal $\mathfrak{P}$ in $\mathcal{O}$. (Are other balls also cosets?) Problem 57 gets translated to the statement that the residue field is finite.

70 Yes it is always the case in the examples we considered, but no, it is not always true. The property of having a principal valuation ideal is really closely related to the fact that in all our examples the valuation is discrete (see the next problem).

71 Checking that $v(x)$ is a valuation is an easy exercise in logs and inequalities. For the other three statements:

(i) If $v_p(x) = n$, then $|x| = p^{-n}$, so that $v(x) = n \log p$. Hence v and v_p differ by multiplication by a constant, $\log p$. The image of v is $\log p \cdot \mathbb{Z}$, i.e., the real numbers which are integral multiples of $\log p$. (It's easy to see that this is a subgroup of $\mathbb{R}$, and that it is isomorphic to $\mathbb{Z}$.)

(ii) If the value group is discrete, look at the element x with smallest nonzero $v(x)$. It's not too hard to prove that it must be a generator of $\mathfrak{P}$. Conversely, if $\mathfrak{P}$ is principal check that the valuation of a generator must be the minimal nonzero element of the value group, which must then be discrete.

(iii) This is difficult. We showed in (ii) that in this case $\mathfrak{P}$ is a principal ideal, but we still need to show that every other ideal is too. See [61, Ch. I §2] for a detailed discussion of what hypotheses are necessary and for the proofs.

72 As the hint suggests, choose any $x_0 \in \Bbbk$, $x_0 \neq 0$, such that $|x_0|_1 < 1$. Then (iii) says that $|x_0|_2$ is also less than 1, so that there exists a positive real number α such that $|x_0|_1 = |x_0|_2^\alpha$. (Just take logs on both sides to find α; why is it important to choose $|x_0|_1 < 1$? What if no such x_0 exists?) This gives us our α.

Now choose any other $x \in \Bbbk$, $x \neq 0$. If $|x|_1 = |x_0|_1$, then we must also have $|x|_2 = |x_0|_2$, because otherwise either x/x_0 or x_0/x would have $|\ |_2$ less than 1 and (iii) would be violated. So in this case the equation $|x|_1 = |x|_2^\alpha$ holds.

If $|x|_1 = 1$, then we must have (by (iii) applied to either x or $1/x$) that $|x|_2 = 1$ also, so that the equation $|x|_1 = |x|_2^\alpha$ holds trivially.

Notice that the equality for some x implies the equality for any power of that x; in particular, we know that $|x_0^n|_1 = |x_0^n|_2^\alpha$ for any integer n.

It remains to consider the case when $|x|_i \neq 1$ and $|x|_i \neq |x_0|_i$ for $i = 1, 2$. As before, choose β such that $|x|_1 = |x|_2^\beta$; again, this means that we also have $|x^n|_1 = |x^n|_2^\beta$ for all integers n. In particular, we can assume that $|x|_1 < 1$ (otherwise replace it with $1/x$), which of course also implies that $|x|_2 < 1$.

What we want to do is show that α and β must be equal. Since we want to use (iii), the natural way to proceed is to show that if they are not equal, then we can manufacture an element that has $|\ |_1 < 1$ but $|\ |_2 > 1$, which would contradict (iii). The reader should fiddle with this idea for a while to convince himself that it is not very easy to carry it through. Faced with that, we have no choice but to take a more roundabout and more devious route.

Let n and m be any two positive integers. Then we have

$$|x|_1^n < |x_0|_1^m \iff \left|\frac{x^n}{x_0^m}\right|_1 < 1 \iff \left|\frac{x^n}{x_0^m}\right|_2 < 1 \iff |x|_2^n < |x_0|_2^m.$$

Taking logs of the first and last equations, we get

$$n \log |x|_1 < m \log |x_0|_1 \iff n \log |x|_2 < m \log |x_0|_2,$$

which we can write as

$$\frac{n}{m} < \frac{\log |x_0|_1}{\log |x|_1} \iff \frac{n}{m} < \frac{\log |x_0|_2}{\log |x|_2}.$$

This says that the set of fractions which is smaller than the first quotient of logs is exactly the same as the set of fractions which is smaller than the other; since there are fractions as close as we like to any real number, this means that the two numbers must be equal (otherwise, some fraction will be bigger than one but smaller than the other). Thus, we get

$$\frac{\log |x_0|_1}{\log |x|_1} = \frac{\log |x_0|_2}{\log |x|_2},$$

and therefore

$$\frac{\log |x_0|_1}{\log |x_0|_2} = \frac{\log |x|_1}{\log |x|_2}.$$

But plugging in $|x_0|_1 = |x_0|_2^\alpha$ shows that the first quotient equals α, and similarly the second quotient equals β. This shows $\alpha = \beta$, and we are finally done!

73 This only requires a straightforward reading of the definition.

74 The point is that saying $|x| = 1$ is equivalent to saying that neither $|x|$ nor $|1/x|$ are < 1. Then do the obvious thing.

75 According to the Proposition, equivalent absolute values differ by raising to a positive power. Since both $1^\alpha = 1$ and $0^\alpha = 0$ for any α, anything equivalent to the trivial absolute value is itself trivial. We would only need to change the definition of "nontrivial" if there were more absolute values that were equivalent to the trivial one. Since there aren't, we can let "trivial" mean "trivial" and have done with it.

76 Easy: $|p|_p < 1$ but $|p|_q = 1$ whenever p and q are two different primes, and $|p|_\infty > 1$ for any prime p.

77 Let A be the image of $\mathbb{Z}$ in $\Bbbk$, i.e., the elements of $\Bbbk$. We showed that $||$ was non-archimedean if and only if we had $|a| \leq 1$ for all $a \in A$. Now use Problem 74.

78 We don't need to worry about $n = 1$, since $|1| = 1 = 1^\alpha$. If $1 < n < n_0$, the estimate $|n| \leq Cn^\alpha$ would still be true, since in this case $|n| = 1$ (remember that we chose n_0 to be the smallest integer with absolute value more than 1). Furthermore, when we go on to consider n^N we will certainly get to integers bigger than n_0, so the proof doesn't need to consider this case separately.

79 Since the balls are nested, if $m > n$ then $a_m \in \overline{B}(a_n, r_n)$, which tells us that $|a_m - a_n| < r_n$. Since $r_n \to 0$, we can find N such that $|r_n| < \varepsilon$ if $n \geq N$, and so if $m > n \geq N$ we have $|a_m - a_n| < \varepsilon$. So the centers a_n form a Cauchy sequence. Since $r_n \to 0$, the intersection will contain at most one point. If it is nonempty, that point will be the limit of the sequence of centers.

Conversely, given a Cauchy sequence we can create such a sequence of nested balls: for each k find N such that $m > n \geq N$ implies $|a_m - a_n| < \frac{1}{k}$, and set $a_k = a_N$, $r_k = \frac{1}{k}$. The intersection of the nested balls will be the limit of the subsequence a_k and hence the limit of the original sequence a_n.

Next question: what if the nested balls had radii $r_n \to r \neq 0$? Would the intersection still be nonempty? A field with this property is called *spherically complete*.

80 If we assume that the real numbers are known, this is easy: just take a sequence of rational numbers like

$$1, \quad 1.4, \quad 1.41, \quad 1.414, \quad 1.4142, \quad \text{etc.}$$

which get closer and closer to $\sqrt{2}$. This is clearly Cauchy and has no limit in $\mathbb{Q}$.

If we want to do this within $\mathbb{Q}$, we just need to be a bit more careful. For example, we might use Newton's method for finding approximate roots of polynomials to generate a sequence of rational numbers which approximate $\sqrt{2}$.

81 Let $\Bbbk = \mathbb{R}$, and let $||$ be the usual absolute value. The famous example is

$$x_n = 1 + \frac{1}{2} + \frac{1}{3} + \cdots + \frac{1}{n}.$$

For this sequence, we have $x_{n+1} - x_n = 1/(n+1)$, which clearly tends to zero as $n \to \infty$. The sequence is increasing, so that if it has a limit its terms must all be bounded (they are all less than any number which is bigger than the limit). However, a standard argument that the reader will very likely have seen in her calculus class shows that $x_{2^k} \geq (k+2)/2$, so that the sequence cannot be bounded, hence cannot have a limit.

The infinite sum

$$1 + \frac{1}{2} + \frac{1}{3} + \cdots + \frac{1}{n} + \cdots$$

is called the *harmonic series*, and it is a staple of calculus courses because it shows that a series can get infinitely large even though its summands get closer and closer to zero.

82 The first one is clear: if a field contains $\mathbb{Q}$ and is complete, it must contain the limit of any Cauchy sequence made up of elements of $\mathbb{Q}$. For the second, you need to show that any real number can be arbitrarily well approximated by rational numbers. Can you prove that?

83 No, because $\mathbb{Q}$ is already complete with respect to the trivial absolute value: since the only possible absolute values are 0 and 1, a sequence will be Cauchy only if $|x_m - x_n| = 0$ for all large enough m and n. But this means that $x_m = x_n$ for all large enough m and n, and of course any such sequence converges (because it just stops).

84 Use the approach we described to construct a sequence tending to a cube root of 3. The main point is to show that one can always get a solution modulo 2^{n+1} from a solution modulo 2^n, and this is done just as in the other case.

85 The sum is easy, since

$$(x_n + y_n) - (x_m + y_m) = (x_n - x_m) + (y_n - y_m).$$

For the product, use the identity

$$x_n y_n - x_m y_m = x_n(y_n - y_m) + y_m(x_n - x_m),$$

plus the fact that x_n and y_m cannot get arbitrarily big as n and m grow.

86 The zero element is the sequence

$$0, 0, 0, 0, 0, 0, \ldots$$

The unit element is

$$1, 1, 1, 1, 1, 1, \ldots$$

A sequence (x_n) is invertible exactly when the x_n are bounded away from zero (i.e., there exists a bound b such that $|x_n| > b$ for all n; in particular $x_n \neq 0$ for all n).

We need to know they are bounded away from 0 rather than simply nonzero, because otherwise the "inverse sequence" might not be Cauchy! (Make sure you understand this one: what's an example of a Cauchy sequence of non-zero rational numbers x_n such that the sequence given by $y_n = 1/x_n$ is not Cauchy?) On the other hand, things are less bad than they seem: you should be able to show that if a Cauchy sequence does not tend to zero, then it is bounded away from zero.

87 Use the inequality

$$|y_n - y_m| \le |y_n - x_n| + |x_n - x_m| + |x_m - y_m|.$$

88 Here's an example: sequence one is

$$0, \, p, \, 0, \, p^2, \, 0, \, p^3, \, 0, \, p^4, \, \ldots$$

and sequence two is

$$p, \, 0, \, p^2, \, 0, \, p^3, \, 0, \, p^4, \, \ldots$$

89 In any Cauchy sequence (x_n), the terms x_n are bounded (if this is not immediately clear, you should write down a proof). Hence, if $y_n \to 0$, then also $x_n y_n \to 0$, which is what we want to prove.

90 As it says, just follow the proof through. The argument you used in Problem 86 to show that the inverse of an invertible Cauchy sequence with terms bounded away from zero is itself a Cauchy sequence will work for "almost inverses" too.

91 If $\lambda = 0$, then $(x_n) \in \mathcal{N}$, so that $x_n \to 0$, so that $|x_n|_p \to 0$, so $|\lambda|_p = 0$, which is only reasonable. On the other hand, if $\lambda \ne 0$, then Lemma 3.2.10 says that the sequence $|x_n|_p$ is constant for sufficiently large n, which means it certainly has a limit.

92 If the difference tends to zero, the absolute value of the difference tends to zero, so that the difference of the absolute values tends to zero.

93 Once you do remember, there is nothing left to prove.

94 Problem 93 handled one part. What remains to be shown are the multiplicativity and the non-archimedean property. Both are easy: just write down the known properties of $|\ |_p$ for the terms of any sequence, and take the limit. For example, if λ is represented by (x_n) and μ is represented by (y_n), then the product $\lambda\mu$ is represented by $(x_n y_n)$. Now, for each n we have $|x_n y_n|_p = |x_n|_p |y_n|_p$. Taking the limit gives $|\lambda\mu|_p = |\lambda|_p |\mu|_p$. Something similar works for the addition.

95 Yes, this is essentially obvious.

96 Lemma 3.2.10 says it all.

97 $<$ becomes $\leq$ because it's perfectly possible for a sequence to tend to a certain value while remaining consistently smaller than that value. We need to decrease ε slightly to guarantee that y doesn't end up in the *closed* ball of radius ε. This is hardly ever a real problem, however, since we can always use the same trick of using a slightly smaller radius.

98 This shouldn't be too hard. The key idea for (i) is that the field operations are continuous functions. Part (ii) is clear. Part (iii) follows from the fact that, as suggested, the absolute value function is (uniformly) continuous, because
$$\big|\,|x| - |y|\,\big| \leq |x - y|.$$

99 For something to be determined "really uniquely," it must not only be unique up to isomorphism, but up to unique isomorphism. Here is an example from linear algebra: all the vector spaces of dimension n are isomorphic, but it is still unwise to simply identify them all, because there are *many* different ways to chose a basis of a vector space, and each choice gives a different isomorphism. Why should we favor any one of them over the others?

   Another example of the same thing is the process of forming an algebraic closure of a field. Given a field $\Bbbk$, its algebraic closure is unique up to isomorphism, but has a great many automorphisms, so that given two algebraic closures there are a great many *different* isomorphisms between them. What this means is that the algebraic closure is not really canonically determined. (One should always speak of *an* algebraic closure, but one *can* speak of *the* completion.)

100 This is easy: since $\mathbb{Z}_p$ is the closed unit ball with center 0, it is an open set containing 0, hence a neighborhood of 0. Since multiplication by p sends open sets to open sets, this means that for every n the set $p^n\mathbb{Z}_p$ is a neighborhood of zero. That
$$\bigcup_{n \in \mathbb{Z}} p^n\mathbb{Z}_p = \mathbb{Q}_p$$
is clear from the first statement in the Corollary; to see that they are a fundamental system of neighborhoods we need to show that any open set containing zero contains a $p^n\mathbb{Z}_p$, and this is clear (because any open ball containing 0 contains a closed ball of smaller radius, which will be one of the $p^n\mathbb{Z}_p$).

101 There are lots, of course; an example would be the family consisting of the open intervals $(-1/n, 1/n)$ plus the open intervals $(-n, n)$, where n ranges through the positive integers.

102 The proof we sketched for Problem 100 pretty much shows this already.

103 This has all been done, albeit in different terms. Multiplication by p^n is injective because $\mathbb{Z}_p$ is contained in $\mathbb{Q}_p$, which is a field in which we can divide by p^n. Now consider the homomorphism $\mathbb{Z} \longrightarrow \mathbb{Z}_p$. Clearly $p^n\mathbb{Z}$ maps into $p^n\mathbb{Z}_p$. Since $p^n\mathbb{Z}$ is the kernel of the projection $\mathbb{Z} \longrightarrow \mathbb{Z}/p^n\mathbb{Z}$, we get an injective homomorphism

$$\mathbb{Z}/p^n\mathbb{Z} \longrightarrow \mathbb{Z}_p/p^n\mathbb{Z}_p.$$

But Proposition 4.2.2 tells us that any element of $\mathbb{Z}_p$ is p^{-n}-close to an integer, which translates to saying that any element of $\mathbb{Z}_p$ is congruent mod $p^n\mathbb{Z}_p$ to an element of $\mathbb{Z}$, which says the isomorphism is onto. Finally, in the discrete topology a singleton set is open, but the inverse image of a point in $\mathbb{Z}/p^n\mathbb{Z}$ is a closed ball with radius p^n, which is an open set in $\mathbb{Z}_p$.

104 You'll probably need to look some of these up. A good reference is [47, §26–§29]. Statement (i) is easy: just use the fact that the inverse image of an open set by a continuous function is again open. It is also easy to see that any compact set is closed and bounded, but the main claim in (ii) is pretty hard; it is known either as the Heine–Borel Theorem or the Bolzano–Weierstrass Theorem, depending on which of the equivalent definitions of compactness you use. Proving (iii) takes some thought to come up with a way to relate covering sets and sequences; notice that we need to be in a metric space because we haven't really defined convergence in general topological spaces (and indeed that is not easy to do). For (iv), to show that any compact set will have these two properties is not too hard (use (iii) for the first one), but the converse takes some work; see [47, §45].

105 A closed interval is compact, and is a neighborhood of any of its interior points, so it's enough to note that any point is in the interior of some closed interval; e.g., $x \in [x - 1, x + 1]$.

106 The hint pretty much proves everything.

107 Because any ball in $\mathbb{Q}_p$ is equal to a ball of radius p^n for some $n \in \mathbb{Z}$, and any ball in $\mathbb{Z}_p$ with radius greater than or equal to one will simply be all of $\mathbb{Z}_p$.

108 To reproduce the argument we gave for $\mathbb{Z}_p$, we only need to check that the other quotients $\mathcal{O}/\mathfrak{P}^n$ are also finite, since $\mathcal{O}$ is always the closed unit ball in $\Bbbk$. To see this, we look at the obvious map $\mathcal{O}/\mathfrak{P}^n \longrightarrow \mathcal{O}/\mathfrak{P}$; its kernel is $\mathfrak{P}^n/\mathfrak{P}$. If we show the kernel is finite, we will be done (because the assumption is that the image is too). Can you do that?

The "do we really need" questions are both pretty hard. (The answer is likely to be "yes" in both cases.)

109 $\mathbb{Z}^\times = \{\pm 1\}$, $\mathsf{F}[t]^\times = \mathsf{F}^\times$, and $\mathbb{C}[[t]]^\times$ is all power series with nonzero initial term, i.e., of the form $a_0 + a_1 t + \cdots$ with $a_0 \neq 0$.

110 Let $x \in \mathbb{Z}_p$. Since we know $\mathbb{Z}$ is dense in $\mathbb{Z}_p$, we can find, for each n, an integer x_n such that $|x - x_n| \leq p^{-n}$. It suffices then to show that for each n we can find $a_n \in A^\pm(a, b)$ such that $|x_n - a_n| \leq p^{-n}$. That amounts to finding an integer k such that $a_n = a + kb \equiv x_n \pmod{p^n}$ and has the right sign. The congruence can always be solved because $\gcd(b, p^n) = 1$ and adding or subtracting multiples of p^n allows us to adjust the sign.

111 Since $\mathbb{Z}_p$ is a closed set, we already know this. But here is a direct proof: if $x_n \in \mathbb{Z}$ for all n, then $|x_n| \leq 1$ for all n. Now if $x_n \to x$, then there is some n such that $|x - x_n| < 1$. But then

$$|x| = |x_n + (x - x_n)| \leq \max\{|x_n|, |x - x_n|\} \leq 1$$

so that $x \in \mathbb{Z}_p$.

112 This is not too hard, but does rely on the reader being comfortable with topology and with infinite products. That the map is an injective homomorphism is not too hard to show, because any element in $\mathbb{Z}_p$ is the limit of its associated coherent sequence. For details on how to construct $\mathbb{Z}_p$ from this point of view, see chapter 2 of [60].

113 Given such a family of maps $f_n : R \longrightarrow A_n$, and given $r \in R$, the sequence $(f_n(r))$ is a coherent sequence. By the previous problem, we can find an element of $\mathbb{Z}_p$ corresponding to this sequence. Taking this as the image of r gives a map $R \longrightarrow \mathbb{Z}_p$ which does what we want.

114 We need $x = a_0 + a_1 p + \cdots$ with $a_0 \neq 0$.

115 Let x_n be a sequence of elements of $\mathbb{Z}_p$. We want to pick out a subsequence that converges. To do this, use the following iterative procedure:

(i) There are only p possible choices for the zeroth coefficient in the p-adic expansion of the x_n. Hence there must be infinitely many x_n all of which have the same initial term a_0. Choose n_0 such that x_{n_0} is one of these.

(ii) For each of the infinitely many x_n whose p-adic expansions start with a_0, there are p choices for the first coefficient. Hence there must be infinitely many x_n all of whose p-adic expansions start with $a_0 + a_1 p$. Choose n_1 so that x_{n_1} is one of these.

Now keep going to get a convergent subsequence. Why does this procedure fail for sequences in $\mathbb{Q}_p$?

116 Since $x = 0 \equiv 6 \pmod 3$, the zero-th digit is 6. Now

$$0 = 6 + (-6) = 6 + (-2) \times 3.$$

Since $-2 \equiv 7 \pmod 3$, the next digit is 7. So

$$0 = 6 + 7 \times 3 - 27 = 6 + 7 \times 3 + (-3) \times 3^2.$$

Since $-3 \equiv 6 \pmod 3$, the next digit is 6 again. So

$$0 = 6 + 7 \times 3 + 6 \times 3^2 - 81 = 6 + 7 \times 3 + 6 \times 3^2 + (-3) \times 3^3.$$

Now we have -3 again, so the process will start repeating. Thus

$$0 = 6 + 7 \times 3 + 6 \times 3^2 + 6 \times 3^3 + \cdots + 6 \times 3^n + \cdots$$

It's easy to check that this is correct by adding up the geometric series.

117 This can be done in many ways. Notice that, however we do it, the Taylor expansion is *finite*, so we don't need to worry about convergence questions.

Here's the jazziest proof I can think of: the field generated by $\mathbb{Q}$ and the coefficients of $F(X)$ can be embedded in $\mathbb{C}$, and the theorem is clearly true for polynomials with complex coefficients. (Talk about overkill...)

A nicer way to prove it is to work in the field $\mathbb{Q}[x, h]$ of polynomials in two variables. Then $F(x+h)$ can be written as a polynomial in h with coefficients in $\mathbb{Q}[x]$:

$$F(x + h) = F(x) + F_1(x)h + F_2(x)h^2 + \ldots,$$

and we can check that $i!F_i(x) = F^{(i)}(x)$, which is an easy exercise in binomial coefficients. Indeed, if $F(x)$ has integer coefficients we can even work in $\mathbb{Z}[x, h]$ to see that $F_i(x) \in \mathbb{Z}[x]$.

Notice, however, that what we need for the proof is even simpler, namely

$$F(x + h) = F(x) + F'(x)h + h^2 R(x, h),$$

which is easily seen to be true in $\mathbb{Z}[x, h]$.

118 Just follow what was done to go from α_1 to α_2.

119 If $F'(\alpha_1)$ is divisible by p, it is not invertible in $\mathbb{Z}_p$, so that we can't pick the b_1 in the computation, and the proof falls through. Indeed, the polynomial $X^2 - 3$ has roots modulo 2 but no roots in $\mathbb{Q}_2$.

120 See [14, Lemma 3.1] for a full proof. Here are the highlights. We want to use the iteration

$$\alpha_{n+1} = \alpha_n - \frac{F(\alpha_n)}{F'(\alpha_n)}.$$

Let $n = 1$ and set $b_1 = -\frac{F(\alpha_1)}{F'(\alpha_1)}$. Since

$$|b_1| = \left| \frac{F(\alpha_1)}{F'(\alpha_1)} \right| < |F'(\alpha_1)| \leq 1,$$

we see that $b_1 \in \mathbb{Z}_p$. Using the Taylor expansion, there is some $K \in \mathbb{Z}_p$ such that

$$F(\alpha_1 + b_1) = F(\alpha_1) + F'(\alpha_1)b_1 + Kb_1^2.$$

But our choice of b_1 makes $F(\alpha_1) + F'(\alpha_1)b_1 = 0$, so

$$|F(\alpha_1 + b_1)| \le |b_1|^2 < |F(\alpha_1)|.$$

On the other hand,

$$|F'(\alpha_1 + b_1) - F'(\alpha_1)| \le |b_1| < |F'(\alpha_1)|$$

implies, by the ultrametric inequality, that

$$|F'(\alpha_1 + b_1)| = |F'(\alpha_1)|.$$

If we now set $\alpha_2 = \alpha_1 + b_1 = \alpha_1 - \frac{F(\alpha_1)}{F'(\alpha_1)}$, we see that $|F(\alpha_2)| < |F(\alpha_1)|$ but $|F'(\alpha_2)| = |F'(\alpha_1)|$.

This is what sets up the iteration: at each step $F(\alpha_n)$ gets smaller and $F'(\alpha_n)$ does not. This allows us to show that the sequence α_n converges to a root α.

Working through it all, we also get that

$$|\alpha - \alpha_1| \le \left| \frac{F(\alpha_1)}{F'(\alpha_1)} \right|$$

and that α is the only root of $F(x)$ satisfying that condition.

An example where the stronger result is necessary is the equation $X^2 - 17$, whose derivative is $2X$ and so will never satisfy the condition $F'(\alpha_1) \not\equiv 0$ (mod 2). But if we take $\alpha_1 = 1$ we see that $|F(\alpha_1)| = 2^{-4}$ and $|F'(\alpha_1)| = 2^{-1}$, and we can conclude that $X^2 - 17$ does have roots in $\mathbb{Q}_2$.

If you know what the discriminant D of a polynomial $F(X)$ is, you can use this to show that if we can find $\alpha_1 \in \mathbb{Z}_p$ such that $|F(\alpha_1)| < |D|^2$ then $F(X)$ has a root in $\mathbb{Z}_p$. See [14, pp. 49–52].

121 If α_1 exists, then its image in $\mathbb{Z}/p\mathbb{Z}$ is an element of order dividing m in the cyclic group $(\mathbb{Z}/p\mathbb{Z})^\times$ of order $p-1$. It follows that $\gcd(m, p-1) \ne 1$ unless $\alpha_1 \equiv 1 \pmod{p}$. Furthermore, the least exponent m with this property must be a divisor of the gcd, and hence must be a divisor of $p-1$. Conversely, in a cyclic group of order $p-1$, if $e|(p-1)$ and x is a generator, $x^{(p-1)/e}$ is of order e. Indeed, the set of elements of order e is a cyclic group generated by $x^{(p-1)/e}$.

122 It's basically straight Hensel's Lemma together with Problem 121. For each m dividing $p-1$ we can find m incongruent roots of $X^m - 1 \equiv 0 \pmod{p}$, and then Hensel's Lemma shows that we have m different roots of $X^m - 1$, which are the mth roots of unity.

The only hard part is to show that there are no other roots of unity. Specifically, we show that if $\zeta^k = 1$ and $p \nmid k$ then in fact $\zeta^m = 1$ for some m dividing $p - 1$. This is where the uniqueness part of Hensel's Lemma is important. Suppose $\zeta^k = 1$. Then also $\zeta^k \equiv 1 \pmod p$ and Problem 121 says that there is an m dividing $p-1$ such that $\zeta^m \equiv 1 \pmod p$. By Hensel's Lemma, there is a unique $\zeta_1 \equiv \zeta \pmod p$ such that $\zeta_1^m = 1$. But then, since m divides k, ζ_1 is a root of $X^k - 1$ as well, and it is congruent mod p to ζ. The uniqueness part of Hensel's Lemma then forces $\zeta_1 = \zeta$.

123 The roots of unity are exactly the elements of $\mathbb{Z}_p^\times$ that satisfy $x^m = 1$ for some power m. It is easy to see that the set of such elements in any abelian group always forms a subgroup. To see that there are $p - 1$ roots, note that the numbers $1, 2, 3, \ldots, p - 1$ are all solutions of $X^{p-1} \equiv 1 \pmod p$, and are all incongruent modulo p. Applying Hensel's Lemma gives $p - 1$ roots which are all incongruent modulo p, and in particular are all different. Since a polynomial can only have as many roots as its degree, these must be all the roots. Since we know that any root of unity must be a root of this polynomial, these must be all the roots of unity in $\mathbb{Q}_p$. Finally, any finite subgroup of any field is cyclic. (To see that the group of roots of unity is cyclic in a more direct way, apply the reasoning above to all polynomials $X^d - 1$ as d ranges through the divisors of $p - 1$, and count to show that some $(p - 1)$-st root of unity must exist which is not a root of any of these. Such a root of unity will be a generator.)

124 The first assertion is an easy application of the stronger form of Hensel's Lemma. For the second, write the 2-adic unit in the form $1 + 2x$, and square. The conclusion follows by considering $\mathbb{Z}/8\mathbb{Z}$.

125 Polynomials that are quite different in $\mathbb{Z}_p[X]$, such as $X + 1$ and $X + (p + 1)$, are identical modulo p, so being relatively prime modulo p is a more restrictive condition than being so over $\mathbb{Z}_p$.

126 We use the notation in the proof, and focus mostly on the question about the final twist. Since $g_1(X) = X$ and $h_1(X) = 1$, the obvious solution for $a(X)g_1(X) + b(X)h_1(X) \equiv 1 \pmod p$ is $a(X) = 0$, $b(X) = 1$, which yields $\tilde{s}_1(X) = 0$ and $\tilde{r}_1(X) = X^2 + 1$. But now if we simply set $g_2(X) = g_1(X) + p\tilde{r}_1(X)$ and $h_2(X) = h_1(X) + p\tilde{s}_1(X)$, we end up with $g_2(X) = 2X^2 + X + 2$ and $h_2(X) = 1$, which yields a factorization, all right, but a rather unsurprising one!

If we do it right, we get $r_1(X) = 1$ (the remainder of dividing $X^2 + 1$ by X) and $s_1(X) = X$, which gives $g_2(X) = X + 2$ and $h_2(X) = 1 + 2X$, and all is well. The reader should go through at least one more iteration herself.

127 Just follow what we did in our proof—carefully.

128 If you did the previous problem, this should be easy.

129 For $p \neq 2, 17$ one can do this by a straight application of Hensel's Lemma, as follows: if neither 2 nor 17 are squares modulo p, then their product must be a square modulo p; then use Hensel's Lemma. (To see why the product of two quadratic non-residues must be a square modulo p, remember that $(\mathbb{Z}/p\mathbb{Z})^{\times}$ is cyclic, so that being a square means being an even power of the generator; the product of two odd powers of the generator must be an even power of the generator!)

For $p = 2$, note that 17 is a square in $\mathbb{Q}_2$. For $p = 17$, note that $6^2 \equiv 2$ (mod 17), so that (Hensel's Lemma!) 2 is a square in $\mathbb{Q}_{17}$. For $p = \infty$, there are clearly six different roots. And there are clearly no rational roots.

130 This isn't too hard if one uses more advanced tools such as biquadratic reciprocity. An elementary (but not easy) proof can be found in [14], page 57, and one using algebraic number theory is outlined in [15], page 88.

131 Hensel's Lemma tells us that all sorts of polynomials are irreducible over $\mathbb{Q}$ and reducible over some $\mathbb{Q}_p$ (think of $X^2 + 1$ for example), so the "only if" part is bunk. The "if" part works, because if a polynomial were reducible over $\mathbb{Q}$ it would certainly be reducible over all the $\mathbb{Q}_p$. Of course, the "if" part is not very interesting...

How about this: is it true that a polynomial will be irreducible over $\mathbb{Q}$ if and only if it is irreducible over *some* $\mathbb{Q}_p$? (In other words, given a polynomial that is irreducible over $\mathbb{Q}$, can I find a prime p such that the polynomial is irreducible over $\mathbb{Q}_p$?) If so, this proves the statement in the exercise with "irreducible" replaced by "reducible."

132 We have

$$ax^2 + by^2 + cz^2 = a'n^2x^2 + by^2 + cz^2 = a'(nx)^2 + by^2 + cz^2,$$

which establishes the correspondence. We are interested in deciding when there are roots in the rational numbers (or p-adic numbers, or integers modulo p for some p). Since these are all fields, we can divide by n, so the correspondence shows that (in each case) the equation with a will have a root if and only if the equation with a' does. So we might as well work with a'. Doing the same for b and c, we see that we can assume that all three coefficients are square-free.

133 By the previous problem, we may assume a, b, and c are square-free and have no common factors, and we do. We want to show that any two coefficients are relatively prime. Suppose $k = \gcd(a, b)$ is greater than 1. Notice that k must be square-free. Then we can set $a = ka'$, $b = kb'$, and we know that k is relatively prime to c. Suppose that $ax^2 + by^2 + cz^2 = 0$. If we look at the last equation, we see that k must divide cz^2, since it divides the other two terms. Since it is prime to c, it divides z^2. Since it is square-free, it must divide z. So write $z = kz'$, plug in, divide by k, and continue from there to get a contradiction.

134 For $n = 0$, we get the sum of p ones, which is p, hence is $\equiv 0 \pmod{p}$. Instead of trying to give a general proof, here's the proof for $n = 1$: choose and fix a number a, $2 \le a \le p-1$ (this is possible, since p is odd). I want to compare the two sums

$$\sum_{x=0}^{p-1} x \quad \text{and} \quad \sum_{x=0}^{p-1} ax.$$

It is not hard to show that the numbers 0, a, $2a$, ... $(p-1)a$ are all non-congruent modulo p. (Do it!) Since there are p of them, they must be congruent, in some order, to 0, 1, ... $p-1$. (In other words, modulo p the list of the ax is just a permutation of the list of x.) This means that the two sums are congruent:

$$\sum_{x=0}^{p-1} x \equiv \sum_{x=0}^{p-1} ax \pmod{p}.$$

We can rewrite this as

$$0 \equiv \sum_{x=0}^{p-1} x - \sum_{x=0}^{p-1} ax \pmod{p}$$

$$\equiv (1-a)\sum_{x=0}^{p-1} x$$

and, since $1 - a \not\equiv 0 \pmod{p}$, the conclusion follows.

Reorganizing this to take care of more general exponents is not too hard: what we need is to show that we can choose our a so that $a^n \not\equiv 1 \pmod{p}$. If so, the same proof will work!

135 What we need to check is that the polynomial $f(X) = aX^2 + by_0^2 + cz_0^2$ satisfies the conditions in Hensel's Lemma. But that's easy: $f(x_0) \equiv 0$ is our assumption, and $f'(x_0) = 2ax_0 \not\equiv 0 \pmod{p}$ because p is odd and does not divide a or x_0.

136 Well, certainly in the application of Hensel's Lemma (we need to know that whichever of x_0, y_0, or z_0 is not divisible by p has a coefficient next to it which is not divisible by p). But presumably also in the Proposition: where? (Hint: suppose one of a, b or c is divisible by p; is the Proposition still true?)

137 Of course, the idea is to use the result in Problem 120, using Hensel's Lemma as we did in Problem 135. The difficulty is that, since $p = 2$, there is no doubt that the derivative will be divisible (once) by p. If we look hard at the conditions in Problem 120, we see that we need to find an initial solution (x_0, y_0, z_0) such that $ax_0^2 + by_0^2 + cz_0^2 \equiv 0 \pmod 8$.

Very well, we know that the sum of two of the coefficients, say a and b, is divisible by 4: $a + b \equiv 0 \pmod 4$. Now there are two possibilities:

- if $a + b \equiv 0 \pmod 8$, then we can choose $x_0 = y_0 = 1$ and $z_0 = 0$, and all is well;

- if not, we will have $a + b \equiv 4 \pmod 8$; choosing $x_0 = y_0 = 1$ and $z_0 = 2$ will then do what we want (check!).

In either case, we are in business, and the stronger form of Hensel's Lemma gives a solution in $\mathbb{Q}_2$.

138 Suppose a is even, b and c are odd, and $ax^2 + by^2 + cz^2 = 0$. As before, we can assume that at least one of x, y and z is a 2-adic unit, and that all three are in $\mathbb{Z}_2$. There are two cases to consider:

- x is in $2\mathbb{Z}_2$. Then clearly ax^2 is divisible by 8, and it is easy to see that y and z must be 2-adic units. Since the square of a 2-adic unit is always in $1 + 8\mathbb{Z}_2$, it follows that

$$0 = ax^2 + by^2 + cz^2 \equiv b + c \pmod 8.$$

- x is a 2-adic unit. Then y and z again must be 2-adic units (if, say, $y \in 2\mathbb{Z}_2$, then $ax^2 + by^2$ would be divisible by 2, and therefore cz^2 would be divisible by 2; but we know c is odd, so z would have to be in $2\mathbb{Z}_2$; but then $by^2 + cz^2$ would be in $4\mathbb{Z}_2$, hence so would $ax^2 \ldots$). Once again, the square of a 2-adic unit is always in $1 + 8\mathbb{Z}_2$, and we get

$$a + b + c \equiv 0 \pmod 8.$$

The converse is once again an application of the generalized form of Hensel's Lemma (but it's actually easier this time, because we have information about a, b, and c modulo 8).

139 Necessity is easy: since $p | a$, we have $by^2 + cz^2 \equiv 0 \pmod p$, and it's not hard to see that both y and z will have to be p-adic units. Hence, we can rewrite the equation as $b + (y/z)^2 c \equiv 0 \pmod p$, and it is now a matter of showing that if a p-adic unit fits into this equation, then we can find an integer that does (and that is easy).

The sufficiency is Hensel's Lemma again, of course.

140 What this suggests is that given information about *all but one* prime $p \leq \infty$ the behavior at the missing prime is determined. This is in fact true. The magic word here is "reciprocity." Have a chat with the local number theorist, who is likely to wax poetic over this one.

141 Remember that powers of p are small, and these are all easy to work out: $n!$ converges to zero, n and $1/n$ diverge, p^n converges to zero, $(1+px)^{p^n}$ converges to 1.

142 We've pretty much already proved this. Look at Lemma 3.2.10.

143 Exactly the same proof that works over $\mathbb{R}$ works here also. Basically, the fact that $|x+y| \leq |x| + |y|$ says that when the sequence of partial sums of $\sum |a_n|$ is Cauchy, then so is the sequence of partial sums of $\sum a_n$.

144 If $\sum a_n = 0$, the inequality is vacuous. If not, for any partial sum, we have

$$\left| \sum_{n=0}^{N} a_n \right| \leq \max_{0 \leq n \leq N} |a_n|$$

by the non-archimedean property. Now note that for large enough N we have

$$\max_{0 \leq n \leq N} |a_n| = \max_{n} |a_n|$$

because the a_n tend to zero, and

$$\left| \sum_{n=0}^{\infty} a_n \right| = \left| \sum_{n=0}^{N} a_n \right|$$

by Lemma 3.2.10.

145 Examples in $\mathbb{R}$:

$$\sum \frac{1}{n} \qquad \sum \frac{1}{n \log n} \qquad \sum_{p \text{ prime}} \frac{1}{p}$$

are all divergent.

146 Almost any example with positive terms works. Say,

$$1 + \frac{1}{2} + \frac{1}{2^2} + \frac{1}{2^3} + \cdots + \frac{1}{2^n} + \cdots = 2,$$

and $\max \left\{ \left| \frac{1}{2^n} \right| \right\} = 1 < 2$.

147 This follows the proof given in [39, Theorem 3.6].

Suppose we have a sequence of terms

$$a_1, a_2, \ldots, a_n, \ldots$$

with $a_n \to 0$. If we take a bijection

$$\sigma : \{1, 2, 3, \ldots\} \longrightarrow \{1, 2, 3, \ldots\},$$

we can produce another sequence $b_n = a_{\sigma(n)}$ which is a reordering of the original.

Notice that $a_n \to 0$ if and only if $b_n \to 0$, because both are equivalent to saying that for any $\varepsilon > 0$ all but finitely many have absolute value $< \varepsilon$. So the series $\sum a_n$ and $\sum b_n$ either both converge or both diverge. We want to show that if they converge they have the same sum.

Since both sequences tend to zero, there is an N_1 such that if $n \geq N_1$ we have both $|a_n| < \varepsilon$ and $|b_n| < \varepsilon$. If $A = \sum a_n$, we can also find N_2 so that for any $k \geq N_2$ we have

$$\left| A - \sum_{n=0}^{k} a_n \right| < \varepsilon.$$

Now take $N = \max\{N_1, N_2\}$ and take $k \geq N$; look at the two sums

$$S = \sum_{n=0}^{k} a_n \qquad \text{and } S' = \sum_{n=0}^{k} b_n.$$

We know that all terms after the k-th are ε-small, but both S and S' involve adding some terms with $|a_n| \geq \varepsilon$ and some terms with $|b_m| \geq \varepsilon$. But notice that these are exactly the same terms, since they are all terms of the original sequence with larger absolute values, just written in a different order. Let S_1 be the sum of all of those terms, so that

$$S = S_1 + \text{ terms } a_n, \ 1 \leq n \leq k, \text{ with absolute value } < \varepsilon$$

and

$$S' = S_1 + \text{ terms } b_n, \ 1 \leq n \leq k, \text{ with absolute value } < \varepsilon.$$

Now the ultrametric inequality gives $|S - S_1| < \varepsilon$ and $|S' - S_1| < \varepsilon$, from which we get $|S - S'| < \varepsilon$. Since we also know (by our choice of k) that $|S - A| < \varepsilon$, we get $|S' - A| < \varepsilon$.

So we have found an N so that if $k \geq N$ we have

$$\left| A - \sum_{n=0}^{k} b_n \right| < \varepsilon,$$

which shows $\sum b_n = A$.

148 We used the ultrametric inequality a number of times, but basically in two ways:

i) to conclude that a series converges when we know that its terms tend to zero,

ii) to conclude that a sum is less than ε when each of the summands is less than ε.

Both uses are crucial, so we don't expect that this result remains true over $\mathbb{R}$. Can you construct a counterexample? There are theorems of this sort that are true over $\mathbb{R}$, but in that case the crucial property is absolute convergence. See, for example, section 8.21 of [4].

149 This is true in the classical setting (i.e., over $\mathbb{R}$), and the same proof works here. But it's not hard to do: work with partial sums. We have

$$\sum_{n=0}^{N} a_n + \sum_{n=0}^{N} b_n = \sum_{n=0}^{N} c_n$$

because these are all *finite* sums; now take the limit.

150 In the classical setting, this is true if one of the two series converges absolutely, but not otherwise; see [55, Example 3.49], where $\sum c_n$ fails to converge. Abel showed, however, that if $\sum c_n$ does converge, the limit will be the product; see [55, Theorem 3.51]. In some sense, then, the point is that in the p-adic case $\sum c_n$ always does converge.

As usual when we work in a non-archimedean setting, absolute convergence is not needed. One can use Proposition 5.1.4 to prove this; see [19, Cor. 2.10], for example. Or one could check that Abel's theorem is true in the p-adic situation and then just prove that $\sum c_n$ converges. But here's a brute-force approach.

Let

$$A = \sum_{n=0}^{\infty} a_n, \qquad B = \sum_{n=0}^{\infty} b_n.$$

The key difficulty is this: if we take

$$A_m = \sum_{n=0}^{m} a_n, \qquad B_m = \sum_{n=0}^{m} b_n,$$

then since A_m converges to A and B_m converges to B it is clear that $A_m B_m$ converges to AB. The problem is that $\sum_{n=0}^{m} c_n \neq A_m B_m$. To show that $\sum c_n$ does converge to AB, we need to estimate the difference. This is a finite sum, and with a little bit of care the ultrametric inequality will save the day.

We have

$$A_m B_m = \sum_{n=0}^{m} \sum_{k=0}^{m} a_n b_k,$$

while

$$\sum_{\ell=0}^{m} c_\ell = \sum_{\ell=0}^{m} \sum_{n+k=\ell} a_n b_k.$$

The first sum includes all the $a_n b_k$ where $0 \le n, k \le m$, while the second includes only the ones where $n + k \le m$. So the difference is the sum of the terms where $0 \le n, k \le m$ and $n + k > m$. Notice that the inequalities imply that neither n nor k is zero and that $\max\{n, k\} > m/2$, so we can split the sum into the part where $m > n/2$ and the part where $k > m/2$:

$$A_m B_m - \left(\sum_{\ell=0}^{m} c_\ell \right) = \sum_{\substack{n > m/2 \\ 0 < n, k \le m \\ n+k>m}} a_n b_k + \sum_{\substack{k > m/2 \\ 0 < n, k \le m \\ n+k>m}} a_n b_k.$$

Since both a_n and b_k tend to zero, we can choose m such that $n > m/2$ implies $|a_n| < \varepsilon$ and $k > m/2$ implies $|b_k| < \varepsilon$. So for each term in the first sum, we have $|a_n b_k| < \varepsilon \max |b_k|$ and for each term in the second sum we have $|a_n b_k| < \varepsilon \max |a_n|$. By the ultrametric inequality, it follows that

$$\left| A_m B_m - \left(\sum_{\ell=0}^{m} c_\ell \right) \right| < \varepsilon (\max |a_n| + \max |b_k|).$$

Since ε is arbitrary, this shows that $\sum_{\ell=0}^{m} c_\ell$ has the same limit as $A_m B_m$, namely AB, and we are done.

151 Well, the basic content of the intermediate value theorem is that the image of an interval under a continuous function is an interval. This is a special case of a general fact, true in any metric space: the image of a connected set by a continuous function is a connected set. This is true in the p-adic context, but is kind of silly, since the only connected sets are those which consist of exactly one point!

A more interesting question is this: suppose $f(X)$ is a polynomial (might as well choose an easy function to work with). What can you say about the range of values of $f(x)$ as x runs through $\mathbb{Z}_p$?

152 It's what you think it should be!

153 The injectivity is clear, since we can recover the p-adic expansion of x from that of $f(x)$. So this function is nowhere near locally constant!

For the derivative, we follow the proof in [56]. Let's estimate $|f(x) - f(y)|$ when $|x - y|$ is known. Since every absolute value is a power of p and we are working on $\mathbb{Z}_p$, we can assume $|x - y| = p^{-k}$ for some $k \geq 0$. If so, the p-adic expansions of x and y must be identical up to the $(k-1)$st digit. Therefore, the expansions of $f(x)$ and $f(y)$ agree up to the $(2k-1)$st digit (they agree up to $2(k-1)$ by the definition of f, and all odd digits are zero). That means $|f(x) - f(y)| \leq p^{-2k}$. So we conclude that for all $x, y \in \mathbb{Z}_p$ we have

$$|f(x) - f(y)| \leq |x - y|^2.$$

But then

$$\left| \frac{f(x) - f(y)}{x - y} \right| \leq |x - y|,$$

which goes to zero as $y \to x$, so $f'(x) = 0$.

154 The chain rule is true because the usual proof works quite well. Once that is known, it's easy to see that one can make more "pseudo-constant" functions by taking one such and composing with any other non-constant differentiable function. That'll yield a great many examples!

155 In each case, one has to look at how the p-adic valuation of the general term changes as $n \to \infty$. For (i), note that $v_p(p^n x^n) = n + n v_p(x) = n(1 + v_p(x))$; if $v_p(x) > -1$, this will tend to infinity with n, so that

$$|p^n x^n|_p = p^{-n(1 + v_p(x))}$$

will tend to zero, and the series will converge. Otherwise, the series will diverge. So the radius of convergence is given by $v_p(x) > -1$, or $|x|_p < p$. (ii) is very similar. The hardest one is (iii): we want to compute $v_p(n! x^n) = v_p(n!) + n v_p(x)$. The difficulty is to estimate $v_p(n!)$. This will be done later in the chapter, but give it a go now. If you can show that $v_p(n!)$ grows faster than linearly in n, then the series will converge for all x. Does it?

156 All that needs to be checked is that the definition of the sum and product power series agrees with the sum and product series in problems 149 and 150.

157 Clearly the formula is

$$c_n = \sum_{m=1}^{n} a_m d_{m,n},$$

where

$$g(X)^m = \sum_{n=m}^{\infty} d_{m,n} X^n.$$

So what's needed is a formula for the coefficient $d_{m,n}$ of degree n in $g(X)^m$. That can be gotten by induction from the definition of the product of two power series. If you can't find it by yourself, look further down in this section!

158 Here's what I get (as usual, line breaks added):

```
gp> funct(x)=exp(2*x^2-2*x)
%1 (x)->exp(2*x^2-2*x)

gp> hseries=truncate(exp(2*x^2-2*x))
%2 92413472/638512875*x^16 - 165592576/638512875*x^15
+ 3824768/8513505*x^14 - 4548032/6081075*x^13
+ 560608/467775*x^12 - 285568/155925*x^11 + 37984/14175*x^10
- 2096/567*x^9 + 1528/315*x^8 - 1856/315*x^7 + 304/45*x^6
- 104/15*x^5 + 20/3*x^4 - 16/3*x^3 + 4*x^2 - 2*x + 1

gp> funct(1+O(2^20))
%3 1 + O(2^21)

gp> subst(hseries,x,1+O(2^20))
%4 1 + 2 + 2^2 + 2^3 + 2^4 + 2^5 + 2^7 + 2^9 + 2^12 + 2^15
    + 2^17 + 2^18 + 2^19 + O(2^21)
```

The answers are actually very far from each other. Plugging into the function gives the expected answer, 1 (up to the 2-adic precision). Plugging into the series gives what looks like a random 2-adic number very close to $-1 = 1 + 2 + 2^2 + 2^3 + \dots$. (Is it actually equal to -1? Our computation may not have enough precision for us to decide.)

Might it be a problem of not enough precision? You can get a longer series by first changing the series precision in GP: \ps 50 will get you a series of degree 50. Does the answer change?

159 The proof is similar to the one we gave for Theorem 5.4.3. To prove this version we just need to check that the condition on r suffices to allow us to reverse the order of summation at the crucial point. See [56, 41.2].

160 For any convergent series we know that

$$|g(x)| \leq \max_{n \geq 1}\{|b_n x^n|\}$$

by the ultrametric inequality. Since we have condition iii as well, it follows that

$$|g(x)| = \max_{n \geq 1}\{|b_n x^n|\} = r.$$

In the theorem we assume $f(g(x))$ converges, so we know that

$$\lim_{n \to \infty} |a_n| r^n = \lim_{n \to \infty} |a_n (g(x))^n| = 0.$$

161 The series $h(X)$ is the case $p = 2$ of a function due to Dwork.

It's easy to see that the extra condition in Theorem 5.4.3 doesn't hold, since $g(1) = 0$ is certain to be smaller than the terms of the series. Trying to use the condition in Problem 159 leads to the same conclusion: in our case $r = \max\{|b_n x^n|\} = 1/2$, but the exponential converges only when $|x| < 1/2$.

The first few terms of $h(X)$ look like this:

$$h(X) = 1 - 2x + 4x^2 - \frac{16}{3}x^3 + \frac{20}{3}x^4 - \frac{104}{15}x^5 + \frac{304}{45}x^6 - \frac{1856}{315}x^7 + \dots$$

Since $g(X)$ is just a binomial, it isn't hard to write out the general term of $h(X)$, but the resulting formula isn't very helpful.

Assuming the estimate on the coefficients, $v_2(a_n) \geq 2$ as soon as $n \geq 4$. Since $a_3 = -16/3$ is also divisible by 4, we see that $h(1) = 1 - 2 + \text{multiples}$ of 4, so that $h(1) \equiv 3 \pmod 4$. So $h(1) \neq 1$.

How hard is it to prove the estimate on the coefficients of $h(x)$? All the proofs I know depend on the theory of the Artin–Hasse exponential function. I first learned it (for general p) from [43, 14.2]. There are alternative accounts (with slightly different estimates) in [17, Section 2] and [53, p. 394]. When reading those, keep in mind that π is used for an element with $v_p(\pi) = 1/(p-1)$, so that when $p = 2$ we can take $\pi = 2$ as well.

The estimate $v_2(a_n) \geq 1 + n/4$ is sharp, because $v_2(a_4) = 2$. Indeed, a computation shows that $v_2(a_{2^k}) = 1 + 2^{k-2}$ for $1 \leq k \leq 10$. Is it always true?

162 This is straightforward manipulation of formal series. (Or can you think of a smarter way to prove these?)

163 Let $f(X) = \sum a_n X^n$ and let $\mathcal{D}$ be its region of convergence, which we know is a ball centered at 0 which might be open or closed. If $f(x)$ converges and $|x| = r$, then we know by Proposition 5.4.1 that the closed ball $\overline{B}(0,r)$ is contained in $\mathcal{D}$. By the Lemma, f is continuous on $\overline{B}(0,r)$, hence continuous at x.

It's also possible to imitate the classical proof, which is based on the theorem that a uniformly convergent series whose terms are continuous functions has a continuous sum. That is true in the p-adic setting as well, which reduces the problem to showing that power series always converge uniformly on closed balls (and not just on compact sets).

164 The phrase assumes that the roles of f and g in the proposition are symmetric, that is, that if we start with g and construct a new series as specified, the result will be f. Can you check that?

165 This is a matter of writing out $g(x)$ and using Proposition 5.1.4 to reorganize it into a power series in x. Say $f(X) = \sum c_n X^n$. Since $|a| = 1$ and $|b| < \rho$, we have

$$|x| < \rho \Longleftrightarrow |ax + b| < \rho$$

and

$$g(x) = f(ax + b) = \sum_{n=0}^{\infty} c_n (ax + b)^n = \sum_{n=0}^{\infty} \sum_{k=0}^{n} c_n \binom{n}{k} a^k x^k b^{n-k}.$$

Now check that we can reorganize this to get

$$g(x) = \sum_{k=0}^{\infty} \left(\sum_{n=k}^{\infty} \binom{n}{k} c_n a^k b^{n-k} \right) x^k.$$

166 First, the region of convergence of a power series is either an open or a closed ball, hence is an open set. Hence, if $x_m \to x$ and $f(x)$ and $g(x)$ converge, we can conclude that $f(x_m)$ and $g(x_m)$ converge for large enough m. Now use Proposition 5.5.3 to reduce the problem to the case where $x = 0$.

167 We give a proof making full use of the ultrametric inequality to simplify things. We assume that the limits exist, and let

$$A = \lim_{h \to 0} f(h), \qquad a_n = \lim_{h \to 0} f_n(h).$$

We want to show $\sum_{n=0}^{\infty} a_n = A$.

Let $\varepsilon > 0$ be given. We want to find an M so that $m \geq M$ implies $\left| A - \sum_{n=0}^{m} a_n \right| < \varepsilon$. By uniformity, we know that we can find an M so that if $m \geq M$ we have $|f_m(h)| < \varepsilon$ for all $|h| \leq r$. It follows that for all $|h| \leq r$ we have

$$\left| f(h) - \sum_{n=0}^{M} f_n(h) \right| = \left| \sum_{n=M+1}^{\infty} f_n(h) \right| \leq \max_{n>M} |f_n(h)| < \varepsilon.$$

Since the limits exist, for each n we can find δ_n such that

$$|h| < \delta_n \implies |f_n(h) - a_n| < \varepsilon.$$

Notice that if $m \geq M$ and $|h| \leq r$ we know $|f_m(h)| < \varepsilon$, so this implies when $m \geq M$ we must have

$$|a_m| \leq \max\{|f_m(h)|, |a_m - f_m(h)|\} < \varepsilon$$

as well. Finally, we can find δ so that

$$|h| < \delta \implies |f(h) - A| < \varepsilon.$$

Now take $|h| < \min\{r, \delta, \delta_0, \delta_1, \dots, \delta_M\}$ and $m \geq M$. We have:

i) $|A - f(h)| < \varepsilon.$

$ii)$ $\left| f(h) - \sum_{n=0}^{M} f_n(h) \right| < \varepsilon.$

$iii)$ $\left| \sum_{n=0}^{M} f_n(h) - \sum_{n=0}^{M} a_n \right| \leq \max_{0 \leq n < M} \{|f_n(h) - a_n|\} < \varepsilon.$

$iv)$ $\left| \sum_{n=0}^{M} a_n - \sum_{n=0}^{m} a_n \right| \leq \max_{M < n \leq m} \{|a_n|\} \leq \varepsilon.$

Using the ultrametric inequality, we get that if $m \geq M$ then

$$\left| A - \sum_{n=0}^{m} a_n \right| \leq \varepsilon,$$

and so $\sum_{n=0}^{m} a_n = A$, as claimed.

168 This is almost identical to the classical proof.

If we just jump into the formula in Proposition 5.4.1, we need to compare

$$\limsup_{n \to \infty} \sqrt[n]{|a_n|} \qquad \text{and} \qquad \limsup_{n \to \infty} \sqrt[n-1]{|na_n|}$$

(both are limits in $\mathbb{R}$). But that n versus $n-1$ thing is confusing, so let's use a trick to avoid that. Notice that $f'(x) = \sum na_n x^{n-1}$ converges if and only if $xf'(x) = \sum na_n x^n$ converges, so we can work with the latter series. So it suffices to prove that

$$\limsup_{n \to \infty} \sqrt[n]{|a_n|} = \limsup_{n \to \infty} \sqrt[n]{|na_n|}.$$

This will be true if

$$\lim_{n \to \infty} |n|^{1/n} = 1.$$

If we were working with the real absolute value, we would have $|n| = n$ and the result is an easy computation. To do it over $\mathbb{Q}_p$, the key thing to notice is that (since we are working with the p-adic absolute value) $1/n \leq |n| \leq 1$, so in $\mathbb{R}$ we have the inequalities

$$n^{-1/n} \leq |n|^{1/n} \leq 1.$$

Since the limit (in $\mathbb{R}$) of $n^{-1/n}$ as $n \to \infty$ is 1, this gives what we want. Thus, the two radii of convergence are the same.

169 Just use the proposition repeatedly (equivalently, use an induction proof where the step is provided by the proposition).

170 Well, the formula for a_k suggests that it's $f^{(k)}(x)/k!$ that's the interesting quantity, and notice that the formula says that if the a_n are in $\mathbb{Z}_p$ then so are the coefficients of $f^{(k)}(x)/k!$. That's kind of neat.

Actually, in certain cases one wants to consider whether the "quasi-derivative" defined by

$$f^{[k]}(x) = \sum_{n \geq k} \binom{n}{k} a_n (x-a)^{n-k}$$

has nice enough properties to replace the derivative. This is relevant, for example, if we are working over a field of characteristic p and we have $k > p$.

171 We need to consider $c_{nj} = a_n x^j a^{n-1-j}$. Since both x and α are in $\mathbb{Z}_p$, we get $|c_{nj}| \leq |a_n| \to 0$, which gives one of the conditions we need to check. For the other, note that $c_{nj} = 0$ if $j \geq n$.

172 If we put $f(X) = \sum a_n X^n$, and assume it converges on $p^m \mathbb{Z}_p$, then, as in the proof of the Corollary, we have to look at the series $\sum a_n p^{mn} X^n$. We need to find N such that

$$|p^{mN} a_N| = \max_n |p^{nm} a_n| \quad \text{and} \quad |p^{mn} a_n| < |p^{mN} a_N| \text{ for } n > N.$$

Then $f(X)$ has at most N zeros on $p^m \mathbb{Z}_p$.

173 The first series converges for $|x| < p$, hence for $x \in \mathbb{Z}_p$, and since $|p^n| = p^{-n}$ is strictly decreasing, we have $N = 0$, so that there are no roots in $\mathbb{Z}_p$. (In fact, we have

$$\sum p^n x^n = \frac{1}{1-px} \qquad \text{when } |x| < p,$$

and this is clearly never equal to zero.) The second series converges on $p^2 \mathbb{Z}_p$, and changing variables as above gives $N = 0$ again. (What is the sum?) The third one is again the hardest; to count the roots in $\mathbb{Z}_p$, one needs to find the last n such that $n!$ is not divisible by p, which gives $N = p - 1$. Thus, the series has at most $p - 1$ roots in the unit disk. If you managed to determine the precise radius of convergence, can you say anything about other possible roots?

174 Since $v_p(n)$ is the largest m such that p^m divides n, it's clear that $v_p(n) \leq \log n / \log p$. But then $v_p(n)/n \leq \log n / n \log p$, which tends to zero as $n \to \infty$, which gives what we want.

175 If $p = 2$, then $-1 = 1 - p \in B$, so that $\log_2(-1)$ makes sense. On the other hand, we must have $2\log_2(-1) = \log_2(-1)^2 = \log_2(1) = 0$, so that $\log(-1) = 0$. Writing out the series for $\log(1 - 2)$ gives

$$-\left(2 + \frac{2^2}{2} + \frac{2^3}{3} + \frac{2^4}{4} + \cdots + \frac{2^n}{n} + \cdots\right)$$

and saying that this converges to zero in $\mathbb{Q}_2$ amounts to saying that its partial sums get more and more divisible by 2. This was the claim in chapter one. For an estimate of the power of 2 dividing a partial sum, we might write

$$\left(2 + \frac{2^2}{2} + \frac{2^3}{3} + \frac{2^4}{4} + \cdots + \frac{2^N}{N}\right) + \left(\frac{2^{N+1}}{N+1} + \cdots\right) = 0,$$

which shows that

$$\left|2 + \frac{2^2}{2} + \frac{2^3}{3} + \frac{2^4}{4} + \cdots + \frac{2^N}{N}\right|_2 = \left|\frac{2^{N+1}}{N+1} + \cdots\right|_2 \leq \max_{n > N}\{|2^n/n|_2\}.$$

Thus we need to estimate $|2^n/n|_2$, or $v_2(2^n/n)$, for large n. Now, $v_2(2^n/n) = n - v_2(n) \geq n - \log n/\log 2$, so a lower bound for the exponent will be given by the least value of $n - \log n/\log 2$ for $n > N$. Now use some calculus.

176 Define a power series by $f(X) = \log(1 + pX)$, which will converge for $x \in \mathbb{Z}_p$. We need to find the last N for which the coefficient a_N has the maximum absolute value. Writing down the series explicitly (do it!), one sees that $N = 1$ if $p \neq 2$ and $N = 2$ if $p = 2$, which gives us the answer we want.

177 Notice first that if $x^p = 1$, then $|x| = 1$, so that any such root must be in $\mathbb{Z}_p$. Reducing modulo p gives an element $\overline{x}$ of $\mathbb{Z}/p\mathbb{Z}$ whose p-th power is one. Since $p \neq 2$, this implies that $\overline{x} = 1$ in $\mathbb{Z}/p\mathbb{Z}$, i.e., that $x \in 1 + p\mathbb{Z}_p$. In a nutshell, a p-th root of unity in $\mathbb{Q}_p$ must be in $1 + p\mathbb{Z}_p$.

Now we can just use the previous problem: if $x \in 1 + p\mathbb{Z}_p$ satisfied $x^p = 1$, then clearly $\log_p(x) = 0$, and this can only happen if $x = 1$. So there are no roots of unity in $1 + p\mathbb{Z}_p$.

Putting it all together, we conclude that there are no nontrivial p-th roots of unity in $\mathbb{Q}_p$.

178 This is very similar to, but easier than, the previous problem.

179 Write out the expression of $n!$ as a product, and work out how many numbers are multiples of p, how many of p^2, etc.

180 What the hint says.

181 Not serious ones!

182 Start from

$$\exp_p(x) - \exp_p(y) = \exp_p(y) \left(\exp_p(x - y) - 1 \right)$$

and note that since $|\exp_p(y) - 1| < 1$ we must have $|\exp_p(y)| = 1$.

183 Since $\log_2(-1) = 0$ and the terms of the series are non-zero, there's no chance that the condition $|g(x)| \geq |a_m x^m|$ is going to be satisfied. This points out a general fact: whenever $g(x) = 0$, we'll have trouble applying Theorem 5.4.3.

184 This is very similar to what we did in the text for the exponential. The regions of convergence will, of course, be the same as those for the exponential function. (Why "of course"?) The "p-adic trig functions" won't be periodic, because of Corollary 5.6.5.

185 The elements of $\mathbb{Z}/n\mathbb{Z}$ can be represented by the integers between 1 and n. It's easy to see that if a is invertible in $\mathbb{Z}/n\mathbb{Z}$, then $\gcd(a, n) = 1$ (can you prove it?). For the converse, use the fact that if $\gcd(a, n) = 1$ then we can find integers r and s such that $ra + sn = 1$, and reduce modulo n.

186 $(1 + qx)(1 + qy) = 1 + q(x + y + qxy)$, and

$$\frac{1}{1 + qx} = 1 - qx + q^2 x^2 - q^3 x^3 + \dots$$

which converges because $x \in \mathbb{Z}_p$. Similarly with p instead of q.

187 Let $x \in 1 + 2\mathbb{Z}_2$. Then $x \equiv \pm 1 \pmod{4}$. In the first case, $x \in 1 + 4\mathbb{Z}_2$ and Theorem 5.7.8 shows that $\log_2(x) \in 4\mathbb{Z}_2$. On the other hand, if $x \equiv -1 \pmod{4}$ then $-x \in 1 + 4\mathbb{Z}_2$ and

$$\log_2(x) = \log_2(-1) + \log_2(-x) = \log_2(-x) \in 4\mathbb{Z}_2.$$

188 We already know that there are exactly $(p - 1)$ roots of unity in $\mathbb{Z}_p$, by a combination of Hensel's Lemma (Problem 123) and Strassman's Theorem (Problem 177). Further, Hensel's Lemma already tells us that no two of the $(p - 1)$-st roots of unity are congruent modulo p. (For $p = 2$, one needs to change this slightly; see problems 124 and 178.) Can you come up with a more direct argument?

189 Since π gives an injective homomorphism between V and $(\mathbb{Z}/q\mathbb{Z})^\times$, and these groups have the same number of elements, π must in fact be an isomorphism. Now let $u \in \mathbb{Z}_p^\times$, and suppose $\pi(u) = \overline{n} \in (\mathbb{Z}/q\mathbb{Z})^\times$. Choose $\zeta \in V$ such that $\pi(\zeta) = \overline{n}$. Then $u_1 = u\zeta^{-1} \in U_1$. The map

$$u \mapsto (\zeta, u_1)$$

gives the isomorphism between $\mathbb{Z}_p^\times$ and $V \times U_1$.

190 The main thing is that when $p = 2$ the Teichmüller character is trivial: $\omega(x) = 1$ if $x \in \mathbb{Z}_2^\times$ and $\omega(x) = 0$ if $x \in 2\mathbb{Z}_2$. For x a unit, then, the factorization $x = \omega(x)x_1$ just says $x = x$. Instead, what we should do is factor $x \in \mathbb{Z}_2^\times$ as $x = \pm x_2$, with $x_2 \in 1 + 4\mathbb{Z}_2$.

191 If $x \in p\mathbb{Z}_p$ it is clear that $x^{p^n} \to 0$, which is the desired answer. If $x \in \mathbb{Z}_p^\times$ and $p \neq 2$, we know $x = \omega(x)\langle x \rangle$.

Since $\omega(x)^{p-1} = 1$, we have $\omega(x)^p = \omega(x)$; taking p-th powers over and over, we see that $\omega(x)^{p^n} = \omega(x)$ for any n.

On the other hand, $\langle x \rangle = 1 + qy$ for some y. Taking p-th powers,

$$\langle x \rangle^p = (1 + qy)^p = 1 + pqy + \text{multiples of } q^2,$$

so that $\langle x \rangle^p \in 1 + p^2\mathbb{Z}_p$. Repeating, we see that $\langle x \rangle^{p^n} \in 1 + p^{n+1}\mathbb{Z}_p$, so that $\langle x \rangle^{p^n}$ tends to 1 as $n \to \infty$. Putting these together gives what we want.

A similar computation handles the case $p = 2$.

192 If $|x| = 1$, we want to look at the sequence $\binom{\alpha}{n}$ as n tends to infinity. If it tends to zero, then the series converges for $|x| = 1$; if not, not, and the radius of convergence is 1. Can you decide? The answer may very well depend on α!

193 Keep in mind that `a^(1/3)` and `a^(1/3+O(7^20))` need not be the same. In GP it seems possible to compute a^b when $a, b \in \mathbb{Z}_p$ without problems. Try to figure out what the program is doing. Sage can also compute a^b in many cases. I have no idea what algorithm(s) are used. Go play.

194 This one is much easier: if $|\alpha| > 1$, then $|\alpha - i| = |\alpha|$ (because "all triangles are isosceles"). Putting this together with our various estimates on $|n!|$ should allow you to get an answer.

195 When α is a *positive* integer, this is obvious, since

$$\mathbf{B}(\alpha, x) = (1 + x)^\alpha = \sum_{i=0}^{\alpha} \binom{\alpha}{i} x^i$$

is actually a polynomial. For negative integers, all we need to notice is that we have $\mathbf{B}(\alpha, x)^{-1} = (1 + x)^{-\alpha}$.

196 This takes quite a bit of work, though some bits aren't hard. For example, we already know that $\mathbf{B}(1/2, x)$ converges if $|x| < 1$, that is, if $x \in p\mathbb{Z}_{(p)}$ (since x is rational). So we need to know for which a/b is it true that $(a/b)^2$ is a one-unit. Now $(a/b)^2 - 1 \in p\mathbb{Z}_p$ means that p divides $a^2 - b^2 = (a + b)(a - b)$ (which is the numerator), and hence that it divides either $a + b$ or $a - b$. This shows the "if" in (i). For the converse, we need to know that the series does not converge if $x \notin p\mathbb{Z}_p$. See how far you can get. There are some hints in [42].

197 See any book on real analysis or general topology. I like [39, Section 4.2], [55, Theorem 4.19] and [47, Theorem 27.6].

198 The easiest way is to exploit the proposition that follows this problem in the text: a function on $\mathbb{Z}$ that cannot be extended to $\mathbb{Z}_p$ will work. Say, choose an element $\alpha \in \mathbb{Z}_p$ which is not in $\mathbb{Z}$ (say, any a/b with $p \nmid b$ and $b > 1$), and define $f(x) = 1/(x - \alpha)$ for any $x \in \mathbb{Z}$. Then f is continuous on $\mathbb{Z}$ but not uniformly continuous (check!).

199 The main point is that $a_n - a_m$ small implies $f(a_n) - f(a_m)$ small by the uniform continuity. It's pretty much a direct check in $\varepsilon - \delta$ style. For more detail on this and the next two problems, see [39, Section 4.2].

200 If b_k is another sequence tending to x, then $a_k - b_k$ tends to zero; by the boundedness and uniform continuity, it follows that $f(a_k) - f(b_k)$ tends to zero, which is what we want.

201 This is very similar to the other two problems: easy $\varepsilon - \delta$ stuff.

202 Basically, all that needs to be done is to run through the argument and check that the only result we needed was that $\mathbb{Z}_p$ was compact and that $\mathbb{Z}$ was a dense subset. Hence the argument works for any compact subset of $\mathbb{Q}_p$ and any dense subset of that. (Even for $\mathbb{Z}_p$, that's an advantage. For example, if $p \neq 2$ the even integers are dense in $\mathbb{Z}_p$, and we can interpolate from them to all of $\mathbb{Z}_p$.)

203 This one's really pretty hard. Here's one way. First check that each term in the series for $\mathbf{B}(\alpha, x)$ is continuous as a function of α. This is easy, since $\binom{\alpha}{n}$ is a polynomial in α. Then check that the series converges uniformly as a series of functions of α. This implies (exactly as in the classical case) that the sum is a continuous function of α.

204 It works if there are no convergence woes (and gives the same result, by continuity). The difficulty is that we need $|\alpha \log_p(n)| < p^{-1/(p-1)}$ to be able to compute the exponential, and $|n - 1| < 1$ to be able to compute the logarithm. The second condition is one we had to impose in our setup anyway ($n \in 1 + p\mathbb{Z}_p$), so it doesn't bother us. We've shown above that it implies $|\log_p(n)| < 1$, which is enough if $\alpha \in \mathbb{Z}_p$ and $p \neq 2$ (because of the remark we made above, that in $\mathbb{Z}_p$ having absolute value less than one implies having absolute value $\leq 1/p$). In a more general situation, this would work less well.

205 This is simpler than it seems. To begin with, in $\mathbb{Z}_2$ there is a good notion of "even," since 2 is not invertible. Hence if we define $(-1)^\alpha = 1$ if 2 divides α and $= -1$ if not, everything works. On the other hand, if $p \neq 2$, then 2 is invertible, and there's clearly no good way to extend the whole function.

If we want to do the interpolation "in pieces," as in the text, then it works. Take $p = 3$, so that $p - 1 = 2$, and the two choices for α_0 are 0 and 1. In fact, f_0 and f_1 are pretty easy to work out: $f_0(\alpha) = 1$ for all α and $f_1(\alpha) = -1$ for all α. Then f_0 interpolates $(-1)^\alpha$ for $\alpha \equiv 0 \pmod 2$, i.e., for even α, and f_1 does the same for odd α. A dumb example, but maybe it sheds some light on what is going on...

What happens if $p = 5$?

206 Easy: do the same as you did in Problem 50.

207 Again, this is a repeat of Problem 51.

208 The hardest property to check is (ii). For the sup-norm, even that one comes easily: we want to check that $\|\mathbf{v} + \mathbf{w}\| \leq \|\mathbf{v}\| + \|\mathbf{w}\|$. Let $\mathbf{v} = a_1\mathbf{v}_1 + \cdots + a_n\mathbf{v}_n$ and $\mathbf{w} = b_1\mathbf{v}_1 + \cdots + b_n\mathbf{v}_n$; the inequality translates into

$$\max_i |a_i + b_i| \leq \max_i |a_i| + \max_i |b_i|.$$

But that follows easily from the fact that $|a_i + b_i| \leq |a_i| + |b_i|$ for each i.

For the r-norms, it's a little harder to get (ii); in fact, it may be worth looking it up in books on functional analysis (where it's done in much greater generality). If you'd like to give it a try, here is an outline of the standard proof.

First of all, it's relatively easy to prove the triangle inequality if $r = 1$ or $r = 2$, so we'll concentrate on providing hints for the rest. (Actually, the proof we sketch works fine for $r = 2$.) Next, for each $r > 1$, let $r' > 1$ be the real number such that

$$\frac{1}{r} + \frac{1}{r'} = 1.$$

We sometimes call r and r' a dual pair.[1] A lot of the proof depends on the duality between the r-norm and the r'-norm. The first lemma is the following:

• Let α and β be positive real numbers, and let r and r' be as above. Then we have

$$\alpha\beta \leq \frac{\alpha^r}{r} + \frac{\beta^{r'}}{r'}.$$

To prove this, plot the function $y = x^{r-1}$, the lines $x = \alpha$ and $y = \beta$, and try to locate in your picture the various quantities that appear in the inequality.

---

[1] Notice that if $r = 2$, then $r' = 2$; this is what makes the case $r = 2$ special. On the other hand, if $r = 1$, the only sensible choice for r' is $+\infty$. What norm would that correspond to?

Now the next step: prove the *Hölder Inequality*. Let $v_1, v_2, \ldots, v_n$ be a basis, and take two elements

$$v = a_1 v_1 + a_2 v_2 + \cdots + a_n v_n$$

and

$$w = b_1 v_1 + b_2 v_2 + \cdots + b_n v_n.$$

Then show that

$$\sum_{i=1}^{n} |a_i|\,|b_i| \leq \|v\|_r \|w\|_{r'}$$

For the proof, apply the previous inequality with $\alpha = |a_i|/\|v\|_r$ and $\beta = |b_i|/\|w\|_{r'}$ for each $i = 1, 2, \ldots, n$, and add the results. (For $r = 2$, this should be a familiar formula—is it?)

Finally, use the Hölder Inequality to prove that $\|v + w\|_r \leq \|v\|_r + \|w\|_r$. Here's the idea: start with the sum whose r-th root is the norm:

$$\sum_{i=1}^{n} |a_i + b_i|^r = \sum_{i=1}^{n} |a_i + b_i|^{r-1}|a_i + b_i|$$

$$\leq \sum_{i=1}^{n} |a_i + b_i|^{r-1}|a_i| + \sum_{i=1}^{n} |a_i + b_i|^{r-1}|b_i|$$

(where we've just used the triangle inequality for the absolute value), and now apply Hölder's inequality to both summands and the pair (r, r').

This was a hard one!

209 See figure B.1.

210 Try a vector space of dimension one.

211 Well, $\|(1, -1)\| = 0$ kind of messes things up. (On the other hand, the other two conditions are satisfied; is that significant?)

212 To prove that equivalent norms define the same topology, it's enough to show that an open ball with respect to one norm is an open set with respect to the other. Since this is a vector space with a norm, it's enough to prove this for one ball, say, the open unit ball. So let $B = \{x \in V : \|x\|_1 < 1\}$. If $x \in B$, then let $r = \|x\|_1$. Choose $R < (1 - r)/C$; it's easy to see that the set $N = \{y \in V : \|y - x\|_2 < R\}$, which is an open ball with respect to $\|\ \|_2$, is contained in B. This shows B is open with respect to $\|\ \|_2$, and, since everything is symmetric, proves what we want.

For the converse, we can be direct or we can be fancy. For a direct approach, show that if the two topologies are equivalent, the closed unit ball with respect to one norm must contain a closed unit ball with respect to the

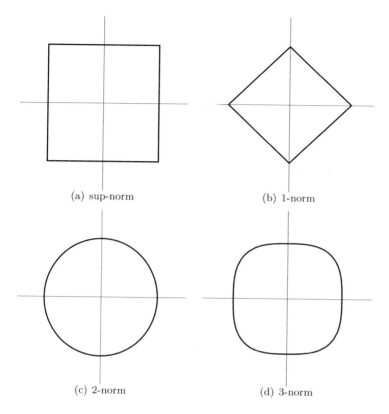

(a) sup-norm   (b) 1-norm

(c) 2-norm   (d) 3-norm

Figure B.1: Unit balls for various norms

other. (For example, argue that the open unit ball for $\| \ \|_1$ is open with respect to $\| \ \|_2$, and hence contains an open $\| \ \|_2$-ball around zero, which contains a closed $\| \ \|_2$-ball—of slightly smaller radius—around zero.) Then look closely at what this means to get one of the inequalities we want.

A fancier approach would be this: consider the identity map $\iota : V \longrightarrow V$, so that $\iota(v) = v$. We give the "first" V the norm $\| \ \|_1$ and we give the "second" V the norm $\| \ \|_2$. Since this yields the same topology on "both" V's, both ι and its inverse are continuous linear transformations. Unwinding the continuity yields the inequalities we want.

213 The two inequalities in the definition of equivalence can be restated as

$$\frac{1}{D}\|\boldsymbol{v}\|_2 \leq \|\boldsymbol{v}\|_1 \leq C\|\boldsymbol{v}\|_2$$

for any $\boldsymbol{v} \in V$. This clearly translates to what we said about closed balls.

214 Easy.

215 The sketches make it clear that any ball with respect to one of the norms, say $\| \ \|_1$, both contains and is contained in balls with respect to the other norms, and this translates directly into the existence of C and D.

216 $\max\{|a|,|b|\} \leq |a| + |b| \leq 2\max\{|a|,|b|\}$.

217 This is straightforward if we do it in the usual "circle of implications" way. To see, for example, that (i) implies (ii), suppose that f is continuous at $\mathbf{0}$. Then given any $\varepsilon > 0$ there exists a $\delta > 0$ such that $\|\mathbf{v}\| \leq \delta$ implies $\|f(\mathbf{v})\| \leq \varepsilon$. Making δ smaller if necessary, we can find an element $x \in \Bbbk$ such that $|x| = \delta$. But then we have

$$\|\mathbf{v}\| \leq 1 \Longrightarrow \|x\mathbf{v}\| \leq |x| = \delta \Longrightarrow \|f(x\mathbf{v})\| \leq \varepsilon$$
$$\Longrightarrow \|xf(\mathbf{v})\| \leq \varepsilon \Longrightarrow \delta\|f(\mathbf{v})\| \leq \varepsilon$$
$$\Longrightarrow \|f(\mathbf{v})\| \leq \frac{\varepsilon}{\delta},$$

so that the sup is finite. The other implications are similar.

For an added challenge, show that all of these conditions are also equivalent to the assertion that there exists some $\mathbf{v} \in V$ and some positive $r \in \mathbb{R}$ such that f is bounded on the closed ball of radius r around $\mathbf{v}$.

218 This is a standard example in functional analysis (these spaces are sometimes known as $\ell_\infty(K)$ and $\ell_1(K)$). If necessary, look it up.

219 See Chapter 7!

220 The picture in two dimensions is like this: the unit ball with respect to the sup-norm is a "square" defined by $|a_1| \leq 1$ and $|a_2| \leq 1$, where a_1 and a_2 are the coordinates with respect to our basis. (In $\mathbb{R}^2$, this is the square given by $-1 \leq x \leq 1$ and $-1 \leq y \leq 1$.) What we are doing is partitioning the sides of the square into pieces of radius less than ε and using this partition to cut the "square" up into lots of "rectangles." Then we show that the rectangles do the job. *Now* draw the picture.

221 Checking that 2 is not a square in $\mathbb{Q}_5$ is just a matter of seeing that it is not a square modulo 5, which is easy. For the norm, we can try the 2-norm with respect to the basis $\{1, \sqrt{2}\}$:

$$\|a + b\sqrt{2}\| = \sqrt{|a|^2 + |b|^2}.$$

This gives the 5-adic norm when $b = 0$, i.e., on $\mathbb{Q}_5$, but is not an absolute value on $\mathbb{Q}_5(\sqrt{2})$—why not?

222 Either $\sigma(\sqrt{D}) = \sqrt{D}$ or $\sigma(\sqrt{D}) = -\sqrt{D}$, because those are the only two roots of $X^2 - D$. But any field that contains $\sqrt{D}$ contains $-\sqrt{D}$.

223 The point is that one of the cube roots of two is real, and the other two are complex. The field obtained by adjoining the real root is contained in $\mathbb{R}$, hence can't be equal to its image under an automorphism mapping $\sqrt[3]{2}$ to a complex cube root.

224 Any σ must map ζ to another root of $X^2 + X + 1$; the roots are ζ and ζ^2, so we're OK. Similarly the image of $\sqrt[3]{2}$ must be a cube root of two; there are three: $\sqrt[3]{2}$, $\zeta\sqrt[3]{2}$, and $\zeta^2\sqrt[3]{2}$, and they are all in K.

225 The point is that $\{1, \alpha, \alpha^2, \ldots, \alpha^{n-1}\}$ is a basis for $K(\alpha)$ over K. With respect to this basis, the matrix of multiplication by α is in rational canonical form and it's easy to check that its determinant is $(-1)^n a_n$.

226 Notice that $\alpha(\alpha^i b_j) = \alpha^{i+1} b_j$, so that with respect to the given basis the matrix of multiplication by α is a diagonal arrangement of r identical $n \times n$ blocks of the form we saw in the previous problem. Hence the determinant is $((-1)^n a_n)^r = (-1)^{nr} a_n^r$.

227 If we remember that $K = F(\alpha)$ is isomorphic to the quotient of $F[X]$ by the ideal generated by $f(X)$, it's not hard. Let $\mathbf{C}$ be an algebraically closed field containing K. For any root α' of $f(X)$, consider the map $K[X] \longrightarrow \mathbf{C}$ mapping X to α'; pass to the quotient to get a map from $K = F(\alpha)$ to $\mathbf{C}$, whose image must be K, by normality. To get the final conclusion, write $f(X)$ as a product of linear factors.

228 If K/F is normal, but K is not equal to $F(\alpha)$, then $K/F(\alpha)$ is normal. Now use the method in Problem 226.

229 Does taking the product in the normal closure work?

230 This is in the same spirit as the solution of Problem 226: choose the basis so that the matrix is in a form that allows one to compute the determinant directly. See, for example, [54, 4.109].

231 A general quadratic extension works exactly like the example in the text. For the second half, it can be easier or harder depending on the elements you choose to work with; I'd try ζ, $\sqrt[3]{2}$, and $\zeta + \sqrt[3]{2}$. The first two are easy; the determinant method is tempting for the last one, but we'll have to compute a six-by-six determinant...

232 No big deal. The only real point: given x and y in the algebraic closure, the field $\mathbb{Q}_p(x, y)$ is a finite extension of $\mathbb{Q}_p$; hence the norm we have defined gives an absolute value on $\mathbb{Q}_p(x, y)$. It follows that $|x + y| \leq \max\{|x|, |y|\}$ and that $|xy| = |x| \, |y|$, which is what we needed to prove.

233 Use the same strategy as in the Lemma, i.e., reduce modulo p after making sure that everything is in $\mathbb{Z}_p$.

234 Suppose $f(X)$ factors in $\mathbb{Q}_p[X]$; by the Lemma, it also factors in $\mathbb{Z}_p[X]$. Since $f(X)$ is monic, the top coefficients of each of the factors must be invertible in $\mathbb{Z}_p$ (yes?), and therefore are non-zero modulo p. If we now reduce modulo p we get a non-trivial factorization in $\mathbb{F}_p[X]$.

235 The argument will still work if we assume that the top coefficient of $f(X)$ is invertible. Otherwise, the reduction modulo p of $f(X)$ will have degree *smaller* than the degree of $f(X)$, and things begin to get weird.

236 Well, modulo p an Eisenstein polynomial looks like X^n. If we factor that as $X^r \cdot X^s$, the factors are not relatively prime, so we can't apply the Lemma. If we factor as $X^n \cdot 1$, we can, but the factorization will be as the product of a polynomial of degree n and a polynomial of degree zero, which means it will be the trivial factorization.

237 Yes, and this is proved in Chapter 6. In fact, proving this first would allow us to simplify many of the proofs in this section.

238 Let $f(X) \in \mathbb{Z}_p[X]$ be a monic polynomial of degree n such that $\bar{f}(X)$ is irreducible in $\mathbb{F}_p[X]$ and whose roots generate the unique extension $\mathbb{F}/\mathbb{F}_p$ of degree n. Let K be the extension of $\mathbb{Q}_p$ obtained by adjoining a root of $f(X)$. We know both K and $\mathbb{F}$ are extensions of degree n, and by uniqueness it's clear that $\mathbb{F}$ is a normal extension. Now: (1) use Hensel's Lemma to show that $f(X)$ has n roots in K, then (2) conclude that K is a normal extension of $\mathbb{Q}_p$. Use the fact that automorphisms preserve absolute values to show (3) that every automorphism of $K/\mathbb{Q}_p$ induces an automorphism of $\mathbb{F}/\mathbb{F}_p$. This gives a map from $\mathrm{Gal}(K/\mathbb{Q}_p)$ to $\mathrm{Gal}(\mathbb{F}/\mathbb{F}_p)$. Then it's a matter of showing this map is injective (and is therefore an isomorphism).

239 Eliminate one prime at a time from the denominator.

240 Yes, of course: a polynomial in $\mathbb{Z}[X]$ which satisfies the conditions in the Eisenstein criterion for some prime p is irreducible in $\mathbb{Q}_p[X]$, and *a fortiori*[2] irreducible in $\mathbb{Q}[X]$.

241 Yes. Can you prove it?

242 Let's do the first one;

$$\mathbf{N}_{F_1/\mathbb{Q}_5}(1 + 3\sqrt{2}) = (1 + 3\sqrt{2})(1 - 3\sqrt{2}) = 1 - 18 = -17,$$

---

[2]Run for the dictionary! What does he mean, *a fortiori*?

so

$$v_5(1 + 3\sqrt{2}) = \frac{1}{2}v_5(-17) = 0.$$

The others are similar, but keep in mind that what we know about valuations still works. For example, it's easy to see that $v_5(\sqrt{2}) = 0$, by computing the norm, and then it follows that $v_5(5\sqrt{2}) = 1$; hence ("all triangles are isosceles") $v_5(1 + 5\sqrt{2}) = 0$.

243 The nicest one of these is $x = \sqrt{5}$:

$$\mathbf{N}_{F_2/\mathbb{Q}_5}(\sqrt{5}) = (\sqrt{5})(-\sqrt{5}) = -5,$$

so $v_5(\sqrt{5}) = 1/2$. As well it should be!

244 Let's try $x = 1 - \zeta$:

$$\mathbf{N}_{F_3/\mathbb{Q}_3}(1 - \zeta) = \mathbf{N}_{\mathbb{Q}_3(\zeta)/\mathbb{Q}_3}\left(\mathbf{N}_{F_3/\mathbb{Q}_3(\zeta)}(1 - \zeta)\right)$$

$$= \mathbf{N}_{\mathbb{Q}_3(\zeta)/\mathbb{Q}_3}(1 - \zeta)^2.$$

(Remember that F_3 is an extension of degree 2 of $\mathbb{Q}_3(\zeta)$.) To compute the norm, take $\{1, \zeta\}$ as a basis for $\mathbb{Q}_3(\zeta)$ over $\mathbb{Q}_3$. The matrix of multiplication by $1 - \zeta$ is

$$\begin{pmatrix} 1 & 1 \\ -1 & 2 \end{pmatrix}$$

(remember that $1 + \zeta + \zeta^2 = 0$ for that one), so the norm (which is the determinant) is 3. It follows that

$$\mathbf{N}_{F_3/\mathbb{Q}_3}(1 - \zeta) = 9$$

and then that

$$v_3(1 - \zeta) = \frac{1}{4}v_3(9) = \frac{1}{2}.$$

Try some of the others.

245 It's not too hard to see that the answers must be $e = 1, 2, 2$, respectively. But how would a proof go?

246 For F_1, $\pi = 5$ will do, and for F_2, $\pi = \sqrt{5}$. For F_3, the computation is problem 244 helps: $\pi = 1 - \zeta$ does the job.

247 Follow the hints; this is mostly straightforward. For example, to show that $\mathfrak{p}_K$ is principal, just note that

$$x \in \mathfrak{p}_K \implies v_p(x) > 0 \implies v_p(x) \geq 1/e$$
$$\implies v_p(\pi^{-1}x) \geq 0 \implies \pi^{-1}x \in \mathcal{O}_K$$
$$\implies x \in \pi\mathcal{O}_K.$$

This is already enough to show π generates $\mathfrak{p}_K$.

The only non-trivial bit in the remainder is showing that the elements of $\mathcal{O}_K$ are exactly the elements of K which are roots of monic polynomials with coefficients in $\mathbb{Z}_p$. In one direction, it's easy: if α is the root of such a polynomial, then its norm is (up to sign) a power of the zeroth coefficient, which is in $\mathbb{Z}_p$. Hence, $v_p(\alpha) = \frac{1}{n} v_p(\mathbf{N}(\alpha)) \geq 0$. For the converse, look at Lemma 6.3.6.

248 For F_1, we get $\mathcal{O} = \mathbb{Z}_5[\sqrt{2}]$, and $k = \mathbb{F}_5[\sqrt{2}]$ is a field of order 25. For F_2, $\mathcal{O} = \mathbb{Z}_5[\sqrt{5}]$ and $k = \mathbb{F}_5$. For F_3, $\mathcal{O} = \mathbb{Z}_3[\zeta, \sqrt{2}]$ and k is a field of order 9.

249 Routine, but important routine. Make sure you understand how both portions of the proof work—for the most part, it's a question of keeping track of what is divisible by what. Can you come up with a more conceptual proof?

250 $X^2 - 5$, of course.

251 We've done all the work already, when we computed the norm of $1 - \zeta$, which is 3. It follows that $v_3(1 - \zeta) = 1/2$, and the extension (which is of degree 2) is totally ramified. The Eisenstein polynomial for $1 - \zeta$ is $X^2 - 3X + 3$ (just square $1 - \zeta$ and see what coefficients work, or use the fact that $\zeta^2 + \zeta + 1 = 0$). Notice that in this case there is another uniformizer, $\sqrt{-3}$, since it's easy to see that

$$\zeta = \frac{-1 + \sqrt{-3}}{2}$$

(since it is a root of $X^2 + X + 1$).

252 Easy: *exactly* the same proof works—just replace all the p's by π's.

253 The obvious reformulation works, and again the proof is the same.

254 Well, if you solved problem 120, then your solution solves this one too.

255 F_1 contains the 24-th roots of unity; for F_2, there's no new information (only that it contains the 4-th roots of unity, which $\mathbb{Q}_5$ already does). F_3 contains the 8-th roots of unity (the degree is 4, and $e = 2$, so $f = 2$). As for other roots of unity, F_3 certainly contains the cube roots of unity, by construction. (Notice that the cube roots of unity are 1-units, since $\zeta - 1$ is a uniformizer. That means they are "invisible" from the Hensel's Lemma side, and therefore not predicted by the Corollary.) What about the other two fields?

256 The first is really easy: $x^m = 1$ implies $|x| = 1$. For the second part, if $x^m = 1$ and m is prime to $p^f - 1$, then look at the image of x in $k^\times$, and remember that this last is a cyclic group of order $p^f - 1$.

257 Write $x = 1 + \pi u$ and raise to the p-th power, remembering that π is a divisor of p. The version for general r requires, of course, an easy induction argument.

258 Use the uniqueness statement in Hensel's Lemma.

259 Expanding $(1 - x_1)^\ell - 1 = 0$ shows that

$$\ell x_1 + \sum_{i=2}^{\ell} \binom{\ell}{i} x_1^i = 0.$$

Dividing by x_1 and rearranging shows that

$$|\ell| = \left| \sum_{i=2}^{\ell} \binom{\ell}{i} x_1^{i-1} \right|.$$

But the left-hand side is 1, since $\ell \neq p$, and every term of the right-hand side is in $\mathfrak{p}_K$, which is a contradiction.

260 The quotients are clearly abelian, and problem 257 shows that $x \in U_n$ implies $x^p \in U_{n+1}$, so that every element of the quotient is of order p. Now: why is the quotient a finite group? (An idea: fix a uniformizer, and consider the map $U_n \to \mathcal{O}_K$ given by $1 + \pi^n x \mapsto x$; what properties does this have?)

261 To see that they are both unramified, it's enough to check that both $X^2 - 2$ and $X^2 - 3$ are irreducible over $\mathbb{F}_5$. That they are the same simply means that one can express $\sqrt{2}$ as $a + b\sqrt{3}$ with a and b in $\mathbb{Q}_5$. That's rather hard to imagine, isn't it? But notice that 6 is a square in $\mathbb{Q}_5$ by Hensel's lemma! So if $\gamma \in \mathbb{Q}_5$ satisfies $\gamma^2 = 6$, we have $\sqrt{2}\sqrt{3} = \pm\gamma$, and there we are:

$$\sqrt{2} = \pm\frac{\gamma}{\sqrt{3}} = \pm\frac{\gamma\sqrt{3}}{3}.$$

As to the p-adic expansion of a 24-th root of unity, we need to choose our "digits" first. Since the residue field is $\mathbb{F}_5[\sqrt{2}]$, we might take coefficients from the nonzero elements of the set $\{a + b\sqrt{2} : 0 \leq a, b \leq 4\}$. Then, to find the expansion, we need to determine, first, the reduction modulo p. That'll have to be an element of order 24 in $\mathbb{F}_5[\sqrt{2}]$. Once you find one, use the procedure in Hensel's Lemma to get closer and closer to the real thing.

262 F_3 is an extension of degree four, and is ramified (in fact, e=2). Hence, the subfield $\mathbb{Q}_3(\sqrt{2})$, which is unramified, must be the maximal unramified subfield.

263 Here's a fancy proof that uses the uniqueness we have just proved. Since $n = 3$ we must have either $e = 1$ or $e = 3$. Suppose $e = 1$, so that the extension is unramified. Then it is equal to the unique unramified extension of degree 3. Now consider the extension $K' = \mathbb{Q}_3(\zeta \sqrt[3]{2})$ obtained by adjoining a different cube root of 2 (here, as before, $\zeta^3 = 1$, $\zeta \neq 1$). If K is unramified, then so is K', since they are clearly isomorphic. If they are both unramified, then they are equal, by the uniqueness. If $K = K'$, then, since both $\sqrt[3]{2}$ and $\zeta\sqrt[3]{2}$ are in K, we must have $\zeta \in K$ and, in fact, $\mathbb{Q}_3(\zeta) \subset K$, which is impossible: extensions of degree 3 can't contain subextensions of degree 2! Hence, K must be ramified, and since e must divide the degree, we must have $e = 3$.

Of course, that's a very fancy chain of reasoning! (It's the one that the author followed, though. . .) Can you exhibit directly an element of K whose minimal polynomial over $\mathbb{Q}_3$ is Eisenstein?

264 Let $\alpha \in K$ be a uniformizer; since K is totally ramified, the minimal polynomial for α is an Eisenstein polynomial, so $\alpha^e + a_{e-1}\alpha^{e-1} + \cdots + a_1\alpha + a_0 = 0$ with $p|a_i$ for all i and $p^2 \nmid a_0$. Now rearrange the equation to get

$$\alpha^e + a_0 = -(a_{e-1}\alpha^{e-1} + \cdots + a_1\alpha).$$

Every term on the right-hand side is divisible by $p\alpha$, so that $v_p(\alpha^e - a_0) \geq 1 + 1/e$. This suggests that the pu in the problem will be $-a_0$, but to make it work we have to show that we can pass from the "approximate root" α to a real root. The obvious way to do this is Hensel's Lemma, but that method doesn't work: if we put $f(X) = X^e - pu$, our estimates give $v_p(f(\alpha)) \geq 1 + 1/e$ and $v_p(f'(\alpha)) = 1 - 1/e$, which isn't enough to use problem 254.

Here's a more direct method (taken from [42]) that avoids that dilemma by using some analysis: we have $|\alpha^e - pu| \leq p^{-1}p^{-1/e} = |pu|p^{-1/e}$. Dividing through by $|pu|$ gives $|(\alpha^e/pu) - 1| \leq p^{-1/e}$. In other words, α^e/pu is a 1-unit. Using the binomial series, we can raise any 1-unit to any p-adic integer. Since $p \nmid e$, we have $1/e \in \mathbb{Z}_p$; use the binomial series to get $(\alpha^e/pu)^{1/e}$, and then use that to get a root.

265 Yes, of course. Just replace p by a uniformizer π everywhere.

266 What we need to do is to define "relative" notions of the ramification index and the residue degree; then everything works. See [34, Ch. 14].

267 The point is that for any such m one can find an r such that m divides $p^r - 1$, and the $(p^r - 1)$-st roots of unity are in $\mathbb{Q}_p^{\mathrm{unr}}$.

268 The image of v_p is still $\mathbb{Z}$, since there has been no ramification. The residue field is the algebraic closure of $\mathbb{F}_p$.

269 The residue field is the same (it can't very well become any bigger), but the image of v_p on $\overline{\mathbb{Q}}_p$ is $\mathbb{Q}$.

270 We know that $|\pi| = p^{-1/e}$, so $|\pi^n| = p^{-n/e}$. For the denominators, remember that $|n| \geq 1/n$ (with equality if and only if n is a prime power). Then $|1/n| \leq n$. So

$$\left| \frac{\pi^n}{n} \right| \leq np^{-n/e},$$

which clearly goes to 0 as $n \to \infty$.

Clearly the best possible bound B is

$$B = \max_{n \geq 1} \left| \frac{\pi^n}{n} \right|.$$

This clearly depends only on the ramification index e. Is there a clean formula for how B depends on e?

271 Our discussion shows that we need K to be a ramified extension. If π is a uniformizer we expect $\log_p(1 + \pi)$ to be as big as possible. Take $p = 2$ for simplicity and let's adjoin the root of an Eisenstein polynomial in Sage:

```
sage: K=Qp(2)
sage: S.<x>=ZZ[]
sage: f=x^3-2
sage: F.<w>=K.ext(f)
```

Remember that w is the uniformizer, so we compute the relevant logarithm:

```
sage: log(1+w)
   w^-2 + w^2 + w^4 + w^6 + w^7 + w^9 + w^11 + w^12
 + w^13 + w^15 + w^16 + w^17 + w^18 + w^19 + w^21
 + w^25 + w^26 + w^27 + w^28 + w^31 + w^32 + w^34
 + w^39 + w^40 + w^41 + w^45 + w^46 + w^47 + w^48
 + w^51 + w^53 + w^55 + w^57 + O(w^58)
```

So in this case, with $e = 3$, $\log_2(1 + \pi) \notin \mathcal{O}_K$. A similar computation shows that $e = 2$ is not large enough to provide an example. It's clear that the larger the ramification index the larger the image will be.

272 This is mostly a matter of time and patience. There should be no difficulty in checking that everything still works as before.

273 This is hard! I don't know how to do it in general. Since $v(\pi) = 1/(p-1)$, the expansion begins with the term in π^{p-1}. Let's see...

We know that π is a root of the polynomial

$$\Phi_p(X + 1) = X^{p-1} + \cdots + p$$

(we've shown that everything in the dots is divisible by p, but we haven't determined the coefficients exactly, though that is doable). Plugging in gives

$$\pi^{p-1} + a_{p-2}\pi^{p-2} + \cdots + a_1\pi + p = 0$$

rearranging,

$$p = -(\pi^{p-1} + a_{p-2}\pi^{p-2} + \cdots + a_1\pi).$$

Since all the a_i are divisible by p, hence divisible by π^{p-1}, we see that

$$p \equiv -\pi^{p-1} \equiv (p-1)\pi^{p-1} \pmod{\pi^p}.$$

Because for integers congruence $\pmod \pi$ is the same as congruence $\pmod p$. This is enough to show that the first coefficient is $(p-1)$, but how do we go on from there?

For any particular prime p, Sage can do it. Here it is with $p = 7$. First of all, we need to create the field extension as explained in the text. We need the polynomial $f(X) = \Phi_7(X+1)$, which is easy to compute:

$$f(X) = X^6 + 7X^5 + 21X^4 + 35X^3 + 35X^2 + 21X + 7.$$

```
sage: K=Qp(7)
sage: S.<x>=ZZ[]
sage: S
sage: f=x^6 + 7*x^5 + 21*x^4 + 35*x^3 + 35*x^2 + 21*x + 7
sage: F.<u>=K.ext(f)
sage: F
  7-adic Eisenstein Extension Field in u defined by
  x^6 + 7*x^5 + 21*x^4 + 35*x^3 + 35*x^2 + 21*x + 7
sage: F(7)
  6*u^6 + 3*u^7 + 3*u^8 + 2*u^9 + 2*u^10 + 3*u^11 + ...
```

That is only the beginning of the expansion. Since K is created with precision 20 by default and F is an extension of degree 6, the relative precision becomes 120 when we write expansions in u. I asked Sage to give it in base 7 notation:

$$7 = \ldots 1050356111013426642232336000000.$$

274 Just use $x_1 = 1$ as your initial root.

275 Well, if I could, I would very likely have put the simpler one in the text... Can you?

276 If $K = \mathbb{Q}_p(\zeta_p)$ contained *both* a $(p-1)$-st root of $-p$ and a $(p-1)$-st root of p, then it would contain an element ξ such that $\xi^{p-1} = -1$. Such a thing would be a $2(p-1)$-st root of unity, and we already know that K (a totally ramified extension of $\mathbb{Q}_p$) contains exactly as many prime-to-p roots of unity as $\mathbb{Q}_p$ does, and hence can't contain ξ.

277 Show first that the minimum will occur at a power of p.

278 This is really an open-ended project, which should be fun to play with. Is the extension still totally ramified? Can we get higher-order roots of $-p$? How about the logarithm function?

279 To use Krasner's Lemma, what we want to do is find a generator x of $\mathbb{Q}_p(\zeta_p)$ and a $(p-1)$-st root a of $-p$ such that $|x-a|$ is less than $|a-a'|$ whenever a' is some other $(p-1)$-st root of $-p$. (The a' are the conjugates of a over $\mathbb{Q}_p$, since the polynomial $X^{p-1}+p$ is irreducible.) Understanding the $|a-a'|$ part isn't hard: any a' must be of the form ξa, where $\xi^{p-1}=1$, and hence, $|a-a'|=|a-\xi a|=|a|\,|1-\xi|=p^{-1/(p-1)}$. (Remember that we proved before that $|1-\xi|=1$ when $\xi\neq 1$ is a root of unity of order prime to p.) The other bit—finding the x—is a little harder. One idea is to repeat the proof in the text to get

$$(1-\zeta)^{p-1}\cdot(\text{a 1-unit})=-p.$$

Use this equation to choose a appropriately so that we get

$$(1-\zeta)\cdot(\text{a 1-unit})=a.$$

(In other words, we want to "take the $(p-1)$-st root of both sides of the equation;" of course, $-p$ has many $(p-1)$-st roots, and what the equation does is tell us how to choose the right one.) This gives

$$|(1-\zeta)-a|<p^{-1/(p-1)},$$

which gives what we want.

   (Notice that the real work of the proof ends up being the same. Krasner's Lemma just replaces Hensel's Lemma as the trump card.)

280 Suppose we can show that the function

$$\phi:(a_0,a_1,\ldots,a_{n-1},b_0,b_1,\ldots,b_{n-1})\mapsto D$$

is continuous. Then notice that $\phi(a_0,a_1,\ldots,a_{n-1},a_0,a_1,\ldots,a_{n-1})=0$ (since in that case the λ's and the μ's are the same). By continuity, it will follow that we can make D as small as we like by choosing the bs close enough to the as, which proves Claim 2.

   It remains to show that ϕ is indeed continuous. In fact, it's not hard to see that ϕ is a *polynomial* in $a_0, a_1,\ldots,a_{n-1}, b_0, b_1, \ldots, b_{n-1}$. This is a very classical fact, and one that you may have met before. (It may help to know that the number D is known as the *resultant* of the polynomials f and g.) See [44, IV.8].

281 Solving this one is, at least at a first stage, a matter of reading through proofs carefully to see what fails if we drop any of the hypotheses. It's rather clear, for example, that characteristic zero plays a minor role (though we might need to add a separability condition if we drop it). Will the theorems work in complete archimedean fields? (That's not much fun, because they will be talking about roots of polynomials with real coefficients, not a very mysterious topic.) Will they work if we drop completeness? That would be quite interesting, since we would then have a choice of absolute values to work with.

282 Well, we've come close to proving this one in the proof of the Corollary 6.8.3, since there we obtained the conclusion by showing that some root of $g(X)$ was very close to some root of $f(X)$. See if you can push it through to get this result. If you can't, check [42, Section III.3].

283 What happens if we add a term $b_m X^m$ to $g(X)$ where m is very large and b_m is very small? How do the roots of

$$b_0 + b_1 X + \cdots + b_{10} X^{10}$$

relate to the roots of

$$b_0 + b_1 X + \cdots + b_{10} X^{10} + p^{100000} X^{100000}?$$

(See the section on Newton polygons, in Chapter 6, for further light on this one.)

284 Yes. (Prove it.)

285 Elements of $\mathbb{Q}_p^{\mathrm{unr}}$ still have p-adic expansions, since p is still a uniformizer. The coefficients in such an expansion will be chosen from a set of lifts of elements of the residue field, and the roots of unity we have used are precisely such a set of lifts. It's not clear that this clarifies anything. Note, however, that there are constructions of transcendental elements in $\mathbb{R}$ which proceed by constructing an appropriate decimal expansion. Is there any analogy?

286 This should be clear. There is nothing in our constructions that depends explicitly on the field being $\mathbb{Q}$: any field with a non-archimedean valuation will clearly do.

287 The definitions follow this problem in the text. The residue field is the algebraic closure of $\mathbb{F}_p$; the valuation ideal is not principal (there is no smallest positive rational number!), and therefore there is no uniformizer.

288 Suppose $x \in \mathbb{C}_p$ and $v_p(x) = r = a/b$. Choose a root π of $X^b - p^a$ in $\overline{\mathbb{Q}}_p$; it's fair to say that π is a "fractional power of p," and it's also clear that $v_p(\pi) = a/b$. Then $y = x/\pi$ is clearly a unit. Its image in $\overline{\mathbb{F}}_p$ lifts to a root of unity $\zeta \in \overline{\mathbb{Q}}_p$, and $\zeta^{-1}y$ is a 1-unit.

289 Follow the outline. (This makes a nice longer project.)

290 One would need to show that the closed unit ball is not compact. To do that, you need to exhibit a covering of the closed unit ball by open sets which has no finite subcovering. Can you find one? (The closed unit ball is just the valuation ring $\mathfrak{O}$. Consider the image of $\mathfrak{O}$ under reduction modulo $\mathfrak{P}$; how many elements are in the image? Now translate back to topological language.)

291 Mostly routine. The point about ρ is simply that there are enough different possible radii for balls in $\mathbb{C}_p$ (any p^r with $r \in \mathbb{Q}$ is allowed, and this is a dense subset of $\mathbb{R}$).

292 It was true in $\mathbb{Q}_p$ because the ideal in question was a principal ideal. That isn't true in $\mathbb{C}_p$.

293 Imitate the proof in Chapter 3.

294 Given the caution about choosing δ appropriately, it's just a matter of repeating the original proof.

295 Since every polynomial with coefficients in $\mathbb{F}$ has a root, having a common factor is the same as having a common root.

296 Yes, because we know that $g(X)$ is monic.

297 Parts (i) and (iv) are clear, (ii) follows by writing out the sum and applying the ultrametric inequality coefficient by coefficient, and (vi) is an instance of Problem 36. Finally, (v) is an easy application of the ultrametric inequality (and was done in the text just before the statement of the theorem).

298 Clear, since the absolute value of each of the coefficients is independent of the field we think it belongs to.

299 The inequality is very easy to get: just use the non-archimedean property directly. Over $\mathbb{C}_p$, the equality holds, but this takes some proving. Let's do it in the special case where $c = 1$. In this case, after multiplying by a constant if necessary, we can assume $\|f(X)\|_1 = 1$, so that all the coefficients are in $\mathfrak{O}$ and at least one is a unit. Then reduce it modulo $\mathfrak{p}$ to get a polynomial with coefficients in $\mathbb{F}$; the fact that $f(X)$ has a coefficient that is a unit means that the reduced polynomial is non-zero. Since $\mathbb{F}$ is an infinite field, there must be an element $\alpha \in \mathfrak{O}$ such that $\bar{f}(\bar{\alpha}) \neq 0$ in $\mathbb{F}$. Then it's clear that $|f(\alpha)| = 1$, which proves the equality.

Can you generalize to arbitrary c? Does it matter whether or not there is an element in K with absolute value equal to c?

300 Basically, we just replace things like $f(X) \equiv g_1(X)h_1(X) \pmod{\mathfrak{P}}$ with their translation (in this case, $\|f(X) - g_1(X)h_1(X)\|_1 < 1$). As to a version for the $\| \ \|_c$, can you decide? (Take a look, for example, at the proof of Lemma 7.2.2 and the statement of Proposition 7.2.3.)

301 Finding α uses a trick we have used before: if $c = p^r$ and $r = a/b$, we choose α to be a root of the polynomial $X^b - p^{-a}$, which exists because $\mathbb{C}_p$ is algebraically closed. Proving that $\|f(X)\|_c = \|\phi(f(X))\|_1$ is a matter of writing out the definitions. What this tells us is that all of these norms should have similar properties, since the equality allows us to transfer theorems about one to the other. In fancier terms, the theorem gives an isometric isomorphism between two normed rings.

302 One would need to be a lot more careful, the problem being that it is no longer clear that an α exists. What we would need to do is to further restrict c. To be precise, the argument still works over a field whose ramification index is e if we restrict c to be a real number of the form p^r where $r \in \frac{1}{e}\mathbb{Z}$.

303 Not immediately, since no α is available. It is true that in this case we can get A_c as a union of A_{p^r} for a sequence of rational powers of p that approach c, but does that yield a proof?

304 If we have a sequence of polynomials of bounded degree, we might as well think of them as being all of the same degree (padding the top terms with zeros if necessary). So let $f_i(X)$ be a sequence of polynomials of degree n. The first requirement is to dig out a candidate for the limit, and the obvious thing works: consider each of the coefficients and note that they form a Cauchy sequence themselves. Then it's a matter of showing that the polynomial just obtained is the limit we want.

As to why the boundedness is essential, the simplest example makes the point: take $c = 1$ and look at the sequence

$$f_0(X) = 1$$
$$f_1(X) = 1 + pX$$
$$f_2(X) = 1 + pX + p^2X^2$$
$$\cdots$$
$$f_i(X) = 1 + pX + p^2X^2 + \cdots + p^iX^i.$$

This is clearly Cauchy, and clearly its limit cannot possibly be a polynomial.

305 Because we know something about its N-th coefficient. Fill in the details.

306 Just notice that the inequality holds at every step of the inductive construction of $g(X)$.

307 No, because it is clear from the proof that $\|g(X)\|_c = |a_N| c^N$, so that the factorization the theorem asserts is the trivial one.

308 Just from the proof we can see that $g_i(X)$ converges to $g(X)$ at least as fast as δ^i converges to zero. That already says that the convergence is quite good. It may be, of course, that a more delicate analysis shows that the convergence is in fact faster than that.

309 If $c = 1$, then we can multiply $f(X)$ by a constant to assume that $\|f(X)\|_1 = 1$. In that case, the assumption reduces to saying that a_N is a p-adic unit and that $a_j \in \mathfrak{p}$ if $j > N$. The reduction of $f(X)$ modulo $\mathfrak{p}$ is then of degree N, and, after multiplying by another (unit) constant if necessary, we can assume the reduction is monic. This gives a congruence $f(X) \equiv g_1(X) \cdot 1 \pmod{\mathfrak{p}}$. Now apply Theorem 7.1.2.

310 What we need to prove is

- the sum of two series in A_c belongs to A_c,

- the product of two series in A_c belongs to A_c (this implies that the product of a series by a scalar does too, of course).

But both statements follow directly from Proposition 5.4.2.

311 This follows from the fact that absolute values are independent of the field in which we place ourselves.

312 As long as we are thinking of the closed ball in $\mathbb{C}_p$, this is clear. First, if $|a_n| c^n \to 0$, then clearly the series converges for any x such that $|x| \le c$. For the converse, we just need to note that in $\mathbb{C}_p$ we can always find an x whose absolute value is exactly equal to c. Convergence at that x implies that $|a_n| c^n \to 0$.

Notice that it is important to work over $\mathbb{C}_p$. It is easy to come up with a series that converges in the closed ball of radius $c = p^{-1/100}$ in $\mathbb{Q}_p$ but which is not in A_c, simply because the closed ball of radius c in $\mathbb{Q}_p$ is exactly the same as the closed ball of radius p^{-1}.

If c is not a rational power of p, then there are no x such that $|x| = c$, so the closed ball of radius c is the same as the open ball of radius c. Can you go on from there?

313 Since $0 \le |a_n|c_2^n < |a_n|c_1^n$, $|a_n|c_1^n \to 0$ implies $|a_n|c_2^n \to 0$.

314 This is identical to problem 297, except for the fact that we've replaced the equality in (iii) with an inequality. But that makes (iii) much easier.

315 Nothing changes, i.e., everything works just the same way. Can you prove it? (Here's a strategy: handle $c = 1$ first, by exactly the same method, which is feasible because the reduction modulo p is still a polynomial. Then use the usual tricks to handle other values of c.)

316 Same as before.

317 Again, the same argument as was used for polynomials works here, and shows that the map is an isometric isomorphism between the two spaces. (It's even a ring isomorphism.)

318 The map will be continuous if we can find a constant M with the property that
$$\|f(X)\|_{c_1} \le 1 \implies \|f(X)\|_{c_2} \le M.$$
Try to decide whether such a constant exists. (Thinking about the norms as sup-norms on appropriate balls may help.)

319 We have $g(X) = b_0 + b_1 X + \cdots + b_N X^N$ and $|b_N| = \max |b_n|$. Dividing through by b_N gives a monic polynomial whose coefficients all have absolute value less than or equal to 1. So what we want to prove is this: if $g(X) = b_0 + b_1 X + \cdots + b_{N-1} X^{N-1} + X^N$ satisfies $|b_i| \le 1$ for all i and α is a root of $g(X)$, then $|\alpha| \le 1$. To prove it, plug α into $g(X)$ to get
$$\alpha^N + b_{N-1}\alpha^{N-1} + \cdots + b_1\alpha + b_0 = 0,$$
and rewrite this as
$$\alpha^N = -(b_{N-1}\alpha^{N-1} + \cdots + b_1\alpha + b_0).$$
It follows that
$$|\alpha|^N \le \max_{0 \le i \le N-1}\{|b_i||\alpha|^i\},$$
and, since $|b_i| \le 1$, it follows that
$$|\alpha|^N \le \max_{0 \le i \le N-1}\{|\alpha|^i\}.$$
But this clearly implies $|\alpha| \le 1$.

320 Games with two indices are always a little tricky, but a careful walk through the proof should convince you that all is well.

321 No. It's also not Cauchy.

322 If $c = p^r$ for some $r \in \mathbb{Q}$, then it's easy to see that the answer is yes: just use the trick we've been using over and over. How would you handle general c?

323 Is this obvious? If $a_n \to 0$, then given $\varepsilon > 0$ we can find N such that $|a_n| < \varepsilon$ if $n > N$. But then $\max_{n>k} |a_n| < \varepsilon$ as soon as $k > N$.

324 It certainly should.

325 Do write out the details. The proof follows blow-by-blow the proof of Proposition 7.2.3, so there should be no difficulty in putting it together. But doing so will help you understand what's going on.

326 If $\|h(X) - 1\|_1 < 1$, then whenever $|x| \leq 1$ we have $|h(x) - 1| < 1$, which implies $h(x) \neq 0$.

327 By Proposition 169, we can rewrite $f(X)$ as a power series in $(X - x)$, and in this case the fact that n exists becomes obvious. To show that the two definitions of the multiplicity agree, write $f(X)$ as a product as in the Weierstrass Preparation Theorem, and then take derivatives. The advantage of Cassels' definition is, of course, that it doesn't depend on the theorem.

328 It's the usual thing: change variables so as to translate from $\|\ \|_c$ to $\|\ \|_1$. For more general values of c, it's a little harder—see below.

329 This particular game should be routine by now. Just follow the usual outline.

330 If ζ is a p^m-th root of unity, then $f(\zeta - 1) = \log_p(\zeta) = 0$. Hence, $f(X)$ has infinitely many zeros in the open unit ball. How can that be?

331 We assumed $g(X) \in \mathbb{C}_p[X]$ but all we really need to know is that we are working in an algebraically closed field. Write $g(X)$ as a product of linear factors and rearrange as necessary. Remember that in a monic polynomial the coefficient of degree 0 is the product of the roots.

332 The roots of $g_0(X)$ (in $\mathbb{C}_p$) are the roots of $f(X)$ in the closed unit ball, counted with multiplicities. The roots of $g_1(pX)$ are the roots of $f(X)$ in the closed ball of radius p, counted with multiplicities. Hence, every root of $g_0(X)$ (in $\mathbb{C}_p$) is also a root of $g_1(pX)$, with the right multiplicities. Therefore, $g_0(X)$ must be a divisor of $g_1(pX)$.

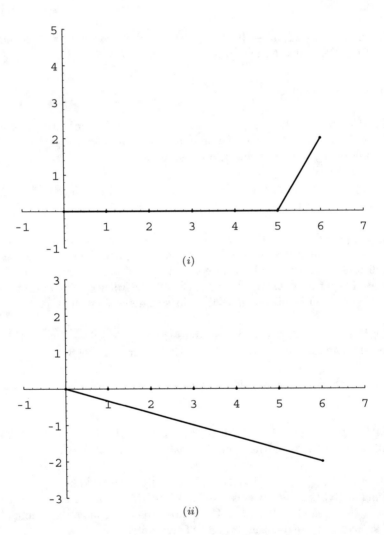

Figure B.2: The first two Newton polygons for problem 338

333 If you are at all hesitant, it might be helpful to write out a detailed proof.

334 We've clearly done enough to prove convergence with respect to $\| \ \|_1$. For $c = p^r$ with $r \in \mathbb{Q}$, we then use the usual dodge: change variables, and note that this transforms entire functions into entire functions. What should you do with other c? (Hint: how does convergence with respect to $\| \ \|_c$ relate to convergence with respect to $\| \ \|_{c'}$ when $c > c'$?)

335 The point is that "convergent in the closed ball of radius c" and "convergent with respect to $\| \ \|_c$" are equivalent.

336 As the next problem suggests, one can make such a function by using an infinite product. Something like

$$f(X) = \prod_{i=1}^{\infty} (1 - p^i X)$$

will work. It's also easy to arrange it directly in a power series, something like

$$\sum_{i=0}^{\infty} p^{n(i)} X^i,$$

by making $n(i)$ grow fast enough.

337 This is basically routine. What we are asking for, in a way, is a p-adic version of the general theory of infinite products. This would make a nice project. For example: we've proved that p-adic infinite series converge whenever their general term tends to zero; is it true that p-adic infinite products converge whenever their general term tends to 1?

338 See figures B.2 and B.3. Note that in (iii), we don't really need to divide by 3, because it is a 5-adic unit (dividing by a unit doesn't change any of the valuations).

339 If the polynomial has degree n, the polygon has only one line segment, of slope $-1/n$. (In the language we'll introduce below, Eisenstein polynomials of degree n over $\mathbb{Q}_p$ are "pure of slope $-1/n$.")

340 We would have to start with a vertical line beginning at $(0, v_p(a_0))$ and use that point as the initial center for the rotation, but otherwise all would be the same. The polygons for $f(X)$ and $af(X)$ will be the same except for a vertical translation.

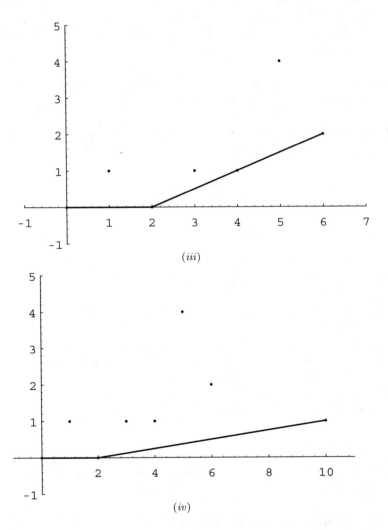

Figure B.3: Two more Newton polygons for problem 338

341 This is just a matter of sorting through the definitions. $h(X)$ will be pure of slope m if and only if $v_p(a_i) \geq mi$ for all i and $v_p(a_n) = mn$; translating this to absolute values gives what we want.

342 Routine, but worth writing up carefully. It's mostly a matter of translating Proposition 7.2.3 to the language of Newton polygons.

343 Use the fact that $\|f(X)g(X)\|_c = \|f(X)\|_c\|g(X)\|_c$ and the characterization of pure polynomials in problem 341.

344 It's just a question of translating inequalities for v_p to the language of absolute values.

345 Write

$$f(X) = 1 + a_1 X + \cdots + a_n X^n = (1 - \lambda_1 X)(1 - \lambda_2 X) \cdots (1 - \lambda_n X)$$

and work out valuations. We can start by noting that $v_p(a_n) = nm = \sum v_p(\lambda_i)$, and go on from there. (This is much trickier than it looks!)
   A more sophisticated method would be to compute $\| \ \|_{p^m}$ of each term in the product, and then to try to use the characterization of pure polynomials in terms of the norm. Does that work?

346 By induction, we can work at the j-th break, assuming that it happens at a point (x_j, y_j) and that the next point is $(x_j + i_{j+1}, y_j + m_{j+1}i_{j+1})$. It's then a matter of translating the assertion that this is the $(j+1)$-th segment of the polygon into valuations and absolute values.

347 Showing that $g_1(X)$ divides $g_2(X)$ is straightforward because we are working in an algebraically closed field: just consider the roots. Yes, $h(X)$ is pure. Can you prove it?

348 The polygon of $h(X)$ is obtained from the polygon of $f(X)$ by removing the segment corresponding to λ, i.e., a segment of slope m and x-length one, and then translating the resulting polygon to the origin. This clearly follows from our analysis of the roots, but it could be used as the starting point for that analysis if we could prove it directly. See [42] for a direct proof.

349 They tell us exactly what sorts of roots each polynomial has:

   i) five unit roots, one root of valuation -2;

   ii) six roots of valuation 2;

   iii) two unit roots, four roots of valuation $-1/2$;

   iv) two unit roots, eight roots of valuation $-1/8$.

Notice that the fact that the last two polynomials are congruent modulo 5 makes them have the same number of unit roots, but that this says very little about the other roots.

350 One polygon has a segment of slope 0 and a segment of slope 3; the other has a segment of slope 0 and a segment of slope 1. So even though the polynomials are "close," their polygons—and therefore their roots—look quite different. In other words, even if $\|f(X) - g(X)\|_1$ is very small, the root distribution of $f(X)$ and $g(X)$ outside the unit ball can be completely different. (*Inside* the unit ball, they will of course have exactly the same number of zeros of each valuation—check!)

351 The obvious thing to try is to require that $f(X)$ and $g(X)$ be close with respect to the c-norm. Does that work?

352 Yes. Find an example.

353 A segment of slope -1 and length 1, followed by an infinite horizontal line. The radius of convergence doesn't change, of course, since we've only made a finite number of changes.

354 Just use the same idea: compare the polygon to a line of slope b, remembering that by assumption the segments in the polygon all have slope less than b. (Does this work when $b = m$ but the sup is not attained? The segments will still "all have slope less than b" in that case... What if the sup *is* attained?)

355 Draw a picture if necessary, and refer to the previous solution.

356 The conclusion is that the radius of convergence is 0. The proofs still work.

357 One can't do much better than the obvious: the series will converge on the closed ball if the points $(n, v_p(a_n))$ get farther and farther above (vertical distance) the last segment of the polygon.

358 The first Newton polygon looks like a parabola; the series defines an entire function. The second polygon is a horizontal line, and so is the third (most coefficients will have valuation zero). The second series converges on the closed ball of radius 1, the third on the open ball. The fourth is tricky: the polygon connects the points $(0,0)$, $(p, -2)$, $(p^2, -4)$,... The series converges on the open ball of radius 1.

359 We'll leave these to the reader, who has certainly got the point by now.

360 It's just a matter of putting together all the information we already have. To show there are no zeros of smaller absolute value, consider a line through the $(k-1)$-st break point of slope smaller than m_k, and so on.

361 Yes it is. Can you prove it?

C   A Brief Glance at the Literature

As we end our promenade, it is important to point out to our reader where to go for further adventures in the p-adic realm. We will limit ourselves to a brief outline of the major sources of information (of which we are aware), and invite the reader to explore at will. The comments are, of course, personal opinion.

C.1   Textbooks

The first category of books are those which are intended as basic textbooks covering the fundamentals of the theory. Most of these are aimed at graduate students, but they should be accessible to anyone who has managed to read this book. Many of these books were major sources of information during the preparation of this book.

p-adic Analysis Compared with Real, by Svetlana Katok, [39], is probably the closest competitor of this book, in the sense that it too is aimed at undergraduates. As the title suggests, the burden of the book is to note the differences and similarities between p-adic analysis and real analysis. It appeared since the publication of the second edition; I found it very much worth reading as I prepared the third edition.

p-adic Numbers, p-adic Analysis, and Zeta-functions, by Neal Koblitz, [42], includes an introduction to p-adic numbers and p-adic analysis, and then goes on to discuss p-adic interpolation, the construction of the p-adic zeta-function, and several other related topics. The culmination of this book is an exposition of Dwork's proof of the rationality of the zeta-function of a hypersurface over a finite field, which is one of the landmarks of modern number theory. While the introductory portion of Koblitz's book has much in common with this book, Koblitz goes much further than we have, and in much less space.

Introduction to p-adic Analytic Number Theory, by Ram Murty, [51], is a bit more advanced than this book or [39], but still introductory. As in Koblitz's book, the main goal is to get to the theory of p-adic zeta and L-functions.

Local Fields, by J. W. S. Cassels, [14], is a much broader book that contains a great deal of interesting material. Its treatment of the fundamentals had a lot of influence on our choices when we were writing this book. A par-

© Springer Nature Switzerland AG 2020
F. Q. Gouvêa, *p-adic Numbers*, Universitext, https://doi.org/10.1007/978-3-030-47295-5

ticularly interesting characteristic of Cassels's book is the large number of examples of honest-to-goodness applications of p-adic methods to the rest of mathematics, especially diophantine equations. In contrast to our "strictly local" approach, Cassels includes a lot of material on how different p-adic completions relate to each other and to $\mathbb{Q}$ and also on the global theory: valuations on finite extensions of $\mathbb{Q}$ and how they relate to their various completions.

A Course in p-adic Analysis, by A. Robert, [53], overlaps with this book at the beginning, but then goes much deeper into p-adic analysis. Robert includes a construction of a spherically complete extension of $\overline{\mathbb{Q}}_p$, extensive coverage of differentiability, Mahler's theory of continuous functions on $\mathbb{Z}_p$ and its generalization, and detailed coverage of p-adic analytic functions. The exposition is terse but clear.

Since [53] appeared just before the second edition of this book, it had little impact on that version. This time around I have tried to give it the attention it deserves. As a result, I have often pointed my readers to Robert's book for more information.

Les Nombres p-adiques, by Y. Amice, [3], is another elementary introduction to p-adic numbers and p-adic methods, a little brief but very useful. Readers who read French might enjoy looking through her book, which is slanted towards functional analysis and rationality theorems.

Ultrametric Calculus, by W. H. Schikof, [56], despite its unprepossessing title, is quite an advanced book. Again, the focus is largely on p-adic analysis, and the author assumes that his reader has a good knowledge of classical analysis as a starting point. Schikhof's book is particularly good on integration,

Keith Conrad has written many expository notes on topics he teaches at the University of Connecticut. I have found the ones on p-adic numbers and p-adic analysis very useful. You can find them (and many more on other topics) at kconrad.math.uconn.edu/blurbs/.

Introduction to p-Adic Numbers and Valuation Theory, by G. Bachman, [6], starts with a basic introduction to p-adic numbers, and then veers off into a discussion of valuation theory in general. It is an interesting book, with very little intersection with this one. It is mentioned here largely because it was my first introduction to the subject.

Introduction to p-adic Numbers and Their Functions, by Kurt Mahler, [45], is rather hard to classify. While it presents itself as an introduction (and does develop the theory from scratch), it is really focused on a rather sophisticated account of continuous functions on $\mathbb{Z}_p$ with a special focus on their interpolation properties. This material is very different in flavor from the topics we have discussed, and the book is well worth the effort.

Finally, there is *Primeiros Passos p-ádicos*, [32], the seed from which this book grew. This one the reader, even if fluent in Portuguese, can safely disregard, since the only things it contains that were not incorporated into this version are the errors, which have been replaced by new and subtler errors.

C.2   Other Books

There are many other books that either deal with specific aspects of the theory or contain material that relates to one or another topic covered in this book. Here are the ones I like best, in no particular order.

I guess I'll mention first the books that started it all: *Theorie der Algebraischen Zahlen*, by Kurt Hensel, [35], and *Zahlentheorie*, [36]. Hensel had introduced *p*-adic numbers in various (often cryptic) journal articles, but these two books were intended to introduce them to a broader audience. The first book focuses on using the *p*-adics in algebraic number theory. The second is more elementary and represents an evolution in Hensel's thought. For example, it starts by constructing the "*g*-adic numbers," which are power series in an arbitrary integer *g*. For people who read German, they are worth a look, particularly to note the differences between Hensel's point of view and the one we have taken. There are no absolute values in Hensel.

Numbers, by a crowd of people headed by H.-D. Ebbinghaus, [25], is a delightful book about number systems in general. Its first part is called "From the Natural Numbers, to the Complex Numbers, to the *p*-adics." It is written in a very compressed style, and the various chapters can only survey the basics of each of the number systems, but reading them still is quite an enjoyable ride. There are two other parts: about real division algebras and about Conway's "surreal numbers." Some sections are more readable than others, but there is much here that is interesting.

Exercises in Number Theory, by the fictitious D. P. Parent, [48] (a translation of the 1978 French original) is a problem book which is really much more ambitious than its title suggests. Each chapter gives a compact introduction to one of the major areas of modern number theory and then presents the reader with problems (full solutions are included). The final chapter is called "*p*-adic Analysis."

Helmut Hasse was Kurt Hensel's student and became one of the great number theorists of the 20th century. His *Number Theory*, [34], is an introduction to number fields and function fields built on a systematic local-global approach. The local part of the book includes a lot of material on the *p*-adic numbers. Despite the old-fashioned notation, it is still a very useful book, but it is not easy to read.

The best elementary treatment of the Hasse–Minkowski theorem is probably the one in J.-P. Serre's *A Course in Arithmetic*, [60]. This book includes a chapter on the basic structure of the p-adic numbers, from the point of view of "coherent sequences," and then goes on to develop the theory of quadratic forms and prove the Hasse–Minkowski theorem over $\mathbb{Q}$. The second half of the book focuses on Analytic Number Theory, and may serve as an introduction to the (classical, rather than p-adic) theories of L-functions and modular forms.

The functional-analytic side of p-adic analysis is the focus of *Analyse Non-Archimedienne*, by A. F. Monna, [46], *Non-Archimedean Functional Analysis*, by A. C. M. van Rooij, [64], and *Nonarchimedean Functional Analysis*, by Peter Schneider, [57]. The newer books [50], [27], [26], and [23] focus on specific subtopics.

p-adic Geometry is a difficult subject, not least because there are many different approaches to the foundations. Perhaps the best place to start is *p-adic Geometry*, by Matthew Baker, Brian Conrad, Samit Dasgupta, Kiran S. Kedlaya, and Jeremy Teitelbaum, [7]. This collects lectures from an Arizona Winter School and so gives a good survey of the subject. Other books on the subject include [11], [10], [28], [29], [1].

These, of course, only scratch the surface, since an enormous amount of research has focused on p-adic methods and their application to number theory and other areas. Rob Benedetto treats arithmetic dynamics in *Dynamics in One Non-Archimedean Variable*, [9]. For Lie groups over the p-adics one can look at *p-adic Lie Groups*, by Peter Schneider, [58] and one of the *Lectures on Profinite Topics in Group Theory*, [41]. For differential equations, look at Kedlaya's *p-adic Differential Equations*, [40].

In the second edition, I mentioned my own[1] *Arithmetic of p-adic Modular Forms*, [31], but it is now outdated; perhaps [8] is now a better place to start.

A good overview of how p-adic numbers are used in number theory is Barry Mazur's article "The theme of p-adic variation" in [5]. Our adventurous reader will have no trouble finding more and more to learn, and may soon be in the position to teach us something herself.

---

[1] Could I resist a chance like this?

Bibliography

[1] Ahmed Abbes. *Éléments de Géometrie Rigide, Volume I.* Birkhäuser, 2011.

[2] Lars V. Ahlfors. *Complex Analysis.* McGraw-Hill, 1979.

[3] Yvette Amice. *Les Nombres p-adiques.* Presses Universitaires de France, 1975.

[4] Tom M. Apostol. *Mathematical Analysis.* Addison-Wesley, 1974.

[5] V. Arnold, M. Atiyah, P. Lax, and B. Mazur, editors. *Mathematics: Frontiers and Perspectives.* American Mathematical Society, 2000.

[6] George Bachman. *Introduction to p-adic Numbers and Valuation Theory.* Academic Press, 1964.

[7] Matthew Baker, Brian Conrad, Samit Dasgupta, Kiran S. Kedlaya, and Jeremy Teitelbaum. *p-adic Geometry.* American Mathematical Society, 2008. Lectures from the 10th Arizona Winter School held at the University of Arizona, Tucson, AZ, March 10–14, 2007, Edited by David Savitt and Dinesh S. Thakur.

[8] Baskar Balasubramanyam, Haruzo Hida, A Raghuram, and Jacques Tilouine. *p-adic Aspects of Modular Forms.* World Scientific, 2016.

[9] Robert L. Benedetto. *Dynamics in One Non-Archimedean Variable.* American Mathematical Society, 2019.

[10] Vladimir G. Berkovich. *Spectral Theory and Analytic Geometry over Non-Archimedean Fields.* American Mathematical Society, Providence, 1990.

[11] S. Bosch, U. Güntzer, and R. Remmert. *Non-Archimedean Analysis.* Springer-Verlag, 1984.

[12] E. B. Burger and T. Struppeck. "Does $\sum_{n=0}^{\infty} \frac{1}{n!}$ really converge? Infinite series and p-adic analysis". *Amer. Math. Monthly*, 103:565–577, 1996.

[13] Henri Cartan. *Elementary Theory of Analytic Functions of One or Several Complex Variables.* Dover Publications, 1995.

© Springer Nature Switzerland AG 2020

F. Q. Gouvêa, *p-adic Numbers*, Universitext, https://doi.org/10.1007/978-3-030-47295-5

[14] J. W. S. Cassels. *Local Fields*. Cambridge University Press, 1986.

[15] J. W. S. Cassels. *Lectures on Elliptic Curves*. Cambridge University Press, 1991.

[16] John Coates and M. J. Taylor, editors. *L-functions and Arithmetic*. Cambridge University Press, 1991.

[17] Keith Conrad. "Artin–Hasse-type series and roots of unity". Search for "Keith Conrad expository" or visit
`https://kconrad.math.uconn.edu/blurbs/gradnumthy/`
`AHrootofunity.pdf`.

[18] Keith Conrad. "Hensel's lemma". Search for "Keith Conrad expository" or visit
`https://kconrad.math.uconn.edu/blurbs/gradnumthy/hensel.`
`pdf`.

[19] Keith Conrad. "Infinite series in p-adic fields". Search for "Keith Conrad expository" or visit
`https://kconrad.math.uconn.edu/blurbs/gradnumthy/`
`infseriespadic.pdf`.

[20] Keith Conrad. "Ostrowski for number fields". Search for "Keith Conrad expository" or visit
`https://kconrad.math.uconn.edu/blurbs/gradnumthy/`
`ostrowskinumbfield.pdf`.

[21] Keith Conrad. "The p-adic expansion of rational numbers". Search for "Keith Conrad expository" or visit
`https://kconrad.math.uconn.edu/blurbs/gradnumthy/`
`rationalsinQp.pdf`.

[22] Albert A. Cuoco. "Visualizing the p-adic integers". *American Mathematical Monthly*, 98(4):355–364, 1991.

[23] Toka Dagana and François Ramaroson. *Non-Archimedean Operator Theory*. Springer, 2016.

[24] Jean Dieudonné. "Sur les founctions continues p-adiques". *Bull. Sci. Math.*, 68:79–95, 1944.

[25] H.-D. Ebbinghaus, H. Hermes, F. Hirzebruch, M. Koecher, K. Mainzer, J. Neukirch, A. Prestel, and R. Remmert. *Numbers*. Springer-Verlag, 1991.

[26] Alain Escassut. *Analytic Elements in p-adic Analysis*. World Scientific, 1995.

[27] Alain Escassut. *Value Distribution in p-adic Analysis*. World Scientific, 2016.

[28] Jean Fresnel and Marius van der Put. *Rigid Analytic Geometry and Its Applications*. Birkhäuser, 2004.

[29] Kazuhiro Fujiwara and Fumiharu Kato. *Foundations of Rigid Geometry I*. European Mathematical Society, 2018.

[30] Lothar Gerritzen and Marius van der Put. *Mumford Groups and Schottky Curves*. Springer, 1980.

[31] Fernando Q. Gouvêa. *Arithmetic of p-adic Modular Forms*. Springer, 1988.

[32] Fernando Q. Gouvêa. *Primeiros Passos P-ádicos*. IMPA–CNPq, 1989.

[33] Fernando Q. Gouvêa. *A Guide to Groups, Rings, and Fields*. MAA Press, 2012.

[34] Helmut Hasse. *Number Theory*. Springer, 1980.

[35] Kurt Hensel. *Theorie der Algebraischen Zahlen*. Teubner, 1908.

[36] Kurt Hensel. *Zahlentheorie*. G. J. Göschen'sche Verlagshandlung, 1913.

[37] Jan E. Holly. "Pictures of ultrametric spaces, the p-adic numbers, and valued fields". *American Mathematical Monthly*, 108(8):721–728, 2001.

[38] Kenkichi Iwasawa. *Lectures on p-adic L-functions*. Princeton University Press, 1972.

[39] Svetlana Katok. *p-adic Analysis Compared with Real*. American Mathematical Society, 2007.

[40] Kiran S. Kedlaya. *p-adic Differential Equations*. Cambridge University Press, 2010.

[41] Benjamin Klopsch, Nikolay Nikolov, and Christopher Voll. *Lectures on Profinite Topics in Group Theory*. Cambridge University Press, 2011.

[42] Neal Koblitz. *p-adic Numbers, p-adic Analysis, and Zeta-Functions*. Springer-Verlag, second edition, 1984.

[43] Serge Lang. *Cyclotomic fields I and II*. Springer-Verlag, second edition, 1990. With an appendix by Karl Rubin.

[44] Serge Lang. *Algebra*. Springer, third edition, 2002.

[45] Kurt Mahler. *p-adic Numbers and their Functions*. Cambridge University Press, second edition, 1981.

[46] A. F. Monna. *Analise Non-Archimedienne*. Springer-Verlag, 1970.

[47] James R. Munkres. *Topology*. Pearson, second edition, 2000.

[48] D. P. Parent. *Exercises in Number Theory*. Springer, 1984.

[49] Sebastian Pauli and Xavier-François Roblot. "On the computation of all extensions of a *p*-adic field of a given degree". *Math. Comp.*, 70:1641–1659, 2001.

[50] C. Perez-Garcia and W. H. Schikhof. *Locally Convex Spaces over Non-Archimedean Valued Fields*. Cambridge University Press, 2010.

[51] M. Ram Murty. *Introduction to p-adic Analytic Number Theory*. American Mathematical Society and International Press, 2002.

[52] Paulo Ribenboim. "The local Fermat problem". In A. Baker and R. Plymen, editors, *p-Adic Methods and Their Applications*. Oxford University Press, 1992.

[53] Alain M. Robert. *A Course in p-adic Analysis*. Springer-Verlag, 2000.

[54] Louis Halle Rowen. *Graduate Algebra: Commutative View*. American Mathematical Society, 2006.

[55] Walter Rudin. *Principles of Mathematical Analysis*. McGraw-Hill, third edition, 1976.

[56] W. H. Schikhof. *Ultrametric Calculus*. Cambridge University Press, 1984.

[57] Peter Schneider. *Nonarchimedean Functional Analysis*. Springer, 2002.

[58] Peter Schneider. *p-adic Lie groups*. Springer, 2011.

[59] Mícheál Ó Searcóid. *Metric Spaces*. Springer, 2007.

[60] Jean-Pierre Serre. *A Course in Arithmetic*. Springer, 1973.

[61] Jean-Pierre Serre. *Local Fields*. Springer-Verlag, 1979.

[62] John Tate. *Collected Works of John Tate*. American Mathematical Society, 2016. Edited by Barry Mazur and Jean-Pierre Serre.

[63] John T. Tate. "Rigid analytic spaces". *Inv. Math.*, 12:257–289, 1971. Reprinted in [62].

[64] A. C. M. van Rooij. *Non-Archimedean Functional Analysis*. Marcel Dekker, Inc., New York, 1978.

[65] A. G. Vitushkin, editor. *Several Complex Variables I*. Springer Verlag, 1989.

[66] Eugene Wigner. "The unreasonable effectiveness of mathematics in the natural sciences". *Comm. Pure Appl. Math.*, 13:1–14, 1960. Reprinted many times.

[67] Oscar Zariski and Pierre Samuel. *Commutative Algebra*. Springer, 1975.

[68] Paul Zimmermann et al. *Computational Mathematics with SageMath*. Society for Industrial and Applied Mathematics, 2019. See `sagebook.gforge.inria.fr/english.html`.

Index

1-units, 151, 201, 202

a fortiori, 221, 338
absolute values, 29, 31–40
    ∞-adic, 32, 53, 99
    archimedean, 32, 40, 56
    at infinity, 32, 53, 54, 61
    discrete, 130, 192
    equivalent, 54–56
    existence of extension, 185, 187
    extensions of, 178, 179
    image of, 68, 73
    independence of ambient field, 180, 184
    non-archimedean, 32–40, 42, 56, 178
    on $\overline{\mathbb{Q}}_p$, 187
    on $\mathbb{Z}[i]$, 36
    on $K(X)$, 231
    on $\mathbb{Q}$, 53–60
    p-adic, 34, 35, 50, 51, 53, 62, 68, 73, 178, 187
    trivial, 32, 53, 56
    uniqueness of extension, 179, 180, 184, 187
A_c, 238
    dependence on c, 239
    dependence on field, 239
    is complete, 241, 243
    polynomials are dense, 243
additive valuations, *see* valuations
algebraic closure of $\mathbb{Q}_p$, 141, 187–190
algebraic functions, 10
algebraic numbers, 10
analytic continuation, 130, 132

analytic functions, 129, 132, 133, 138, 274, 277
    periodic, 140, 275
    zeros of, 137, 138, 140
Anselm of Canterbury, 178
anti-derivatives, 118, 119, 137
archimedean, *see* absolute values, archimedean
Archimedean Property, 40
Artin–Hasse exponential, 273
automorphisms, 181

$\overline{B}(a, r)$, 44
$B(a, r)$, 44
Baire property, 274
balls, 44
balls in a non-archimedean valued field, 45, 46
base p expansion, 10, 11, 16
binomial series, 156–159, 208, 227
boundary points, 44–47

Cantor sets, 86, 87
Cauchy sequences, 60–62, 64, 65, 67, 74, 80, 110, 161, 171, 225
center of a ball, 44, 45
chain rule, 118
characteristic of a field, 169
Chevalley–Warning Theorem, 103
clopen sets, 47, 48, 77
closed balls, 44, 74
    in $\mathbb{Q}_p$, 77, 78
    in vector spaces, 171
closed sets, 44
closure of a set, 47

© Springer Nature Switzerland AG 2020
F. Q. Gouvêa, *p-adic Numbers*, Universitext, https://doi.org/10.1007/978-3-030-47295-5

Printed in the United States
By Bookmasters